Ultra-Cold Neutrons

The ultra-cold neutron (UCN) platform at the Institut Laue-Langevin, Grenoble.

Neutrons produced in the reactor core at a lower level enter the neutron turbine (A) through the curved guide tube (B). After reflection from the moving turbine blades, the neutrons (now slowed to UCN energies) can be distributed to various experimental installations: (C) UCN microscope; (D) fluid walled bottle for measurement of the neutron lifetime; (E) apparatus for a neutron lifetime measurement using magnetic storage; (F) the electric dipole moment experiment is seen with one end of the five-layer magnetic shield removed.

Ultra-Cold Neutrons

Robert Golub
*Technical University of Berlin**

David J Richardson
University of Southampton

Steve K Lamoreaux
University of Washington, Seattle

Long-term Visiting Scientists
Institut Laue–Langevin, Grenoble

* Now at Hahn Meitner Institute, Berlin

CRC Press
Taylor & Francis Group
Boca Raton London New York

CRC Press is an imprint of the
Taylor & Francis Group, an **informa** business

CRC Press
Taylor & Francis Group
6000 Broken Sound Parkway NW, Suite 300
Boca Raton, FL 33487-2742

First issued in paperback 2019

© 1991 by Taylor & Francis Group, LLC
CRC Press is an imprint of Taylor & Francis Group, an Informa business

No claim to original U.S. Government works

ISBN-13: 978-0-367-40304-1

Library of Congress Cataloging-in-Publication Data

Catalog record is available from the Library of Congress

**Visit the Taylor & Francis Web site at
http://www.taylorandfrancis.com**

**and the CRC Press Web site at
http://www.crcpress.com**

Contents

Preface

Maier-Leibnitz once said that he was always surprised to see how a simple little idea could grow and grow until it resulted in something large and complex with implications for other fields of research. This book is about only a part of the outgrowth of an idea of Fermi—that neutrons can undergo total reflection at a material surface. The applications of this idea, which led to neutron guides and their contribution to the flourishing of neutron scattering in recent decades, as well as its applications to neutron optics, are not covered. We are concerned mainly with neutrons which are totally reflected at all angles of incidence and, hence, are capable of being stored in closed vessels for considerable periods of time. However, we will not hold rigidly to this restriction and from time to time will discuss neutrons which do not satisfy this definition of Ultra-Cold Neutrons (UCN).

The development of a new research field, such as UCN research, raises many issues which are usually not discussed in a book of this nature which is intended to be an introduction to the development and present status of the field for new workers, as well as a useful compendium for workers in the field.

As budgets for scientific research are generally fixed at a national, and often institutional level, a new field of research can only obtain support at the expense of existing fields. While the scientific research establishment is reasonably flexible on these questions—most of its members taking real pleasure from, and being willing to support promising new ideas—a new field can only gain support if other fields take less. The field of UCN research had particular difficulties in the early 1970s. Although many important innovations in UCN research have been made at quite weak neutron sources, the really interesting applications require the highest available UCN densities and hence access to a strong neutron source. These rather costly sources have generally been built for, and are supported through, the community of researchers using neutron scattering to study condensed matter. However, even though UCN offer several interesting possibilities for extending the use of neutron scattering studies of condensed matter, a large body of UCN applications is concerned with 'fundamental physics'—the study of the neutron as an elementary particle. Thus it was unclear how UCN research should be supported. The problem can be highlighted by reference to the American Institute of Physics' Physics Classification Scheme (PACS) for classifying physics research. There is really no place for a new field with possible applications across a broad range of existing fields. The problem was essentially only solved by the flexibility of approach taken by the Institut Laue–Langevin and its then directors, Professors Mössbauer and White with the support of the Munich group, Professors Maier-Leibnitz and Steyerl, as well as Professor N Ramsey. In the United States the problem has apparently not yet been solved.

The reader of this book will be struck by the amount of work in this field carried out in the Soviet Union, and it is not clear how much of the previous comments apply to the Soviet system of research support. To a distant observer this often seemed to be somewhat more flexible, but in both systems these things frequently depend on the vagaries of support from certain well-placed individuals.

Having mentioned the large contribution to the field from Soviet colleagues we must now apologize for not giving a complete coverage of this work in the book. This is not possible for reasons of language and the penchant for much valuable work to be published as internal reports. We have endeavoured to give a comprehensive coverage of the work published in journals accessible to Western readers. The reader should be aware that this is only the tip of the iceberg.

Although many workers (ourselves included) come to UCN research without a background in conventional neutron scattering, a good knowledge of this topic is essential to an understanding of many aspects of UCN interactions with matter. It is one of the goals of this book to acquaint the newcomer with the rich physics associated with conventional neutron scattering so as to help him work more fruitfully with UCN.

We hope that the reader of the book will gain an impression of the enjoyment we have had working in this field. This has been increased by the opportunity to interact with co-workers from many countries around the world, and we would like to thank all those with whom we have met and exchanged information and results for sharing their enthusiasms and ideas.

R Golub
D J Richardson
S K Lamoreaux
December 1990

Acknowledgments

The authors are grateful to the following for granting permission to reproduce figures included in this book.

The authors of all figures not originated by ourselves.

Springer-Verlag for figures 2.7, 3.15, 3.17, 3.18, 5.3, 5.6, 5.7, 5.10, 5.11, 5.12, 6.19, 6.20, 8.5, 8.6, 8.11 and table 5.1.

Elsevier Science Publishers for figures 3.1, 3.6(a), 3.7–3.12, 3.16, 4.7, 5.1, 5.17, 6.1–6.3, 6.6, 6.8–6.11 and 7.2–7.16.

American Institute of Physics for figures 3.2, 3.4, 3.5, 5.13, 5.14, 5.18, 5.19, 6.27–6.30, 7.17 and 7.19.

Pergamon Press for figure 4.5.

Les Editions de Physique for figure 7.11.

John Wiley and Sons Inc for figures 8.2 and 8.3.

Les Editions de Physique for figure 8.4.

Akademie-Verlag for figure 8.7.

World Scientific Publishing for figure 8.8.

The authors and IOP Publishing Ltd have attempted to trace the copyright holder of all the figures and tables reproduced in this publication and apologize to copyright holders if permission to publish in this form has not been obtained.

1

Research with ultra-cold neutrons

Following their discovery in 1932 by Chadwick, free neutrons could only be studied under conditions where they spent no more than a brief moment within the experimental apparatus. Even 'long-wavelength' neutrons (long from the point of view of classical neutron scattering) produced in a cold source (see Chapter 3) with $\lambda \sim 10$ Å, $v = 400$ m s^{-1} take only 2.5 ms to travel 1 m. When diffusing through matter neutrons have average lifetimes of, for example, 0.2 ms (H_2O) or 130 ms (D_2O). However, the development, in the last two decades, of the technology of ultra-cold neutrons, has now reached the point where neutrons can be stored in material and magnetic bottles, for times which are essentially limited only by the neutron's β-decay lifetime ($\tau_\beta \sim 900$ s). This has opened up the possibility of a wide range of new applications, some of which have reached a comparatively advanced stage of development while others are only taking their first tentative steps.

It was Fermi who first realized that the coherent scattering of slow neutrons would result in an index of refraction, or effective interaction potential V for slow neutrons travelling through matter, and that this potential would be positive (index of refraction $n < 1$) for most materials (see Chapter 2). This effective potential is crucial for many of the effects grouped together under the topic of Neutron Optics. Sears (1989) provides a comprehensive discussion of this field in a book which in many ways is complementary to this one.

Fermi also realized that the result of this was that those neutrons (with energy E) incident on a surface at a glancing angle θ which satisfied

$$E \sin^2 \theta \leq V \qquad \sin \theta \leq \sin \theta_c = \left(\frac{V}{E}\right)^{1/2} \qquad (1.1)$$

would be totally reflected just as light can be totally reflected on aproaching a glass–air boundary from the glass side. Fermi and Zinn (1946) and Fermi and Marshall (1947) performed the first experimental demonstration of this effect. Total reflection of neutrons has provided the basis for the

1

highly successful technique of neutron guide tubes, in which neutrons whose angles satisfy (1.1) can be transported large distances through guides whose surfaces are smooth enough so that non-specular reflections (reflections for which the angle of incidence is not equal to the angle of reflection) are negligible, as first suggested by Maier-Leibnitz and Springer (1963). The neutron guide technique has virtually transformed slow neutron scattering from a somewhat esoteric technique to one of much wider applications. Bée (1988) and ILL (1988) give a picture of the guide tube installations at the Institut Laue–Langevin in Grenoble while Arif et al (1989) describe the installations at the NIST in Maryland. See also Serebrov (1989) for a description of the impressive facility under construction at Gatchina, south of Leningrad. There are many other installations of which we mention the Rutherford laboratory, Saclay, Julich and Geesthacht.

The observation of the total reflection of neutrons led to the speculation that if neutrons with energies

$$E \lesssim V \tag{1.2}$$

could be obtained—this is not obvious as typical materials have $V \sim 10^{-7}$ eV while thermal neutrons have energies of 2.5×10^{-2} eV—they would undergo total reflection at any angle of incidence and hence could be stored in closed vessels. We refer to such neutrons as Ultra-Cold Neutrons (UCN). Although this book is devoted to UCN we will, from time to time, discuss work with faster neutrons. While many workers in the field of neutron physics attribute the idea of neutron storage to Fermi, the first person to take the idea seriously enough to put it into print was Zeldovich (1959).

He pointed out that, although the lifetime of a neutron in, for example, graphite is only 10^{-2} s (independent of velocity, see Chapter 2), because of the small penetration depth of a UCN during total reflection ($\sim 10^2$ Å $= 10^{-6}$ cm) the fraction of the time that stored UCN would in fact spend in contact with the walls is quite small ($\sim 10^{-7}$) and so one could expect an absorption time of approximately 10^5 s for stored UCN. This is in good agreement with more detailed calculations (Chapter 4). Zeldovich also estimated that a thermal flux of 10^{12} n cm^{-2} s^{-1} cooled to 3 K in liquid helium would produce a UCN density of 50 cm^{-3}. It is interesting to note that such densities have now been achieved at the Institut Laue–Langevin, Grenoble, using a reactor with a thermal flux of 10^{15} n cm^{-2} s^{-1} cooled to 20 K in a deuterium-filled cold source (Chapter 3). Zeldovich suggested it would be interesting to study the interactions of the stored UCN with substances introduced into the cavity e.g. (n,γ) absorbers.

Shortly afterwards Vladimirskii (1961), suggested the use of magnetic field gradients to produce focused beams of polarized neutrons and 'magnetic mirrors' to confine UCN. For a reactor with a thermal flux of 2×10^{13} n cm^{-2} s^{-1} he estimates a UCN density of 10^{-2} cm^{-3} for the small effective potential $V_e \sim 3 \times 10^{-8}$ eV for $B = 5 \times 10^3$ G. Vladimirskii also suggested extraction of the UCN from the reactor by means of a vertical channel with the magnetic bottle located above the reactor, and pointed out the widening in solid angle as the neutrons travel up the guide (see Chapter 3). He emphasized the importance of the potential in the moderator material and that its effects can be countered by vertical extraction.

This was followed by Doroshkevich (1963) who suggested beryllium as a material for a storage vessel and estimated the temperature dependence of the loss rate due to inelastic scattering of the UCN by the walls. He estimated the loss rate due to wall vibrations as less than 10^{-7} s^{-1}.

Foldy (1966) published some speculations concerning the storage of UCN. He suggested a bottle whose walls were coated with liquid helium ($V = 1.1 \times 10^{-8}$ eV). A degenerate Fermi gas of neutrons up to this potential would have a staggering density of 10^{14} UCN/cm^3 but he made no suggestion as to how such densities could be achieved.

In 1968, Shapiro published a review article on the electric dipole moment (EDM) of elementary particles. In this article he pointed out the advantages of UCN for the search for a neutron EDM (see Chapter 7), especially the greatly increased observation time and the reduction of the '$v \times E$' effect (a magnetic field, produced in the frame of the moving neutron by the applied electric field, interacting with the neutron's magnetic moment and mimicking an EDM). See also Golub and Pendlebury (1972) for a more detailed discussion of this point.

Given the fact that the energy V (1.2) is some 10^5 times smaller than the thermal energy of neutrons in the reactor moderator and that the Maxwellian energy spectrum for neutron flux is proportional to E for low energies, it is remarkable that two groups independently had the courage to invest the time and effort to construct the necessary installations on the chance that neutrons so far from the peak of the Maxwell distribution did indeed exist inside the reactor, and that they could be extracted without crippling losses of intensity. That both groups were successful almost simultaneously is one of those coincidences which seem to be so common in the history of physics.

The Dubna group under F L Shapiro (Luschikov et al 1968, 1969) extracted UCN from a very low power pulsed reactor by means of a curved horizontal channnel, 9.4 cm ID, 10.5 m long. Counting rates of 0.8 counts/10^2 s (background $\sim$ 0.4 counts/10^2 s) were obtained. By admitting helium gas to the extraction pipe the authors attempted to estimate the storage time in the pipe. The idea is that when the average lifetime of a neutron for collisions with the helium

$$\tau_{He} = [N_{He}\sigma_{He}\bar{v}_{He}]^{-1} \qquad (1.3)$$

is equal to the average lifetime for wall losses the counting rate should be reduced by a half with respect to that in the absence of helium. This first attempt to measure storage times gave a result of 200 s which is to be compared with the 12 s obtained from later more detailed measurements (Groshev *et al* 1971), the discrepancy being attributed to possible impurities in the helium.

Working at Munich, Steyerl (1969) obtained UCN by vertical extraction from a steady state reactor. The beam was pulsed by a rotating chopper constructed out of 13 boron silicate glass plates located deep within the reactor swimming pool 2 m above the core, allowing time of flight measurements of neutron spectra. The counting rate showed a steep drop below 10 m s^{-1}, probably due to absorption in the aluminium windows, to reflection losses and the limited acceptance angle of the detector. However total cross sections were measured for neutron velocities down to 7 m s^{-1} for gold and 5 m s^{-1} for aluminium (Chapters 6 and 8).

It is noteworthy that both these initial attempts were made at relatively low intensity sources, an average thermal flux of 1.6×10^{10} n cm^{-2} s^{-1} in the Dubna experiment and 10^{13} n cm^{-2} s^{-1} in the Munich experiment, thus demonstrating the ability to carry out really new and important innovations at weak sources.

Following these first experiments Okun (1969) called attention to Shapiro's point that UCN offered a promising method for improving the sensitivity of the search for a neutron EDM emphasizing the potential improvement in observation times—10^3 s for UCN compared with 10^{-2} s in a typical beam experiment. He also mentioned the attraction of UCN for measuring the neutron lifetime.

The neutron lifetime is much less known than that of the μ and π mesons. This is because the neutrality of the neutron means that its trajectories cannot be measured by ionization—the only way to detect a slow neutron is to have it absorbed into a nucleus. Thus, to determine the neutron lifetime in a beam experiment it is necessary to know the absolute efficiencies of both a neutron and a charged particle detector to detect one of the decay products (proton or electron). With stored UCN the measurements can be carried out with a single detector for either the neutrons or the decay products. However with stored neutrons one must be sure that there are no processes other than the β-decay which can result in UCN leaving the storage vessel. For example wall losses must be kept to a minimum or be very well understood. See Erozolimskii (1975) for an early review of this topic and the series of papers dealing with the neutron lifetime in Dubbers (1989), e.g. Erozolimskii (1989). The possibilities of using the decay of polarized UCN to study parity and time-reversal violation in neutron β-decay have not yet been put into practice.

Both these fundamental applications of UCN (EDM and neutron lifetime) have now achieved a first level of success reporting results considerably more precise than those achieved with classical neutron beam methods. The reader is invited to look back over the series of conferences concerned with the fundamental physics of reactor neutrons (von Egidy 1978, Desplanques *et al* 1984, Greene 1986, Dubbers 1989) to see the fascinating development of the applications of UCN to these questions. Naturally this is one of the main themes of this book. Chapter 7 will discuss these topics in detail.

After the initial developments described here the field of UCN research began to expand rapidly. We would like to mention the work of Robson and Winfield (1972) and Bates and Roy (1974). The ensuing work will be discussed in the appropriate following chapters. Further details of the development of the field can be found in the early reviews of Steyerl (1977), Golub and Pendlebury (1979), Luschikov (1977), Golub *et al* (1979). The very comprehensive book by Ignatovich (1986) *Physics of Ultra-Cold Neutrons*, which recently appeared in an English translation (Ignatovich 1990), gives a more comprehensive coverage of the Russian literature. The current work is directed more to the experimentalist beginning work in this field or to the interested reader from another field and so is in many ways complementary to the book by Ignatovich.

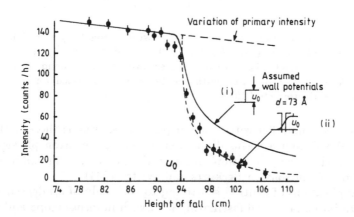

Figure 1.1 The measured intensity reflected from a glass mirror (points) compared with theoretical curves for (i) a step function and (ii) a smoothed step function for the wall scattering potential: broken curve, calculation for monoenergetic neutrons; full curve, calculation for the instrumental resolution. Assumption (ii) may be a model for a hydrogenous surface contamination. Scheckenhofer and Steyerl (1977).

As discussed in detail in Chapter 2 UCN afford the possibility of studying experimentally the one-dimensional square-well potentials, so familiar to all students of quantum mechanics. In figure 1.1 we show some examples of these. Figure 1.1, from Scheckenhofer and Steyerl (1977), shows the reflection from a potential step plotted against perpendicular neutron energy for a glass mirror whose surface potential step is believed to have been smeared out by a hydrogenous surface layer (inset) and figure 1.2 shows the transmission resonances observed in the double-layer structure shown in the inset (Steinhauser *et al* 1980), see Chapter 6.

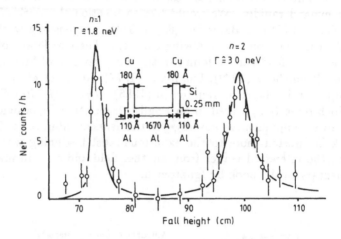

Figure 1.2 Transmission data for a target with nominal layer thicknesses: Al (110 Å), Cu (180 Å), Al (1670 Å), Cu (180 Å) and Al (110 Å). The substrate is silicon 0.25 mm thick. The two resonances observed correspond to $n = 1$ and $n = 2$. The data are compared with the full curve calculated for a multi-step potential. (Steinhauser *et al* 1980.)

The reader unfamiliar with neutron physics (as well as those with more experience) will find a useful discussion of classical neutron physics techniques, including the production and detection, as well as the diffusion and moderation of neutrons, in Beckurts and Wirtz (1964).

The applications of neutron scattering to the study of condensed matter has grown into a large and fascinating field with immense scope and power. We have already mentioned the growth in the number of guide tube installations, here we call attention to the books by Windsor (1981), Lovesey (1984), Marshall and Lovesey (1971) and Egelstaff (1965). The book by Turchin (1965) is recommended for newcomers to the field. Two treasure chests are the books by Champeney (1973) giving many lovely physical insights into aspects of neutron scattering and much else, and that by Lekner (1987) which gives many additional methods for treating the problems discussed in Chapter 2.

2

Interaction of UCN with matter

2.1 INTERACTIONS OF SLOW NEUTRONS

As massive, neutral, nuclear particles (hadrons) neutrons are affected by
all known types of force field except for electric fields; that is neutrons are
affected by the gravitational, magnetic, weak and strong interactions. We
begin this chapter with some introductory comments on each interaction and
then proceed to show how the interaction of UCN with matter follows from
the interaction with single nuclei.

2.1.1 The gravitational interaction

The gravitational interaction with the earth's gravitational field, given by

$$V_g = mgh \qquad (2.1)$$

is equivalent to 102×10^{-9} eV m^{-1} so that UCN of energy ≤ 200 neV can rise
by at most 2 m in the earth's gravitational field. By falling through a distance
of approximately 1 m a UCN will undergo a significant energy increase which
can be useful for improving the transmission probability through detector
windows or polarizing foils.

2.1.2 The weak interaction

The main effect of the weak interaction is that the free neutron is not stable
but decays according to the reaction

$$n \longrightarrow p + e^- + \bar{\nu} = 782 \text{ keV} \qquad (2.2)$$

with a decay time given by $\tau_\beta = 889 \pm 3$ s according to the most recent
measurement (Mampe *et al* 1989). Since the time τ_β is much shorter than

7

the age of the universe it is reasonable to ask how it is that neutrons still exist. The answer is that when neutrons are bound together with protons in a nucleus, for example the deuterium nucleus ^{2}D consists of a proton and neutron bound together with a binding energy of $E_D = 2.23$ MeV, it costs more energy to create a proton in the nucleus than is available from reaction (2.2), 782 keV, so the deuteron is stable. Thus all the neutrons we work with have been preserved inside atomic nuclei and in order to experiment with free neutrons we must arrange to extract them from their nuclear 'package'. This can be done by means of various nuclear reactions and, in the case of most work with UCN, by means of nuclear fission in a reactor.

The fact that the weak interaction violates parity conservation is demonstrated by the observations that the angular distribution of the decay products in reaction (2.2) is not isotropic and that the interaction of a neutron with unpolarized, isotropic materials shows a spin dependence. This latter effect was demonstrated in a remarkable series of experiments (Heckel *et al* 1982).

2.1.3   The magnetic interaction

Although electrically neutral, the neutron possesses a magnetic dipole moment and a spin $\frac{1}{2}\hbar$. The neutron's magnetic moment interacts with any external magnetic field $\boldsymbol{B}(\boldsymbol{r})$ by the interaction

$$V_m(\text{eV}) = -\boldsymbol{\mu} \cdot \boldsymbol{B}(\boldsymbol{r}) = \pm 6 \times 10^{-12} B_G \qquad (2.3)$$

where the sign is determined by the relative direction of the spin and the field.

An inhomogeneous magnetic field will exert a force on a neutron given by

$$\boldsymbol{F}_m = -\boldsymbol{\nabla} V_m = \boldsymbol{\nabla}[\boldsymbol{\mu} \cdot \boldsymbol{B}(\boldsymbol{r})] = \pm |\boldsymbol{\mu}|\, \boldsymbol{\nabla}\, |\boldsymbol{B}(\boldsymbol{r})| \qquad (2.4)$$

where the last equation assumes that the neutron's motion is so slow that the magnetic moment always keeps the same orientation with respect to the field $\boldsymbol{B}$ (adiabatic case). The condition for this is that the time dependence of the field seen by the neutron as it moves through the inhomogeneous field is much less than the precession frequency (Larmor frequency) of the neutron moment in the field

$$\frac{1}{\tau} = \frac{1}{|\boldsymbol{B}|} \cdot \left|\frac{\mathrm{d}\boldsymbol{B}}{\mathrm{d}t}\right| \ll \frac{\boldsymbol{\mu} \cdot \boldsymbol{B}}{\hbar} = \omega_L. \qquad (2.5)$$

In this limit the transition probability $\lesssim 1/(\omega_L \tau)^2$. When (2.5) is badly violated $\omega_L \tau \ll 1$ we find the neutron's dipole moment tends to keep its direction fixed in space (Rabi 1936, Vladimirski 1961) (the sudden approximation).

In addition to applied magnetic fields the neutron interacts with the internal magnetic fields produced by atomic electrons in many materials. As a result of this, neutron scattering can be used to study the distribution of atomic magnetism in ferromagnetic and anti-ferromagnetic materials as well as spin waves. The reader is referred to Lovesey (1984) for an extended discussion of these topics but we will not discuss them further in this book.

2.1.4   The strong interaction

Neutrons and protons are tightly bound together in atomic nuclei. The force responsible for this is called the strong interaction. The force between a neutron and a proton can be represented (at low energies) as a spherical square-well potential with a depth of about $V_0 = 40$ MeV and a radius of $R \sim 2$ fermi (1 fermi $= 10^{-15}$ m), as shown in figure 2.1(a). A better representation is given by a form with rounded corners (Fermi potential) which is also indicated in figure 2.1(a). The force between a neutron and a heavier nucleus has much the same form, the well depth remaining nearly constant from nucleus to nucleus and the radius R growing as approximately $R_0 A^{1/3}$ (constant nuclear density) as one goes to heavier nuclei.

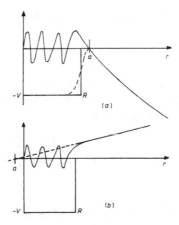

Figure 2.1   (a) The spherical square-well approximation to the neutron–nucleus interaction potential and the more accurate Fermi form (broken curve) $V_{\mathrm{F}} = V_0/(1 + e^{(r-R)/w})$. The wavefunction for a slow neutron ($\lambda \gg R$) moving in such a potential (equations (2.7) and (2.11)) is shown for the case of positive scattering length, a. (b) The wavefunction for the rare case when the strength V and size R of the potential result in a negative a.

In addition to being scattered, neutrons can also be captured by nuclei. This process, called absorption, results in the emission of a γ-ray ((n,γ) reaction) or charged particle (e.g. (n,p) or (n,α) reactions). The cross section for these reactions is proportional to $\left|M^2\right|/$(incident flux) where the matrix element

$$M = \langle \psi_\text{f} | V_\text{int} | \psi_\text{i} \rangle$$

where $|\psi_\text{i,f}\rangle$ are the initial and final neutron states and the incident flux is proportional to v_i, the incident neutron velocity. Since the strong interaction is much larger than the incident neutron energy—although very short range—the wavefunction $\psi_\text{i}(r)$ will be very different from the unperturbed initial wavefunction but only over a very short distance. However, because the neutron energies in which we are interested are so low in comparison with the strength of the interaction, the perturbed wavefunction and the matrix element M will be practically independent of the neutron's energy. Thus the absorption cross section σ_abs will be proportional to $1/v_\text{i}$ (see section 2.4.2) (Landau and Lifschitz 1958). As we will see later, the interaction of a neutron with a magnetic field of 16 kG or a change in height of 1 m or the interaction with a typical material are all of the order of 10^2 neV.

2.2   SCATTERING FROM A SINGLE NUCLEUS

2.2.1   The scattering length

For slow neutrons with de Broglie wavelength λ satisfying

$$kR = \frac{2\pi R}{\lambda} \ll 1 \tag{2.6}$$

the wavefunction for $r > R$, i.e. the region where $V(r) = 0$, has the form of a spherical wave $\sim \mathrm{e}^{\mathrm{i}kr}/r$. For $r < R$, the wavefunction $u = r\psi$ has the form (for the spherical square-well potential)

$$u \sim A \sin Kr \qquad K = \sqrt{\frac{2m(E + V_0)}{\hbar^2}} \tag{2.7}$$

satisfying the boundary condition $u = 0$ at $r = 0$.

In general $KR \gg 1$ so that u makes several oscillations inside the potential well before joining the external wavefunction at $r = R$. Thus we can write for the total wavefunction outside the well (incident plane wave $\mathrm{e}^{\mathrm{i}\boldsymbol{k}\cdot\boldsymbol{r}}$ plus scattered wave)

$$\psi = \mathrm{e}^{\mathrm{i}\boldsymbol{k}\cdot\boldsymbol{r}} + f(\theta)\frac{\mathrm{e}^{\mathrm{i}kr}}{r} \tag{2.8}$$

where $f(\theta)$ is determined by the boundary conditions at $r = R$.

With condition (2.6) the scattering is predominantly s wave (orbital angular momentum, $l = 0$). This follows from the small range of the strong interaction with respect to the neutron wavelength. Classically, we can argue that a neutron travelling a distance d from a nucleus will have angular momentum (mvd), (m is the neutron mass and v its velocity) and for this to be equal to $\hbar$ $(l = 1)$ we require

$$d \sim \hbar/mv = \lambda_\mathrm{n} \tag{2.9}$$

where λ_n is the neutron wavelength. Since λ_n is much greater than the range of the nuclear forces (condition (2.6)) a neutron travelling a distance λ_n from the nucleus will not be scattered. Thus we have

$$f(\theta) = \text{constant} = -a \tag{2.10}$$

because any angular dependence of $f(\theta)$ would imply a non-zero angular momentum. Thus for $R < r < 1/k$ we can write (2.8) as

$$\psi \approx 1 - a/r = (r - a)/r \tag{2.11}$$

so that u has the form of a straight line (figure 2.1).

We see that $\psi = 0$ at $r = a$, and so a can be interpreted as the radius of a hard sphere which would produce the same wavefunction at $r \gg R$ as the actual scattering potential. (The boundary condition for a hard sphere of radius a is $\psi(r = a) = 0$.) When the sign of a is defined as in (2.10) as was first done by Fermi and Marshall (1947), positive a means that the wavefunction is pushed away from the origin by the interaction and this looks like a repulsion if we observe the wavefunction at great distances. From figure 2.1(b) we see that it is possible for a to be negative, but as $KR \gg 1$ the range of V_0, R where this occurs is limited.

By applying the boundary condition that u'/u is continuous at $r = R$ we obtain from (2.7) and (2.11)

$$a = R - \frac{\tan KR}{K} = R \left(1 - \frac{\tan KR}{KR} \right). \tag{2.12}$$

In figure 2.2 we plot $\tan KR$ and KR separately as functions of KR and we see that the shaded regions, where $\tan KR > KR$ are the only places where a can be negative.

Thus except for the first quadrant $(KR < \pi/2)$, the regions where $a < 0$ are very small. For all nuclei except hydrogen it turns out that $KR > 3$ so we expect negative scattering lengths to be rare. This is found to be the case in practice as there are only three or four known cases of negative scattering length. Furthermore, the proton, which is the smallest nucleus, is found to have a negative scattering length. In fact, if we assume that the values of KR are randomly distributed in a uniform manner over the different isotopes we can predict a distribution of values of a which is in good agreement with the observed distribution (Peshkin and Ringo 1971).

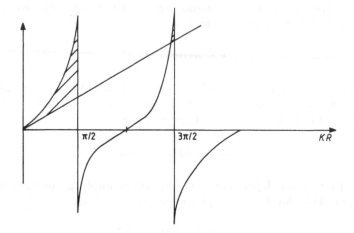

Figure 2.2   Functions relevant to equation (2.12): $\tan KR$ and KR. Shaded regions indicate negative scattering length a.

In terms of the wavefunction (2.8) we calculate the probability of scattering into an element of solid angle $d\Omega$ from

$$dP = (\boldsymbol{j}\cdot\hat{\boldsymbol{r}})r^2\,d\Omega \qquad (2.13)$$

where $\hat{\boldsymbol{r}}$ is a unit vector in the radial direction (the scatterer is assumed to be at $\boldsymbol{r}=0$) and $\boldsymbol{j}$ is the current density associated with ψ

$$\boldsymbol{j} = \frac{\hbar}{2\mathrm{i}m}(\psi^*\boldsymbol{\nabla}\psi - \psi\boldsymbol{\nabla}\psi^*). \qquad (2.14)$$

We assume that the incident beam is defined by an aperture and that we will measure the scattered wave at positions $\boldsymbol{r}$ outside the incident beam so that we need consider only the second term in (2.8) when applying (2.14). We thus find that

$$dP = v[f(\theta)]^2\,d\Omega \qquad (2.15)$$

where $v = \hbar k/m$ is the incident neutron velocity. Dividing by the incident current density $\boldsymbol{j}_{\mathrm{inc}} = \boldsymbol{v}$ as can be seen by applying (2.14) to the first term of (2.8) we obtain the differential cross section per unit solid angle

$$\frac{d\sigma}{d\Omega} = |f(\theta)|^2 = a^2 \qquad (2.16)$$

which shows us the physical meaning of $f(\theta)$ which is called the scattering amplitude. The scattering length a could be calculated for each isotope if a good model for the interaction potential was available. In practice one uses a as an experimentally determined parameter.

As the strong interaction is spin-dependent a can also be spin-dependent although with many nuclei this spin dependence is somewhat weak.

2.2.2 The Fermi potential

Since the interaction $V(r - r_n)$, (r_n is the position of the scattering nucleus) is much larger than the neutron energy, the neutron wavefunction in the presence of the interaction will be very different from the wavefunction in the absence of the interaction (see figure 2.1) and it is not possible to use perturbation theory to describe the scattering. However, since the large changes in the wavefunction only occur over very small distances and the wavefunction outside the range of interaction is only slightly changed by the interaction (see figure 2.1), Fermi realized that it is possible to introduce an equivalent potential which can be used to calculate the small changes in the wavefunction outside the range of the interaction by perturbation theory (Fermi 1936).

We consider the Schrödinger equation for the relative motion of the neutron and nucleus

$$-\frac{\hbar^2}{2\mu}\nabla_\eta^2\psi + [E - V(\eta)]\,\psi = 0 \qquad (2.17)$$

where $\eta = r - r_n$ and μ is the reduced mass. Outside the small region where V is unequal to zero the wavefunction $\psi(\eta)$ is given by (2.11) for small k. We introduce an equivalent potential $U(\eta)$ so that (2.17) becomes

$$-\frac{\hbar^2}{2\mu}\nabla_\eta^2\psi + [E - U(\eta)]\psi = 0 \qquad (2.18)$$

where we assume

$$U(\eta) = -U_0 \qquad \eta < \rho$$
$$= 0 \qquad \eta > \rho$$

where ρ is some distance chosen so that $\rho \ll \lambda_n$, $\rho \gg a$ and $\rho \gg R$.

We hope to find a U so that the solution to (2.17) for $r > R$ will be given by the solution to (2.18) which can be expressed as (2.8) with $f(\theta) = f = $ constant ($l = 0$) and with $f(\theta)$ given by the Born approximation applied to U

$$f(\theta) = -\frac{\mu}{2\pi\hbar^2}\langle k_f|U|k_i\rangle = -\frac{\mu}{2\pi\hbar^2}\int d^3\eta\, U(\eta)e^{i(k_i - k_f)\cdot\eta}. \qquad (2.19)$$

In order for (2.19) to be independent of θ, the range ρ of the equivalent potential $U(\eta)$ must satisfy

$$k\rho \ll 1. \qquad (2.20)$$

If this is the case then (2.19) reduces to

$$f(\theta) = f = \frac{2\mu}{3\hbar^2}U_0\rho^3. \qquad (2.21)$$

In order for the Born approximation to be valid for the effective potential U we require

$$U_0 \ll E \tag{2.22}$$

where $E = \hbar^2 k^2/2\mu$ is the neutron energy. Combining (2.20), (2.21) and (2.22) we find

$$f = \frac{2\mu U_0}{3\hbar^2}\rho^3 \ll \frac{2\mu E}{3\hbar^2}\rho^3 \ll \frac{2\mu E}{3\hbar^2 k^3} \ll \frac{1}{3k} \tag{2.23}$$

which is generally the case $(kf \ll 1)$. Thus it is possible to introduce a potential U, for which the Born approximation will give the correct scattering amplitude (Bethe 1937).

We can rewrite (2.21) as (remember $f = -a$, (2.10))

$$U_0 = \frac{-3\hbar^2 a}{2\mu\rho^3}$$

or

$$\int U(\eta)\,\mathrm{d}^3\eta = -U_0\frac{4\pi}{3}\rho^3 = 2\pi\hbar^2\frac{a}{\mu} \tag{2.24}$$

so that for calculating the matrix elements in (2.19) we can use an effective potential

$$U_\mathrm{F}(\eta) = \frac{2\pi\hbar^2 a}{\mu}\delta^{(3)}(\eta) \tag{2.25}$$

where we understand that U_F is only to be used in the first Born approximation.

We see that for $a > 0$ the effective potential U_F is also positive corresponding to a repulsive interaction. This is in spite of the fact that the actual interaction $V(\eta)$ in (2.17) is negative. The reason that an attractive interaction can produce an effective repulsion is evident in figure 2.1 where we see that the interaction displaces the wavefunction outside the range of interaction in the same sense as would be the case with a repulsive hard sphere.

To summarize, a nucleus located at $\boldsymbol{r}_\mathrm{n}$, produces a scattered wave

$$\psi_\mathrm{sc}(\boldsymbol{r}) = -\frac{a}{|\boldsymbol{r} - \boldsymbol{r}_\mathrm{n}|}\mathrm{e}^{\mathrm{i}k|\boldsymbol{r}-\boldsymbol{r}_\mathrm{n}|}\mathrm{e}^{\mathrm{i}\boldsymbol{k}\cdot\boldsymbol{r}_\mathrm{n}} \tag{2.26}$$

which is the result obtained by substituting (2.25) into (2.19). We have included the amplitude of the incident wave which did not appear in (2.8) because that equation was written for the case where the scattering nucleus is located at the origin $(\boldsymbol{r}_\mathrm{n} = 0)$.

As there are uncertainties in our knowledge of the actual interaction $V(\boldsymbol{r} - \boldsymbol{r}_\mathrm{n})$ it is customary to use (2.25) with experimentally measured values of *a*. In the following we will present values of *a* for some typical materials (table 2.1).

2.3 NEUTRON SCATTERING FROM A COLLECTION OF NUCLEI

2.3.1 Coherent and incoherent scattering

The subject of neutron scattering from condensed matter, which we will discuss in this and the following sections, is a vast topic to which many excellent books are devoted (e.g. Turchin 1965, Lovesey 1984). In this section and the appendices we can only touch on the highlights of this subject. We shall try to explain the physical principles involved with emphasis on those aspects which are important for UCN. For a more detailed discussion the reader is referred to one of the books already mentioned.

We started the previous section by writing equation (2.17) for the relative motion of the neutron and nucleus using the reduced mass, μ. This is correct if the nucleus is free to recoil, i.e. if the only forces on the nucleus come from the neutron–nucleus interaction. From equation (2.19) we see that the scattering length is proportional to the reduced mass. If the nucleus were rigidly bound the reduced mass would be replaced by the neutron mass, m, in the corresponding Schrödinger equation. We introduce the 'bound nucleus scattering length'

$$a_{\mathrm{B}} = \frac{m}{\mu} a \qquad (2.27)$$

so that we can always use the neutron mass m in equations like (2.25), which we can rewrite for a collection of atoms with location specified by $\boldsymbol{R}_i$

$$U_{\mathrm{F}}(\boldsymbol{r}) = \frac{2\pi\hbar^2}{m} \sum_i a_{\mathrm{B}}^{(i)} \delta^{(3)}(\boldsymbol{r} - \boldsymbol{R}_i) \qquad (2.28)$$

where $a_{\mathrm{B}}^{(i)}$ is the 'bound nucleus scattering length' for the ith nucleus.

We now calculate the matrix element of (2.28) between neutron states with initial momentum $\hbar k_0$ and final momentum $\hbar k_{\mathrm{f}}$ for a rigid array of nuclei

$$\langle k_{\mathrm{f}} | U_{\mathrm{F}} | k_0 \rangle = \frac{2\pi\hbar^2}{m} \sum_i a_i \mathrm{e}^{\mathrm{i}\boldsymbol{Q}\cdot\boldsymbol{R}_i} \qquad (2.29)$$

where a_i is the bound atom scattering length of the ith nucleus (we have dropped the subscript B as we will always use bound atom scattering lengths from now on) and

$$\boldsymbol{Q} = \boldsymbol{k}_0 - \boldsymbol{k}_{\mathrm{f}} \qquad (2.30)$$

is the momentum transfer during the scattering (divided by $\hbar$).

In general the scattering lengths a_i will vary from one isotope to another and will depend on the relative spin state of the neutron and nucleus. For scattering from an array of nuclei we use (2.19) substituted into (2.16) with the matrix element (2.29). However the cross section (2.16) must be averaged over the isotopic and spin distributions of the nuclei and (for unpolarized neutrons) the spin of the neutrons. If we denote this average by a bar we obtain

$$\frac{d\sigma}{d\Omega} = \overline{\left| \sum_i a_i e^{i\boldsymbol{Q} \cdot \boldsymbol{R}_i} \right|^2} \tag{2.31}$$

which can be rewritten as

$$\frac{d\sigma}{d\Omega} = \sum_{i,j} e^{i\boldsymbol{Q} \cdot (\boldsymbol{R}_j - \boldsymbol{R}_i)} \overline{(a_i^* a_j)}. \tag{2.32}$$

In most cases the spins and isotopes will be randomly distributed and there will be no correlation between a_i and a_j for different atoms. For the terms in which $i \neq j$ we then have in these cases

$$\overline{a_i^* a_j} = (\overline{a_i^*})(\overline{a_j}) = |\bar{a}|^2 \qquad (i \neq j). \tag{2.33}$$

However in some cases, such as when spin waves are present or in scattering from polarized scatterers there will be a correlation between the scattering lengths of neighbouring atoms and the present discussion will not apply.

For the terms in (2.32) with $i = j$ we have

$$\overline{|a_i^* a_j|} = \overline{|a|^2} \qquad (i = j). \tag{2.34}$$

Thus

$$\overline{a_i^* a_j} = |\bar{a}|^2 + \delta_{ij} \left(\overline{|a|^2} - |\bar{a}|^2 \right) \tag{2.35}$$

and substituting back into (2.32) we obtain

$$\begin{aligned}
\frac{d\sigma}{d\Omega} &= |\bar{a}|^2 \sum_{i \neq j} e^{i\boldsymbol{Q} \cdot (\boldsymbol{R}_j - \boldsymbol{R}_i)} + \sum_{i=1}^{N} \left(\overline{|a|^2} \right) \\
&= |\bar{a}|^2 \left| \sum_i e^{i\boldsymbol{Q} \cdot \boldsymbol{R}_i} \right|^2 + N \overline{|a - \bar{a}|^2} \\
&= a_{\text{coh}}^2 \left| \sum_i e^{i\boldsymbol{Q} \cdot \boldsymbol{R}_i} \right|^2 + N a_{\text{inc}}^2 \tag{2.36}
\end{aligned}$$

where N is the total number of atoms in the scatterer and (2.36) defines the coherent and incoherent scattering lengths a_{coh} and a_{inc}. The first term in

(2.36) is referred to as the coherent scattering cross-section and $\bar{a}$ is called the coherent scattering length a_{coh}. If the atoms are regularly spaced, as in a crystal, there exist some values of Q for which

$$Q \cdot R_i = 2\pi n \qquad (2.37)$$

for all i so that the sum in (2.36) is equal to N and the first term in (2.36) is proportional to N^2. This results in narrow peaks because the condition (2.37) only holds for special values of Q. This sort of scattering is called diffraction and while it forms a very important and interesting subject (Sears 1989) we will not discuss it further in this book, as the wavelength of UCN is too large for (2.37) to hold for any material.

The second term in (2.36) is referred to as the incoherent scattering cross section. Although coherent and incoherent scattering are often discussed as if they were different types of scattering, incoherent scattering really arises from the fact that individual nuclei scatter neutrons differently, because they may be different isotopes, different elements or may be in different spin states. In the sense of the normal usage in physics where incoherent scattering usually refers to scattering with a random phase shift, all slow neutron scattering is coherent, since the phase shift is nearly always the same (very small imaginary part of the scattering length for most materials). As seen from (2.36) the incoherent scattering is given by the mean square fluctuations of the scattering length, due to these causes. In a system where all scattering nuclei have the same scattering length there is no incoherent scattering.

2.3.2 Effective potential and index of refraction

According to equation (2.25) a neutron incident on a solid or liquid would see a 'forest' of δ-function potentials

$$V(r) = \frac{2\pi\hbar^2}{m} \sum_i a_i \delta(r - r_i) \qquad (2.38)$$

where r_i is the position of the ith nucleus and the sum over i includes all nuclei in the system. In other words the wavefunction of the neutron consists of the incident wave and a sum of the spherical waves (2.26) scattered by each nucleus

$$\psi(r) = e^{ik \cdot r} - \sum_i a_i \frac{e^{ik|r - r_i|}}{|r - r_i|} e^{ik \cdot r_i}. \qquad (2.39a)$$

Equation (2.39a) is written under the assumption that the wave incident on each nucleus is the same as the wave incident on the entire sample $e^{ik \cdot r_i}$, i.e. that the scattering does not make any significant change to the incident wave. This assumption is not, in general, correct and (2.39a) should be replaced by

$$\psi\left(\boldsymbol{r}\right) = e^{i\boldsymbol{k_0}\cdot\boldsymbol{r}} - \sum_i a_i\psi_i\left(\boldsymbol{r}_i\right)\frac{e^{ik_0|\boldsymbol{r}-\boldsymbol{r}_i|}}{|\boldsymbol{r}-\boldsymbol{r}_i|} \tag{2.39b}$$

where $\psi_j\left(\boldsymbol{r}_j\right)$ is the wave incident on the jth nucleus, i.e. the total wavefunction $\psi\left(\boldsymbol{r}\right)$ less the contribution of the scattered wave produced by nucleus j, and satisfies an equation identical to (2.39a) except that the sum over i excludes the value $i = j$.

The treatment of multiple scattering according to equation (2.39) is a complicated subject and has been discussed by many authors. A good discussion has recently been given by Sears (1989) (see also Lax 1951, 1952). For our purposes we can follow Foldy (1945) by making the approximation that

$$\psi_i\left(\boldsymbol{r}_i\right) = \psi\left(\boldsymbol{r}_i\right) \tag{2.40}$$

and rewriting (2.39b) as

$$\psi\left(\boldsymbol{r}\right) = e^{i\boldsymbol{k_0}\cdot\boldsymbol{r}} - \int \psi\left(\boldsymbol{r}'\right)\frac{e^{ik|\boldsymbol{r}-\boldsymbol{r}'|}}{|\boldsymbol{r}-\boldsymbol{r}'|}\left[na\left(\boldsymbol{r}'\right)\right]\mathrm{d}^3r' \tag{2.41}$$

where $\left[na\left(\boldsymbol{r}'\right)\right]\mathrm{d}^3r'$ represents the sum of the scattering lengths of those nuclei located in the volume element d^3r' around $\boldsymbol{r}'$.

Operating on both sides of (2.41) with $\left(\nabla^2 + k_0^2\right)$ and using

$$\left(\nabla^2 + k_0^2\right)\frac{e^{ik|\boldsymbol{r}-\boldsymbol{r}'|}}{|\boldsymbol{r}-\boldsymbol{r}'|} = -4\pi\delta^{(3)}\left(\boldsymbol{r}-\boldsymbol{r}'\right) \tag{2.42}$$

we obtain

$$\left(\nabla^2 + k_0^2\right)\psi\left(\boldsymbol{r}\right) = 4\pi\left[an\left(\boldsymbol{r}\right)\right]\psi\left(\boldsymbol{r}\right) \tag{2.43}$$

or the wave equation for a medium with an index of refraction $n_\mathrm{R}\left(\boldsymbol{r}\right)$

$$n_\mathrm{R}\left(\boldsymbol{r}\right) = \sqrt{1 - \frac{4\pi\left[an\left(\boldsymbol{r}\right)\right]}{k_0^2}}. \tag{2.44}$$

Multiplying (2.43) by $-\hbar^2/2m$ we obtain

$$-\frac{\hbar^2}{2m}\left(\nabla^2 + k_0^2\right)\psi\left(\boldsymbol{r}\right) = \frac{-2\pi\hbar^2}{m}\left[an\left(\boldsymbol{r}\right)\right]$$

$$\frac{-\hbar^2}{2m}\nabla^2\psi + \frac{2\pi\hbar^2}{m}\left[an\left(\boldsymbol{r}\right)\right]\psi = \frac{\hbar^2}{2m}k_0^2\psi\left(\boldsymbol{r}\right) = E\psi\left(\boldsymbol{r}\right) \tag{2.45}$$

or the Schrödinger equation for a particle of mass m moving in an effective potential

$$V\left(\boldsymbol{r}\right) = \frac{2\pi\hbar^2}{m}\left[an\left(\boldsymbol{r}\right)\right] \tag{2.46}$$

where a is the bound atom coherent scattering length.

This effective potential is just the volume average of the Fermi potentials (2.38) but the previous discussion, although somewhat simplified, shows the role played by multiple scattering and the self-consistency condition of equations (2.39b) and (2.41) in the exact form of the index of refraction (2.44). The reader may be amused to see the same argument applied to the scattering of sound waves from a distribution of air bubbles in water (Morse and Feshbach 1953, section 11.3). Since the potential (2.46) is independent of neutron energy while the index of refraction is energy-dependent we continue the discussion in terms of the potential function. We return to the index of refraction in Chapter 6.

In the case of magnetic materials the magnetic interaction (2.3) adds to the nuclear interaction (2.47) to produce a total spin-dependent interaction. This serves as the basis for UCN polarizers, (see section 6.3.4).

2.4   REFLECTION OF UCN FROM SURFACES

2.4.1   An ideal lossless surface

According to equation (2.46) we can consider the surface of a material with positive scattering length a as constituting a potential step (figure 2.3), of height

$$V = \frac{2\pi\hbar^2}{m}Na \qquad (2.47)$$

where N is the number density in the material, which is assumed to be homogeneous. Table 2.1 gives values of V for some common materials.

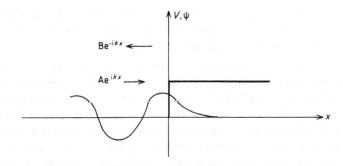

Figure 2.3   Wavefunction of a particle reflected from a material surface represented by a step in potential energy.

Table 2.1   UCN properties of selected materials.

Element	$\rho_{g/cc}$	$N_{form/cc}$ $\times 10^{22}$	$\Sigma_{form} a_{coh}^{bound}$ $\times 10^{-13}$ cm	V_{neV}	σ_{tot} barns	$f = W/V$ $\times 10^{-5}$
Ni^{58}	8.8	9.0	14.4	335	44	8.6
BeO	3.0	7.25	13.6	261	6.6	1.35
Ni	8.8	9.0	10.6	252	48	12.5
Be	1.83	12.3	7.75	252	1.4^a	0.5^a
					0.22^b	0.08^b
Cu^{65}	8.5	8.93	11.0	244	28	7.0
Fe	7.9	8.5	9.7	210	30	8.5
C	2.0	10.0	6.6	180	1.4	0.6
Cu	8.5	8.93	7.6	168	43.5	15.5
PTFE (Teflon)	2.2	2.65	17.6	123	—	—
Pb	11.3	3.29	9.6	83	2.0	0.6
Al	2.7	6.02	3.45	54	2.8	2.25
Perspex $(CH_2H_3O)_n$	1.18	1.65	7.88	33.9	—	—
V	6.11	7.1	−0.382	−7.2	50	—
Polyethylene $(CH_2)_n$	0.92	3.9	−0.84	−8.7	—	—
H$_2$O	1.0	3.34	−1.68	−14.7	—	—
Ti	4.54	5.6	−3.34	−48	58	—

^{a} 300 K.
^{b} 100 K.

$\rho_{g/cc}$—material density; $N_{form/cc}$—molecular density; $\Sigma_{form} a_{coh}^{bound}$—bound coherent scattering length summed over molecular formula; V_{neV}—effective UCN potential, neV; σ_{tot}—total cross-section at neutron wavelength $\lambda = 18$ Å (1 barn $= 10^{-24}$ cm); f—UCN loss factor, eqn (2.66).

A classical particle incident on the step with energy $E_\perp < V$ ($E_\perp$ is the kinetic energy in the direction perpendicular to the surface) will be reflected, while if $E_\perp > V$ it will be transmitted.

Quantum mechanically, particles with $E_\perp < V$ will also be reflected but there is some probability of penetrating the step before reflection. There will also be a reflection probability for particles with $E_\perp > V$, see figure 1.1. The quantum-mechanical treatment of the problem of a reflection from a plane surface of infinite extent can be treated as a one-dimensional problem, and is well known from all introductory texts on quantum mechanics. In fact this field can be considered as one-dimensional quantum mechanics (with square potentials) coming to life.

We consider a one-dimensional potential step located at $x = 0$. For

$x < 0$ $(V = 0)$ we have the wavefunction

$$\psi_I = e^{ik_x x} + R e^{-ik_x x} \qquad (x < 0) \qquad (2.48)$$

representing an incident wave of unit amplitude travelling to the right and a reflected wave (amplitude R) travelling to the left. The energy of the incident particle in the x-direction is given by

$$E_\perp = \frac{\hbar^2 k_x^2}{2m}.$$

Note that the y and z components of the momentum are unchanged by the potential barrier so the one-dimensional wavefunction (2.48) is simply multiplied by $e^{ik_z z} e^{ik_y y}$ in all regions of space. To the right of the potential step $(x > 0)$ we have

$$\psi_{II} = T e^{ik' x} \qquad \frac{\hbar^2 k'^2}{2m} = E_\perp - V \qquad (2.49)$$

and at the boundary at $x = 0$

$$\frac{1}{\psi} \left(\frac{d\psi}{dx} \right) = \text{continuous} = \frac{ik(1 - R)}{1 + R} = ik' \qquad (2.50)$$

thus the amplitude of the reflected wave

$$R = \frac{k - k'}{k + k'} = \frac{(E_\perp)^{1/2} - (E_\perp - V)^{1/2}}{(E_\perp)^{1/2} + (E_\perp - V)^{1/2}}. \qquad (2.51)$$

For $E_\perp > V$ k' is real and the reflection probability $|R|^2 < 1$. For

$$E_\perp < V \qquad k' = i\kappa = i\sqrt{\frac{2m}{\hbar^2}(V - E_\perp)} \qquad (2.52)$$

(if V is real) and $|R|^2 = 1$. In this case the transmitted wave has the form

$$T e^{ik' x} = T e^{-\kappa x} = \left(\frac{2k}{k + i\kappa} \right) e^{-\kappa x} \qquad (2.53)$$

i.e. the classically forbidden region is penetrated by an exponentially decaying wave. The penetration length $1/\kappa$ varies from infinity for $E = V$ to $(\lambda_c/2\pi)$ for $E = 0$, where λ_c is the wavelength of a neutron with $E_\perp = V$, i.e. a penetration on the order of 100 Å for the good neutron reflectors listed in table 2.1. Figure 2.3 shows the general behaviour of the wavefunction.

Those neutrons with total energy E satisfying $E < V$ will be totally reflected for any angle of incidence.

2.4.2 Loss of UCN on reflection

We have seen that the wavefunction of a particle undergoing total reflection penetrates a small distance into the reflecting surface. It is thus possible for the neutron to interact with the nuclei inside the surface. The most important interactions are:

(i) neutron absorption which occurs when a neutron, during the time it is in the deep potential well produced by the nucleus (figure 2.1), emits a γ-ray and makes a transition to a bound state in the nucleus, thus being 'captured' by the nucleus; and

(ii) inelastic up-scattering.

Since the reflecting material is in thermal equilibrium at some temperature T, the atoms of the material are continuously vibrating in a random fashion around their equilibrium positions. Some of the energy of this thermal excitation can be transferred to the reflecting neutron, so that after the reflection the neutron will have an energy $E > V$ and will not be reflected on subsequent encounters with a material surface.

Both these processes should rightly be treated as ordinary quantum-mechanical transition processes:

(i) a neutron in a plane wave state makes a transition to a bound state with emission of a γ-photon; and

(ii) a neutron in a plane wave state of momentum $\hbar k_i$ makes a transition to a state $\hbar k_f$ ($k_f \gg k_i$) while the system of atoms in the material makes a corresponding transition to another state.

However in the case where the probability of the reverse transition is negligible (irreversible transition) the effect of these processes on the incident neutron state can be treated as though it was produced by an effective imaginary potential. The probability of the reverse transition is always negligible when the final state contains a continuum of energy levels e.g. the γ-ray states in (i) or the states of excitation of the solid in (ii).

We consider the time-dependent Schrödinger equation for a complex potential $U = V - iW$

$$i\hbar \frac{\partial \psi}{\partial t} = -\frac{\hbar^2}{2m}\nabla^2\psi + U\psi \tag{2.54}$$

and take its complex conjugate

$$-i\hbar \frac{\partial \psi^*}{\partial t} = -\frac{\hbar^2}{2m}\nabla^2\psi^* + U^*\psi^*. \tag{2.55}$$

Multiplying (2.54) by ψ^* and (2.55) by ψ and subtracting we find

$$i\hbar\frac{\partial|\psi|^2}{\partial t} = -\frac{\hbar^2}{2m}\left(\psi^*\nabla^2\psi - \psi\nabla^2\psi^*\right) + (U - U^*)|\psi|^2. \tag{2.56}$$

The first term on the right-hand side can be rewritten as

$$-\frac{\hbar^2}{2m}\nabla\cdot\left(\psi^*\nabla\psi - \psi\nabla\psi^*\right) = -i\hbar\nabla\cdot\boldsymbol{j} \tag{2.57}$$

where $\boldsymbol{j}$ is the usual quantum-mechanical current density (2.14).

If U is real, equation (2.56) represents the conservation of probability, $|\psi|^2$, which is equal to the probability density, ρ, where

$$\frac{\partial\rho}{\partial t} + \nabla\cdot\boldsymbol{j} = 0 \tag{2.58}$$

i.e. according to (2.58) the decrease in probability density at a point $(\partial\rho/\partial t)$ is equal to the flow of probability away from the point. When $U - U^* = -2iW \neq 0$ we have an additional loss rate of probability density given by

$$\left(\frac{\partial\rho}{\partial t}\right)_{\text{add}} = -\frac{2W}{\hbar}\rho \tag{2.59}$$

which results in a decay of probability density

$$\rho = \rho_0 e^{-2Wt/\hbar}. \tag{2.60}$$

Thus by setting

$$W = \frac{\hbar}{2}\sum_i N_i\sigma_l^{(i)}v \tag{2.61}$$

we get a correct description of the absorption

$$1/\tau_{\text{abs}} = 2W/\hbar = \sum_i N_i\sigma_l^{(i)}v \tag{2.62}$$

where N_i is the density and $\sigma_l^{(i)}$ is the loss cross section (absorption plus inelastic scattering) of nuclear species i. Goldberger and Seitz (1947) present a more rigorous calculation.

Comparing (2.47) and (2.61) we can write

$$U = V - iW = \frac{2\pi\hbar^2}{m}N\left(a_{\text{r}} - ia_{\text{i}}\right) \tag{2.63}$$

where

$$a_i = \left(\frac{\sigma_l k}{4\pi}\right). \tag{2.64}$$

This is the application of the optical theorem to the present problem. According to this theorem

$$\text{Im } f(0) = \frac{k}{4\pi}\sigma_{\text{tot}} \tag{2.65}$$

where $f(0)$ is the forward scattering amplitude (equation (2.8)) and σ_{tot} is the total cross section (see e.g. Schiff 1968). Physically Im $f(0)$ represents the attenuation of the incident beam which, in general, is proportional to the total cross section. In our case of propagation in a uniform medium the coherent wave scattered by each atom adds to the incident wave to produce a phase-shifted wave of unattenuated amplitude so that the coherent scattering cross section does not contribute to the attenuation of the propagating wave or to the right-hand side of (2.64). Goldberger and Seitz (1947) and Lax (1952) show that this is true by treating the self-consistency of the multiple scattering equation (2.39b).

We have seen that V, the real part of the material potential (2.47), is independent of neutron velocity, whereas the imaginary part W (equation (2.61)) appears to depend on the velocity. However the absorption and inelastic scattering cross sections are both proportional to $1/v$ so that $W \propto \sigma_l v$ is velocity independent.

We can see this by considering the inelastic scattering and absorption from the point of view of time-dependent perturbation theory. The rate of transition is proportional to the square of the matrix element of some transition operator between the initial and final states. As the interaction region (the nucleus) is much smaller than the neutron wavelength this matrix element squared will be proportional to $|\psi(0)|^2$ where $|\psi(0)|^2$ is the magnitude squared of the incident particle wavefunction at the position of the nucleus (Landau and Lifschitz 1958). To obtain the cross section we divide by the incident flux (Schiff 1968) and if we normalize ψ to unit current density we have $\sigma \propto |\psi(0)|^2 \propto 1/v$. In other words the cross section for absorption or inelastic scattering depends on the time the incident neutron spends in the vicinity of the nucleus. Note that this argument implies that, at low energies when V (equation (2.47)) is not negligible with respect to E, the absorption and inelastic scattering satisfy $\sigma \propto 1/v'$ where $v' = [(2m)^{-1}(E - V)]^{1/2}$. See Gurevich and Nemirovskii (1962) for a more detailed discussion of this point.

The ratio W/V is given by (2.63) and (2.64)

$$f = \frac{W}{V} = \frac{\sigma_l k}{4\pi a} = \frac{\sigma_l}{2a\lambda} \tag{2.66}$$

which is much less than 1 unless $\sigma_l \sim 2a_r\lambda$ or $\sigma_l \sim 4\times10^4$ b $(1 \text{ b} = 10^{-24} \text{ cm}^2)$ for thermal neutrons $(\lambda \sim 2 \text{ Å})$ and a typical value of $a_r = 10^{-12}$ cm. Only a few very strong absorbers e.g.

$$Cd^{113} \ (a = -1.5 + 1.2i, \sigma_{abs} = 2 \times 10^4 \text{ b at } \lambda = 1.8 \text{ Å})$$

or

$$Gd^{157} \ (a = 4.3 + 4i, \sigma_{abs} = 25.4 \times 10^4 \text{ b at } \lambda = 1.8 \text{ Å})$$

have W on the same order as V. For most materials the ratio W/V is much smaller attaining values of approximately 10^{-5} for some materials used for UCN storage (table 2.1).

The calculation of the reflection coefficient (2.51) can be repeated with the complex potential U replacing V. The result is again given by (2.51) with this replacement. For $W \ll V$, the usual case, we can expand the square roots keeping only first-order terms in W. We thus obtain

$$|R|^2 = 1 - 2f \left(\frac{E_\perp}{V - E_\perp} \right)^{1/2} \equiv 1 - \mu(E, \theta) \qquad (2.67)$$

where $f = W/V$ is given by (2.66) and $\mu(E, \theta)$ is defined as the wall loss probability per bounce. Rewriting (2.67) in terms of the total energy E and the angle of incidence with respect to the surface normal, θ, we have

$$\mu(E, \theta) = 2f \left(\frac{E \cos^2 \theta}{V - E \cos^2 \theta} \right)^{1/2}. \qquad (2.68)$$

If we are dealing with UCN stored in a closed container, we can consider the UCN as forming a perfect gas (see Chapter 4) except for the facts that the density will decay with time due to the β-decay of the neutron ($\tau_\beta \sim 900$ s) and that there are wall losses. In the case of a container whose surfaces are not perfectly flat there will be some fraction of the wall collisions which result in diffuse reflection ($\sim 1\%$ in the case of a good polished surface) so that after a sufficient number (~ 100 in this case) of collisions the directions of the UCN will be randomized and the UCN gas will be isotropic. Thus we can average μ over the angle of incidence for a given energy using the usual kinetic theory rule (Kennard 1938) that the rate of particles incident on a surface is proportional to

$$\cos \theta \, d\Omega = 2\pi \cos \theta \sin \theta \, d\theta = 2\pi \cos \theta \, d(\cos \theta) \qquad (2.69)$$

so that the average loss probability per bounce for UCN of energy E is given by

$$\bar{\mu}(E) = 2 \int_0^1 \mu(E, \theta) \cos \theta \, d(\cos \theta)$$

$$= 2f \left[\frac{V}{E} \sin^{-1} \left(\frac{E}{V} \right)^{1/2} - \left(\frac{V}{E} - 1 \right)^{1/2} \right]. \qquad (2.70)$$

Equation (2.70) is plotted in figure 2.4 as a function of E/V. We see that the maximum value of $\bar{\mu} = \pi f$ at $E = V$ and that at $E/V = 0.5$, $\mu = f$. For completeness we give the exact result for the reflection probability obtained from (2.51) by substituting U for V (Antonov *et al* 1969a).

$$|R|^2 = \frac{E_\perp - \sqrt{E_\perp}\,[2\alpha - 2\,(V - E_\perp)]^{1/2} + \alpha}{E_\perp + \sqrt{E_\perp}\,[2\alpha - 2\,(V - E_\perp)]^{1/2} + \alpha} \qquad (2.71)$$

with

$$\alpha = \sqrt{(V - E_\perp)^2 + W^2}.$$

The reader can verify that expanding (2.71) for small W yields (2.67).

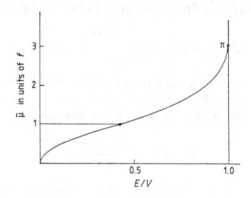

Figure 2.4   The probability of loss per collision with a wall, averaged over angle, $\bar{\mu}$, against UCN energy, E.

Another approach to the problem of the loss of UCN on reflection from a surface, which is useful in the case of more complex surfaces, is the following. We will consider the imaginary part of the potential W as a perturbation, solving the problem for a pure real potential and then using perturbation theory to calculate the loss rate due to W. This approach is only valid for the case $f \ll 1$, but this is the case for most materials of interest. When the neutrons are inside a material with loss cross section $\sigma_l(v)$ and number density N the probability of loss per second is given by

$$1/\tau_l = N\sigma_l(v)v \qquad (2.72)$$

where τ_l is the mean loss time. If the unperturbed wavefunction is $\psi(x)$, then the probability density of the neutrons at the point x is $|\psi(x)|^2$ and the probability of loss per second will be given by

$$P_l' = \int \frac{|\psi(x)|^2}{\tau_l(x)}\,\mathrm{d}^3 x = \int v\,[N\sigma_l(v)]_x\,|\psi(x)|^2\,\mathrm{d}^3 x \qquad (2.73)$$

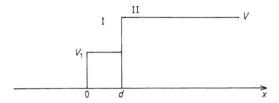

Figure 2.5   A model for a surface of a material with potential V, covered by a film with potential V_1, thickness d.

where we have allowed $N\sigma_l(v)$ to be a function of position. If we normalize $\psi(x)$ to correspond to one incident particle/s/cm^2 then P_l will represent the loss probability per bounce. Applying this to the incident wave (e^{ikx}) in (2.48) we have the incident flux $= v|\psi_{\text{inc}}|^2$ so we must multiply the wavefunction $|\psi|^2$ by $1/v$. Then

$$P_l = \int [N\sigma_l(v)]_x |\psi(x)|^2 \, \mathrm{d}x \qquad (2.74)$$

with ψ normalized as in (2.48). Applying this to the region $x > 0$ in the case of the potential step and using (2.53) for $\psi(x)$ inside the step we have

$$P_l = N\sigma_l|T|^2 \int_0^\infty \mathrm{d}x \, e^{-2\kappa x} = \frac{2N\sigma_l}{\kappa\,[1+(\kappa/k)^2]} = \frac{2W}{V}\frac{\sqrt{E_\perp}}{\sqrt{V-E_\perp}} \qquad (2.75)$$

in agreement with (2.67).

Note that the loss probability per bounce (2.74) is given by $P_l = \tau_D/\tau_l$ where $\tau_D = \int_\Omega |\psi(x)|^2 \mathrm{d}x/v$ is the 'dwell time' of the neutrons in the region Ω, which has been the subject of much discussion in connection with the question of tunnelling (Hauge and Støvneng 1989, Golub *et al* 1990).

2.4.3   Reflection from a surface covered with a film

We consider the surface of a material with potential V covered with a layer of material with potential V_1 and thickness d (figure 2.5). This is a much more realistic model of a surface than that of the previous section. V_1 can be less than V, e.g. if the surface layer is a hydrogenous substance, (remember hydrogen has a negative scattering length) or a layer of hydroxide formed by the reaction of an oxide with atmospheric water. On the other hand V_1 can be greater than V, e.g. some pure metal oxides have potentials higher than the pure metal because oxygen has a large positive scattering length. As before we have

$$\begin{aligned}
\psi(x) &= e^{ikx} + Re^{-ikx} & x &< 0 \\
&= Te^{-\kappa(x-d)} & x &> d.
\end{aligned} \qquad (2.76)$$

In region I $(0 < x < d)$ we have

$$\psi(x) = Ae^{ik'x} + Be^{-ik'x} \tag{2.77}$$

in the case of $E > V_1$. For $E < V_1$ we replace k' by $i\beta$ where

$$\beta = \sqrt{\frac{2m}{\hbar^2}(V_1 - E)} \quad \text{and} \quad k' = \sqrt{\frac{2m}{\hbar^2}(E - V_1)} \tag{2.78}$$

and κ in (2.76) is given by (2.52). The boundary conditions at $x = 0$ and $x = d$ $(\psi, \psi'$ continuous) yield

$$\begin{aligned}
1 + R &= A + B \\
k(1 - R) &= k'(A - B) \\
Ae^{ik'd} + Be^{-ik'd} &= T \\
\left(Ae^{ik'd} - Be^{-ik'd}\right) &= \frac{-\kappa T}{ik'}
\end{aligned} \tag{2.79}$$

with the solutions

$$T = \frac{2}{D} \qquad A = \frac{e^{-ik'd}}{D}(1 + i\kappa/k') \qquad B = \frac{e^{ik'd}}{D}(1 - i\kappa/k') \tag{2.80}$$

where

$$D = \cos k'd[1 + i\kappa/k] + \sin k'd\left(\frac{\kappa}{k'} - i\frac{k'}{k}\right). \tag{2.81}$$

We now apply the integration technique of (2.74) to calculate the losses on reflection. The contribution to the integral from region II $(x > d)$ follows from (2.75) and (2.80)

$$\eta_{II}(E_\perp) = (N\sigma_l)_{II} \frac{2}{\kappa|D|^2}. \tag{2.82}$$

In region I $(0 < x < d)$

$$\int_0^d |\psi_I(x)|^2 \, dx = d\left(|A|^2 + |B|^2\right) + \text{Re}\left[\frac{AB^*}{ik'}\left(e^{2ik'd} - 1\right)\right] \tag{2.83}$$

which yields for the contribution to the loss probability due to region I

$$\begin{aligned}
\eta_I(E_\perp > V_1) = \frac{(N\sigma_l)_I}{|D|^2 k'}\Bigg\{ &2k'd\left(1 + \frac{\kappa}{k'}\right)^2 + 2\frac{\kappa}{k'} \\
&+ \left[1 - \left(\frac{\kappa}{k'}\right)^2\right]\sin 2k'd - 2\frac{\kappa}{k'}\cos 2k'd\Bigg\}
\end{aligned} \tag{2.84}$$

for $E_\perp > V_1$. For $E_\perp < V_1$ we replace k' by $i\beta$ (2.78) in equations (2.80) and (2.81).

2.4.4   Reflection from more complicated surfaces

In the previous section we have treated reflection from surfaces that could be represented by a single step potential as well as a surface covered by a film which can be represented by two steps. More complex surfaces which can be represented by several steps can also be treated by the method described in the previous section. However it is possible that real surfaces are still more complex. For example a continously varying impurity density will cause both the real and imaginary parts of the potential to vary with depth into the material.

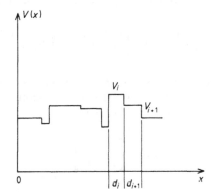

Figure 2.6   Complex surface treated as a series of layers, each with a potential V_i and thickness d_i.

We start with the problem of a series of layers each with a constant potential V_i and thickness d_i (series of steps), as shown in figure 2.6. The particles are incident from the left and the index i increases to the right. Inside the ith layer the wavefunction $\psi_i(x)$ satisfies the Schrödinger equation for a constant potential V_i with solution (2.77, 2.78). The derivative of the wavefunction is

$$\psi'(x) = ik_i \left(Ae^{ik_i x} - Be^{-ik_i x} \right) \tag{2.85}$$

$\psi(x)$ and $\psi'(x)$ can be expressed in terms of their value at the left-hand boundary of the ith layer (taken momentarily as $x = 0$)

$$\psi(x) = \psi_i(0)\cos k_i x + \frac{\psi_i'(0)}{k_i}\sin k_i x \qquad 0 < x < d_i$$

$$\psi'(x) = -k_i \psi_i(0)\sin k_i x + \psi_i'(0)\cos k_i x \tag{2.86}$$

or, expressed in matrix form, we can write

$$
\begin{bmatrix} \psi_i(x) \\ \psi_i'(x) \end{bmatrix} = \begin{bmatrix} \cos k_i x & \sin k_i x / k_i \\ -k_i \sin k_i x & \cos k_i x \end{bmatrix} \cdot \begin{bmatrix} \psi_i(0) \\ \psi_i'(0) \end{bmatrix}
$$

$$
\equiv [M_i(x)] \cdot \begin{bmatrix} \psi_i(0) \\ \psi_i'(0) \end{bmatrix} \tag{2.87}
$$

where we will evaluate ψ and ψ' at $x = d_i$, the right-hand boundary of the ith layer. Since ψ and ψ' are continous at the boundaries we have

$$
\begin{bmatrix} \psi_{i+1}(0) \\ \psi_{i+1}'(0) \end{bmatrix} = \begin{bmatrix} \psi_i(d_i) \\ \psi_i'(d_i) \end{bmatrix} = [M_i(d_i)] \cdot \begin{bmatrix} \psi_i(0) \\ \psi_i'(0) \end{bmatrix}. \tag{2.88}
$$

For $E < V_i$, k_i in (2.87) is replaced by $i\beta_i$ (2.78). For N layers we have

$$
\begin{bmatrix} \psi_{N+1}(0) \\ \psi_{N+1}'(0) \end{bmatrix} = \begin{bmatrix} T \\ ik_2 T \end{bmatrix} = M \cdot \begin{bmatrix} \psi_1(0) \\ \psi_1'(0) \end{bmatrix} = M \cdot \begin{bmatrix} 1 + R \\ ik_1(1 - R) \end{bmatrix} \tag{2.89}
$$

with k_1 and k_2 the wavevectors before and after the potential, and $M = M_N \cdots \cdots M_2 \cdot M_1$, a product of 2×2 matrices. Equations (2.89) can be solved for R and T (Yamada *et al* 1978). Writing

$$
M = \begin{bmatrix} A & B \\ C & D \end{bmatrix} \tag{2.90}
$$

we have

$$
R = \frac{(k_1 k_2 B + C) + i(k_1 D - k_2 A)}{E}
$$

$$
T = \frac{i2k_1}{E} \tag{2.91}
$$

$$
E = k_1 k_2 B - C + i(k_2 A + k_1 D).
$$

This procedure can be carried out for the case of lossy materials using complex k_i where necessary and in this way the loss probability for any given surface model can be calculated. Another method is to neglect the absorption and use (2.88) and (2.87) to calculate the unperturbed wavefunction in each layer. The loss probability can then be calculated using (2.74). Surfaces with continuously variable properties can be treated as a series of thin homogeneous layers by the same methods.

The matrices M_i (2.87) each represent a complete layer. In Appendix A4 we show an alternative method, based on the two linearly independent solutions of the Schrödinger equation where each matrix represents the boundary

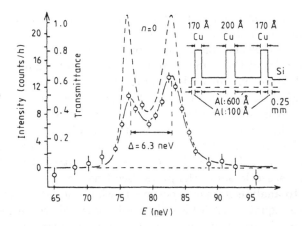

Figure 2.7 Level splitting observed in UCN transmission through a sample with the indicated coupled resonator structure (Steyerl *et al* (1981)). See also Steyerl *et al* (1988a).

between two layers. This method allows a simple derivation of the relation between R and T for incidence from left or right on a given barrier.

The problem of wave transmission by inhomogeneous films has many important practical applications (e.g. Rose 1942) and has been discussed by many authors (Born and Wolf 1970, Jacobsson 1965, Kossel 1948, Lekner 1987). Steyerl (1972a) and Steyerl *et al* (1988a) show some examples of measurements made with UCN on some complex films. Figure 2.7 (taken from Steyerl *et al* (1988a)) shows the measured transmission as a function of UCN energy for the multilayer film whose potential model is shown in the inset.

2.4.5 Reflection from rough surfaces

This is another topic with a wide range of important applications, (ranging from ocean bottoms to distant planets and from sound waves to γ-rays) and a vast literature (Beckmann and Spizzichino 1963, Davies 1954, Rice 1951).

Calculations relevant to UCN have been presented by Steyerl (1972b), Ignatovich and Luschikov (1975), Ignatovich (1973). Schmelev (1972) gives a concise review of the field.

For UCN, rough surfaces are important for two reasons.

(i) Non-specular reflection. For a perfectly smooth plane surface the reflection of UCN would be completely specular, the component of momentum parallel to the surface being conserved and the perpendicular component being reversed. If the surface is not perfectly planar but has some roughness, then the reflection will not be completely specular and there will be some non-specular reflection which will spread further and further from the direction of specular reflection as the roughness increases. This effect is important in the case of neutron guides. For neutrons of total energy $E > V_0$ (V_0 is the potential of the guide material), the possible change in direction with each wall collision will lead to some neutrons approaching the guide wall at an angle higher than the critical angle θ_c ($E \cos^2 \theta_c = V_0$) so that total reflection will not take place and the neutrons will be lost. For UCN ($E < V_0$) this effect does not occur but the spread in angles will, after a number of collisions, lead to the UCN being reversed in direction. This will lead to diffusion-like behaviour when UCN are transmitted through pipes and will lead to increased losses in the pipes. See Chapter 4.

It is easier to analyse storage experiments when the UCN gas being stored is isotropic, e.g. equation (2.70) is only valid under this condition. However UCN storage vessels are filled from a relatively small entrance in one surface so that pure specular reflection on the walls would be unlikely to produce an isotropic velocity distribution.

For this reason some surface roughness is desirable (Mampe *et al* 1989a) in storage experiments, where it is necessary to have good control over the wall losses.

(ii) Increased UCN loss probability. When UCN are reflected from a rough surface the loss probability can be changed with respect to that on a perfectly plane surface of the same composition. If the coherence length of the surface roughness is large compared with the UCN wavelength (wavy surface) this effect will simply be due to the increase in surface area relative to a flat surface. If, however, the roughness correlation length is smaller than the UCN wavelength the effect of roughness will be to smooth out the potential step thus altering the UCN loss probability.

We begin our discussion with a simple model calculation which shows the main physics involved in roughness scattering. We consider a model of a rough surface where the height of the surface above its average plane (taken as $z = 0$ is given by $z = \xi(\rho)$ where ρ is the position vector in the plane of the surface. We take $\xi(\rho)$ as being a random function with average value $\langle \xi(\rho) \rangle = 0$ and autocorrelation function

$$g(\delta) = \lim_{A \to \infty} \frac{1}{A} \int_A d^2\rho \langle \xi(\rho) \xi(\rho') \rangle = \sigma^2 e^{-\delta^2/2w^2} \qquad (2.92)$$

where $\delta = \rho - \rho'$. We will ignore for the moment the reflection caused by the plane surface and treat the scattering due to the rough surface layer as a perturbation. This will miss out some important features of the scattering. Using (2.46), for a continuous medium, and substituting in (2.19) we write the scattering amplitude as the sum of waves scattered by infinitesimal volumes d^3r

$$f(\theta) = a \int n(r) e^{iQ \cdot r} d^3r \qquad (2.93)$$

where $n(\mathbf{r})$ is the number density distribution in the scattering layer taken as a continuum and $\hbar Q = \hbar(\mathbf{k}_1 - \mathbf{k}_2)$ is the momentum transfer. This continuum assumption is valid if the wavelength of the neutrons is much greater than the distance between the atoms. In the case of our rough surface layer $d^3r = \xi(\rho)\,d^2\rho$ so

$$f(\theta) = an_0 \int \xi(\rho)\,e^{i\mathbf{Q}\cdot\rho}d^2\rho$$

and the scattering probability is

$$|f(\theta)|^2 = a^2 n_0^2 \int \int \langle \xi(\rho)\xi(\rho')\rangle\,e^{i\mathbf{Q}\cdot(\rho-\rho')}d^2\rho\,d^2\rho' \qquad (2.94)$$

where we take an average denoted by the angular brackets over the randomly rough surface. If the surface is stationary (i.e. statistically everywhere the same) this average can only be a function of $\delta = \rho - \rho'$ and is equal to the autocorrelation function (2.92) of the rough surface. Thus

$$|f(\theta)|^2 = a^2 n_0^2 \int \int g(\delta)\,e^{i\mathbf{Q}\cdot\delta}d^2\delta\,d^2\rho$$

$$= a^2 n_0^2 S\sigma^2 \int e^{-\delta^2/2w^2}e^{i\mathbf{Q}\cdot\delta}d^2\delta \qquad (2.95)$$

where S is the total area of the surface. The integral can be evaluated to give

$$|f(\theta)|^2 = 2\pi a^2 n_0^2 S\sigma^2 w^2 e^{-w^2 q^2/2} \qquad (2.96)$$

where q is the component of momentum transfer, Q, parallel to the surface. The relation (2.95) shows that the scattering probability or cross section is given by the Fourier transform of the autocorrelation of the density fluctuation of the scattering system. This result is very general (Appendix A1 and A2) and holds over a very wide range of scattering processes and energies from UCN scattering to the scattering of relativistic particles on nuclei (Van Hove 1954, 1958). As is evident from this derivation, it is valid for any wave satisfying a linear wave equation (superposition of scattered waves (2.93)) which undergoes weak scattering.

Keeping in mind the total reflection from the unperturbed flat surface (not accounted for in this calculation) we see that $q = 0$ corresponds to the direction of specular reflection and that the 'roughness scattering' (2.96) is peaked about this direction with an angular width given by $\delta\theta \sim 1/kw$, i.e. sharply peaked for $w \gg \lambda$. As w decreases the angular width of the non-specular peak increases. All these features are shown by the result of a more rigorous treatment (Steyerl 1972b, Ignatovich 1973).

Steyerl treats the rough surface as a perturbation and the flat plane as the unperturbed problem. He calculates the scattered wavefunction from

$$\psi_{\rm s}\,(r) = -[Na]\int_{V_1} G\,(r\,|r')\,\psi_0\,(r')\,{\rm d}^3 r' \tag{2.97}$$

where V_1 is the volume of the surface layer whose 'thickness' can be negative. $\psi_0\,(r)$ is the unperturbed wavefunction for the smooth plane problem and has been given in (2.48) and (2.49). We are now taking the z-axis as perpendicular to the plane surface so z should replace x in those equations. The Green function is the solution of the equation

$$\nabla^2 G\,(r\,|r') + K^{*2}\,(r)\,G\,(r\,|r') = -4\pi\delta\,(r - r') \tag{2.98}$$

where

$$K^*\,(r) = k \qquad z > 0$$
$$= k' \qquad z < 0 \qquad \text{(inside material).}$$

The Green functions are $(z' \geq 0,\ z > 0)$

$$G\,(r\,|\rho', z) = \left[e^{-ikz'\cos\theta} + R(\theta)e^{ikz'\cos\theta}\right]e^{-ik\cdot\rho'}\frac{e^{ikr}}{r} \tag{2.99}$$

and $(z' \geq 0,\ z < 0)$

$$G\,(r\,|\rho', z) = T'(\theta')e^{ikz'\cos\theta}e^{-ik'\cdot\rho'}\frac{e^{ik'r}}{r} \tag{2.100}$$

where R and T are given by (2.51)–(2.53) and

$$T'(\theta') = \frac{k'\cos\theta'}{k\cos\theta}T(\theta) \tag{2.101}$$

where θ' is the angle between k' and the z axis. For $z' < 0$ we have equivalent expressions. The scattered intensity is obtained by dividing the outgoing intensity

$$\frac{\hbar k}{m}r^2\,|\psi_{\rm s}\,(r)|^2$$

by the incident intensity $(\hbar k/m)\cos\theta$ with the result

$$I_{\rm s}(\theta, \varphi) = (2\pi)^2 \frac{[Na]^2}{\cos\theta_i}\,|T\,(\theta_i)|^2\,|T(\theta)|^2 F\,(q) \tag{2.102}$$

where $F\,(q)$ is the Fourier transform of the roughness autocorrelation function (2.92) that appeared in (2.95).

This result gives the scattering due to surface roughness from the surface of a non-lossy material and is important in those cases, previously mentioned, where non-specular reflection is significant. Steyerl has applied these results to the consideration of losses of neutrons propagating in guide tubes as have Brown *et al* (1975), see Chapter 4.

If, however, the material is lossy and the wall losses are significant, e.g. UCN storage in a closed vessel, we can ask about the effect of the surface roughness on the loss probability per bounce, μ. This problem can be approached by means of Green functions as shown previously; only it is necessary to consider second-order corrections to (2.97). This has been done by Ignatovich (1973) and Ignatovich and Luschikov (1975) who found that

$$\mu = \mu_0 \left(1 + \frac{2\sigma^2 k_c^2}{1 + 0.85 k_c w + 2 k_c^2 w^2} \right)^{1/2} \qquad (2.103)$$

where $k_c = (2mV/\hbar^2)$ is the critical wavenumber for total reflection and μ_0 is the loss rate for a perfectly flat surface. For typical parameters $\sigma \sim 50$ Å, $k_c \sim 10^{-2}$ Å^{-1} and taking the worst case $w \longrightarrow 0$ we obtain an increase in loss due to surface roughness of only 25% .

2.4.6   UCN loss due to inelastic scattering

We see from (2.68) and (2.66) that the UCN loss probability per bounce is proportional to $v_{UCN}\sigma(v_{UCN})$ which is independent of the UCN velocity v_{UCN} as discussed in section 2.4.2. This means that the extremely large cross sections for UCN do not lead to unbearable losses. In addition the fact that the UCN only penetrate a small distance into the material (this effect is contained in equation (2.68)) further limits the loss rate.

As previously mentioned both absorption and inelastic scattering contribute to UCN loss on reflection from a material surface. Inelastic scattering, i.e. scattering where the final neutron energy is different (and in the case of UCN usually greater) than the incident energy leads to a loss of UCN because after the scattering the neutrons have energies greater than the potential of the wall material and hence are lost from the system after the next wall collision.

As discussed in section 2.3.1 neutron scattering can be divided into coherent and incoherent scattering. This is the result of the fact that different nuclei in a material, because they are different elements, different isotopes or have different spins, scatter neutrons with different scattering lengths. In Appendices A1, A2 and A3 we give an introduction to the theory of neutron scattering. In Appendix A3 we derive expressions for both coherent and incoherent inelastic scattering cross sections. From equation (A3.27), the expression for the coherent inelastic cross section, we see that both the momentum and energy are conserved in this case. In Appendix A3 we discuss the momentum and energy conservation conditions in connection with figure A3.1. This figure is drawn for a case where the neutron velocity v_n is greater than the sound velocity c_s. For UCN we have $v_{UCN} \ll c_s$ and in this case figure A3.1 will look like figure 2.8. Down-scattering (emisson of a phonon) is forbidden and up-scattering can only occur for the small region of k_f where the 'free neutron' parabola crosses the band around the dispersion curve. For a crystalline material the latter is periodic with a period $2\pi/a$ (a = the lattice constant), and behaves as is shown in figure 2.8. To find the total cross section for coherent inelastic scattering we must integrate (A3.27) over solid angle, $d\Omega$, and energy transfer $d\omega$.

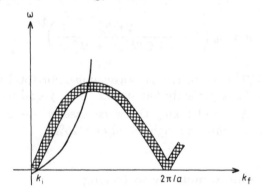

Figure 2.8   Conservation of momentum and energy for coherent scattering of a UCN.

Changing the sum over q in (A3.27) to an integral over d^3q

$$\sum_q \longrightarrow \frac{NB}{(2\pi)^3} \int d^3q \tag{2.104}$$

and carrying out this integration for phonon absorption we obtain

$$\frac{d^2\sigma}{d\Omega\, d\omega} = \frac{k_f}{k_i} a_{coh}^2 e^{-2W(Q)} \frac{\hbar}{2M} \frac{|Q\cdot\gamma(s,q)|^2}{\omega_{s,q}\left(e^{\hbar\omega_{s,q}/k_B T}-1\right)} \delta\left(\hbar\omega_s(q)+\hbar\omega\right)\big|_{(q=\tau-Q)} \tag{2.105}$$

since for UCN scattering $\omega = \hbar k_f^2/2m$ we put

$$d\omega = \frac{\hbar k_f}{m} dk_f = v_n dk_f \tag{2.106}$$

and integrate over dk_f. Then expanding $[\omega_s(q)+\omega]$ around the value of k_f for which this quantity is zero (note $\omega < 0$ for neutron energy gain, equation (A2.7)) and using $\delta(ax) = |a^{-1}|\delta(x)$ we obtain

$$\frac{d^2\sigma}{d\Omega\, d\omega} = \frac{k_f}{k_i} a_{coh}^2 \frac{e^{-2W(Q)}}{|v_n - \partial\omega(q)/\partial q|} \frac{\hbar}{2M} \sum_{s,q} \frac{|Q\cdot\gamma(s,q)|^2}{\omega_s(q)\left(e^{\hbar\omega_{s,q}/k_B T}-1\right)} \tag{2.107}$$

where the sum over s, q is for those phonons satisfying momentum and energy conservation. Since $Q = k_i - k_f$ (2.30)

$$Q^2 = k_i^2 + k_f^2 - 2k_ik_f \cos \theta_{if}$$
$$Q \, dQ = k_i k_f \sin \theta \, d\theta \tag{2.108}$$

we can take

$$d\Omega = 2\pi \sin \theta \, d\theta = \left(\frac{2\pi Q \, dQ}{k_i k_f}\right) \tag{2.109}$$

and since $k_f - k_i < Q < k_f + k_i$ (figure 2.8) and $k_i = k_{UCN} \ll k_f$ we can set $Q = k_f$ and $dQ = 2dk_i$ and $\hbar\omega = \hbar^2 k_f^2/2m = \epsilon_f$. Then

$$\sigma_{tot}^{(coh)} = 4\pi a_{coh}^2 \left(\frac{k_f}{k_i}\right) \frac{e^{-2W(k_f)}}{|v_n - \partial\omega(k_f)/\partial q|} \frac{\hbar^2}{2M\epsilon_f} \sum_{s,q} \frac{|k_f \cdot \gamma|^2}{(e^{\epsilon_f/k_BT} - 1)} \tag{2.110}$$

where we sum over the allowed phonons given by the crossings in figure 2.8. These are relatively few in number.

For incoherent inelastic scattering we obtain from (A3.36) taking $|Q \cdot \gamma|_{ave}^2 = \frac{1}{3}Q^2|\gamma|^2$ for a cubic symmetric crystal, and replacing ω by $-\omega$

$$\sigma_{tot}^{(inc)} = \frac{4\pi a_{inc}^2}{k_i} \frac{m}{M} \int e^{-2W(k_f)} \frac{|\gamma|^2 g(\omega)\sqrt{2m\omega/\hbar}}{(e^{\hbar\omega/k_BT} - 1)} \, d\omega. \tag{2.111}$$

The loss parameter for UCN is obtained by substituting (2.110) and (2.111) into (2.66).

The scattering of UCN by phonons in solids has been discussed by Blokhintsev and Plakida (1977) and Ignatovich and Luschikov (1975) taking into account the decaying exponential nature of the incident wavefunction but the results are unaffected by this.

Most materials used for UCN storage are good coherent scatterers so $a_{inc} \ll a_{coh}$ and the contribution of (2.111) is negligible. However for the case of impurities with light mass M and large incoherent scattering the contribution of (2.111) can be significant. Hydrogen satisfies both these requirements and, because of its chemical properties, it is found on most technical surfaces, e.g. as hydroxides on oxidized metal surfaces. Blokhintsev and Plakida (1977) and Ignatovich and Satarov (1977) have considered the case of strongly bound hydrogen so that the vibration spectrum $g(\omega)$ is given by

$$g(\omega) = \delta(\omega - \omega_0) \tag{2.112}$$

and (2.111) yields

$$\sigma_{tot} \propto \frac{\sqrt{\omega_0}}{(e^{\hbar\omega_0/k_BT} - 1)} \tag{2.113}$$

which yields $\sigma_{tot} \propto T/\sqrt{\omega_0}$ for $\hbar\omega_0 \ll k_B T$ and $\sigma_{tot} \propto \sqrt{\omega_0}e^{-\omega_0/T}$ for $\hbar\omega_0 \gg k_B T$ as possible temperature dependences.

Blokhintsev and Plakida (1977) also considered the case where the hydrogen has a high mobility inside the wall material and moves as an isotropic classical gas. Starting from (A2.12) we can calculate $S_{inc}(\mathbf{Q}, \omega)$ by Fourier transforming (A2.19)

$$S_{inc}(\mathbf{Q}, \omega) \propto \int dt \left\langle e^{-i\mathbf{Q} \cdot \mathbf{R}(0)} e^{i\mathbf{Q} \cdot \mathbf{R}(t)} \right\rangle e^{-i\omega t}. \qquad (2.114)$$

Now from (A2.12), remembering that for UCN $\omega = \hbar Q^2/2m$, $d\omega = \hbar Q \, dQ/m$, $k_f = Q$

$$\frac{d\sigma_{in}}{d\Omega} \sim \frac{a_{inc}^2}{k_{UCN}} \int Q^2 dQ \int dt \left\langle e^{-i\mathbf{Q} \cdot \mathbf{R}(0)} e^{i\mathbf{Q} \cdot \mathbf{R}(t)} \right\rangle e^{-i\omega t}. \qquad (2.115)$$

Treating $\mathbf{R}(t)$ as a classical variable $\mathbf{R}(t) = \mathbf{R}(0) + \mathbf{v}t$ and averaging over the velocity distribution, $P(v)\,dv$, in the gas

$$\frac{d\sigma_{inc}}{d\Omega} = \int dv \, P(v) \int Q^2 \, dQ \int dt \, e^{-i\omega t} e^{i\mathbf{Q} \cdot \mathbf{v}t}$$

$$\sim \int dv \, P(v) \int Q^2 \, dQ \, \delta(\omega - \mathbf{Q} \cdot \mathbf{v}) \qquad (2.116)$$

using $d\Omega = 2\pi \sin\theta \, d\theta = 2\pi \, dx$ $(x = \cos\theta)$

$$\sigma_{inc} \sim \int dv \, P(v) \int Q^2 \, dQ \int_{-1}^{1} dx \, \delta\left(\frac{\hbar Q^2}{2m} - Qvx\right)$$

$$\sim \int dv \, P(v) \int_{-2mv/\hbar}^{2mv/\hbar} \frac{Q \, dQ}{v} \sim \langle v \rangle \sim \sqrt{T}. \qquad (2.117)$$

For pure substances it is possible to estimate the UCN loss factor from measured curves of total cross section at low energy. The calculations are mainly important for estimating the effects of impurities and we can reproduce a range of temperature dependences with reasonable models.

2.5 QUANTUM MECHANICS OF A PARTICLE IN A BOX WITH ABSORBING WALLS

In this section we consider the problem of a particle in a box with absorbing walls and show how one can construct a wave packet which shows the classical behaviour for this case.

While the calculation is not important for UCN stored in a box with weakly absorbing walls which are described perfectly adequately by our previous considerations—particles following classical trajectories and with quantum-mechanical reflection probabilities—the calculation is interesting in its own right and has some relevance to the problem of using UCN to search for neutron–antineutron oscillations (Golub and Yoshiki 1989). See Chapter 7.

We consider a one-dimensional problem, a potential well with steps of height V_1 located at $x = \pm a$. We begin by reviewing the results for the case without absorption (Schiff 1968) and relate the eigenvalues of the particle in the box to the reflection coefficient at the potential steps. For simplicity we discuss only even-parity states, the odd-parity states can be treated in the same way. The wavefunction is given by

$$\psi_k(x) = A \cos kx \qquad |x| \le a$$
$$\psi_k(x) = Be^{-\kappa x} \qquad |x| > a \qquad (2.118)$$

and the boundary condition

$$\frac{1}{\psi} \left[\frac{d\psi}{dx} \right]_{x=\pm a} = \text{continuous} = \frac{-k \sin kx}{\cos kx} = -\kappa \qquad (2.119)$$

leads to

$$(ka) \tan ka = \kappa a \qquad (2.120)$$

with

$$k = \sqrt{\frac{2mE}{\hbar^2}} \qquad \kappa = \sqrt{\frac{2m(V - E)}{\hbar^2}}.$$

We have

$$k^2 + \kappa^2 = 2mV/\hbar^2. \qquad (2.121)$$

The simultaneous solution of equations (2.120) and (2.121) can be carried out graphically. The solutions can be represented as

$$k_n a = n\pi + \epsilon \qquad \text{with } 0 < \epsilon < \pi/2.$$

In the semi-classical treatment we calculate the reflection of an incident plane wave, the wavefunction being given by (2.48) and (2.49).

The reflection probability is, from (2.50) and (2.51)

$$\frac{1+R}{1-R} = \frac{k}{i\kappa}$$
$$R = \frac{1 - i\kappa/k}{1 + i\kappa/k} \equiv \rho(k)e^{i\varphi(k)}. \qquad (2.122)$$

Thus we can rewrite (2.119) as

$$\frac{e^{ika} + e^{-ika}}{e^{ika} - e^{-ika}} = \frac{1 + e^{-2ika}}{1 - e^{-2ika}} = \frac{k}{i\kappa}. \tag{2.123}$$

Comparing (2.122) and (2.123) we see that the eigenvalues can be determined from

$$e^{-2ik_n a} = R(k_n) = \rho(k_n) e^{i\varphi(k_n)}. \tag{2.124}$$

Since we are neglecting the absorption we have $\rho = 1$ and $\varphi(k)$ varying from $\varphi = -\pi$ at $k = 0$ to $\varphi = 0$. From (2.124) we have

$$k_n a = \frac{\varphi(k_n)}{2} + m\pi \tag{2.125}$$

which is equivalent to the previous eigenvalue condition. Using the eigenfunctions (2.118) we can construct a wave packet using (2.124)

$$\psi(x, t) = \sum_m A(k_m - k_0) \left[e^{i(k_m x - \omega t)} + e^{-i[k_m(x - 2a) + \omega t]} R(k_m) \right] \tag{2.126}$$

where $A(k_m - k_0)$ is peaked around $k_m = k_0$.

As is well known, the peak of the wave packet is located where the derivative of the phase with respect to k_m is equal to zero. Thus the first term (2.126) represents a packet centred at $x = 0$ at $t = 0$ and moving to the right with the group velocity $v_g = \partial \omega / \partial k$ and the second term located at $x = a$ at $t_a = a / v_g$ is moving to the left with velocity v_g. For times $t < t_a$ the second term is centred in the non-physical region $x > a$ and for $t > a$ the first term is in this region, so that the incident term disappears after $t = t_a$ and the reflected wave appears after this time (see e.g. Bohm 1951). The wave packet will have a spatial extension l if it contains a band of eigenstates in a region $\Delta k \approx 1/l$ around k_0. The number of eigenstates in the packet is L/l which can be quite large even for relatively macroscopic wave packets (we consider $L \approx 1$ m). Since $(\partial \rho / \partial k)\Delta k \ll \rho$ and $(\partial \varphi / \partial k)\Delta k \ll \varphi$ the behaviour of the wave packet (2.126) will be in good agreement with the semi-classical picture according to which the UCN is considered as following a classical trajectory with the phase of its wavefunction changing by $e^{i\varphi(k)} = R(k)$ at each wall collision.

Turning now to the case where the walls are absorbing we replace the potential V by a complex potential $\tilde{V} = V - iW$ (where the tilde indicates a complex quantity) and κ becomes complex ($\tilde{\kappa}$). The occupation probability of the eigenstate is no longer conserved, both the energy $\tilde{E}$ and wavenumber $\tilde{k}$ becoming complex quantities. The reflection coefficient calculated from (2.122) now has a magnitude $\rho < 1$ (2.71) and again by comparing (2.122) and (2.123) we see that we can set

$$e^{-i2\tilde{k}a} \equiv e^{-i2ka} e^{-2\gamma a} = \rho\left(\tilde{k}\right) e^{i\varphi(\tilde{k})} = R\left(\tilde{k}\right) \tag{2.127}$$

where we have set $\tilde{k} = k - \mathrm{i}\gamma$.

Now $\gamma \ll k$ so we can consider $\tilde{k}$ as being real on the right-hand side of (2.127) and thus obtain

$$-2k_M a = \varphi(k) - 2\pi M \qquad \text{or} \qquad k_M a = \pi M - \varphi(k)/2 \quad (2.128)$$

$$\rho(k) = \mathrm{e}^{-2\gamma a} \qquad\qquad\qquad \gamma = -\ln\rho/2a. \quad (2.129)$$

Hence, we have $\gamma/k = -\ln\rho/\pi M \ll 1$ for $a \approx 1$ m.

By considering the Schrödinger equation in the region $|x| \leq a$ $\left(\tilde{V} = 0\right)$ we find

$$\frac{2m}{\hbar^2}(E - \mathrm{i}\Gamma) = \left(\tilde{k}\right)^2 = k^2 - \gamma^2 - 2\mathrm{i}k\gamma$$

$$\frac{\Gamma}{\hbar} = \frac{\hbar k}{m}\gamma = \frac{v_g}{2a}(-\ln\rho) \approx \frac{v_g}{2a}(1 - \rho) \quad (2.130)$$

where $\hbar k/m = v_g$ (particle velocity). The final result [valid when $(1 - \rho) \ll 1$] is just what one expects classically, i.e. the storage time is equal to the collision time divided by the absorption probability. When $\rho \ll 1$ the classical estimate for the storage time will break down, because, if a single collision gives a large loss probability, the starting position of the particle in the box will be important.

We now construct a wave packet following the form (2.126), taking a band of k values of width Δk around k_0 and using $\tilde{k}$ as previously given (2.127)

$$\psi(x, t) = \sum_M B\left(k_M - k_0\right)\mathrm{e}^{-\Gamma t/\hbar}$$

$$\times \left(\mathrm{e}^{\mathrm{i}(k_M x - \omega t)}\mathrm{e}^{\gamma x} + \mathrm{e}^{-\mathrm{i}[k_M(x-2a)+\omega t]}\mathrm{e}^{-\gamma(x-2a)}\tilde{R}\right). \quad (2.131)$$

As in the case without absorption the first term represents a wave packet moving to the right with velocity v_g which is located at $x = 0$ at $t = 0$ and the second term represents a particle moving to the left located at $x = 2a$ at $t = 0$ (a non-physical region).

What is noteworthy is that for $x = v_g t$ (right travelling packet) or $x = 2a - v_g t$ (left travelling or reflected packet) the terms $\mathrm{e}^{-\Gamma t/\hbar}$ and $\mathrm{e}^{\gamma x}$, or $\mathrm{e}^{\gamma(2a-x)}$, respectively cancel out (using (2.130)) so that the wave packet (2.131), although constructed out of decaying eigenstates, can be written (in the neighbourhood of the peak)

$$\psi(x, t) \approx \sum_M B\left(k_M - k_0\right)\left(\mathrm{e}^{\mathrm{i}(k_M x - \omega t)} + \tilde{R}\mathrm{e}^{-\mathrm{i}[k_M(x-2a)+\omega t]}\right) \quad (2.132)$$

and keeps a constant amplitude in the free-space region in accordance with the classical picture; the amplitude of the packet only changes as a result of a wall collision. Again if $\left(\partial\tilde{R}/\partial k\Delta k \ll \tilde{R}\right)$ all components of the wave packet will have essentially the same reflection coefficient and the behaviour of the wave packet can be described by the classical picture.

3

Production of UCN

3.1 INTRODUCTION

As we discussed in section 2.1.2, neutrons are prevented from β-decaying by the forces binding them in nuclei so they are only available to us in these nuclear 'packages'. In order to do experiments with free neutrons we must 'produce' them i.e. free them from these 'packages'. This can be done by means of a nuclear reaction or fission process. In nuclear reactions a beam of fast particles produced in an accelerator is made to collide with a collection of nuclei. There are many combinations of beam and target particles which result in the release of neutrons. The (d,t) reaction (deuterium beam incident on a tritium target) produces practical quantities of neutrons at the lowest beam energy ($\leq$ 250 keV) (Beckurts and Wirtz 1964, Hanson *et al* 1949). When protons of energies $\sim$ 800–1200 MeV collide with heavy nuclei such as lead or uranium the incident particle is absorbed into the nucleus and its energy produces a general excitation of the nucleus which results in several neutrons being evaporated from each struck nucleus (Carpenter 1977, Windsor 1981). This process, spallation, is the basis of several operating neutron sources at Los Alamos (PSR), (Eckert 1983), Argonne (IPNS) (Lander 1983), KEK Japan (KENS) (Ishikawa 1983) and the Rutherford Appleton Laboratory (SNS). All these sources operate in the pulsed mode, i.e. they produce pulses of neutrons with relatively short duty cycles, and peak thermal fluxes of up to 10^{16} n cm^{-2} s^{-1} (Los Alamos).

Another process producing (freeing) neutrons is fission in which a nucleus splits into two, approximately equal, pieces with the release of a number of neutrons. Fission can occur either spontaneously or following the absorption of a particle. The most commonly used fission reaction is that of uranium 235 following the capture of a thermal neutron. This releases an average 2.5 n/fission. In a nuclear reactor one of these neutrons produces a further fission (chain reaction) and the rest eventually leave the system by diffusion or absorption. Fission reactors used for research come in a variety of sizes ranging from the TRIGA reactors, with powers from 250 kW to 1 MW to the high flux beam reactor (HFBR), power 50 MW, thermal flux 10^{15} n cm^{-2} s^{-1} (ILL 1988).

An interesting hybrid is the pulsed reactor at Dubna, USSR (Frank and Pacher 1983) which is pulsed by a neutron reflector mounted on a rotating wheel. When the reflector is located at a distance from the core, too many neutrons escape and there are not enough neutrons remaining in the core to sustain the fission chain reaction. During the short time that the reflector is close to the core, enough neutrons are reflected back into the core so that the chain reaction is sustained for the short time that the reflector is in this critical region (Windsor 1981, Frank 1979). The peak power is about 8 GW for 50 pulse/s and a pulse duration of about 100 μs. The peak thermal flux is 10^{16} n cm^{-2} s^{-1}.

Nuclear reactions and fission produce neutrons of relatively high energy ($\geq$ 2 MeV). In order to obtain neutrons with thermal or lower energies it is usual to surround the source with some material (moderator) which scatters the neutrons, gradually absorbing the neutron energy by recoil of the scattering nuclei. As this process is fundamental to the operation of nuclear reactors an enormous amount of work has been put into studying it (Amaldi 1959, Beckurts and Wirtz 1964, Egelstaff and Poole 1969). An early elementary treatment was given by Fermi (1950) (age theory).

3.2   PRODUCTION OF UCN

3.2.1   Steady state moderation

If the moderator consisted of materials whose nuclei did not absorb neutrons and was of an infinite size the neutrons would eventually come into thermal equilibrium with the moderator and the energy spectrum would be Maxwellian, corresponding to the temperature of the moderator. In any real system both absorption and boundaries are present and so the thermalization is not complete. It is often a useful approximation to consider the actual neutron spectrum to be Maxwellian, corresponding to a temperature slightly higher than the temperature of the moderator. If one wants to obtain an increase in the flux of neutrons at energies below that corresponding to room temperature it is possible to cool the moderator (Butterworth et al 1957). It is usual to use liquid hydrogen or deuterium at 20 K (Ageron et al 1969).

At thermal equilibrium the density of neutrons with velocities between v and $v + dv$ is given by

$$\rho(v)dv = \frac{2\Phi_0}{\alpha}\frac{v^2}{\alpha^2}\exp\left(-v^2/\alpha^2\right)\frac{dv}{\alpha} \tag{3.1}$$

where Φ_0 is the total thermal flux, used to characterize the strength of neutron sources, $\alpha = (2k_BT_n/m)^{1/2}$ and T_n is the neutron temperature. To determine the density of neutrons available for trapping in a cavity whose walls have a potential V we integrate (3.1) up to an energy $E = V$ and find that for $v \ll \alpha$ the density of neutrons with $E \leq V$ is given by

$$\rho_{\text{UCN}} = \frac{2}{3} \frac{\Phi_0}{\alpha} \left(\frac{V}{k_{\text{B}}T} \right)^{3/2}. \qquad (3.2)$$

We should note that equation (3.2) represents the maximum UCN gas density that can be obtained from a given source at temperature T if the UCN are in thermal equilibrium with the moderator. For $T = 300$ K, $\alpha = 2.2 \times 10^5$ cm s^{-1} and $V = 2.5 \times 10^{-7}$ eV (Be cut-off)

$$\rho_{\text{UCN}} = 10^{-13}\Phi_0 \text{ cm}^{-3}$$

where Φ_0 is in units of n cm^{-2} s^{-1}.

The largest steady thermal fluxes presently available (at Brookhaven National Laboratory and Institut Laue–Langevin) are of the order of $\Phi_0 = 10^{15}$ n cm^{-2} s^{-1}, corresponding to a maximum UCN gas density of 10^2 cm^{-3}. The use of a moderator cooled to liquid-hydrogen temperature (20 K) (Butterworth et al 1957, Webb 1963) may increase this by a factor of 20 so the maximum density available from thermal reactors is about 2×10^3 cm^{-3}. All the various schemes for extracting UCN from reactors involve some considerable degree of loss so the available densities will always be much less than this.

We are interested in the spatial density of UCN in a fixed region of momentum space

$$p^2 \leq 2mV. \qquad (3.3)$$

To increase this density would correspond to increasing the density in phase space. Now the Maxwell distribution (3.1) corresponds to a constant phase space density for momenta such that $(v/\alpha)^2 \ll 1$ and a smaller phase space density outside this region. Therefore systems which use some potential distribution to alter the energy of the neutrons, e.g. extraction of the neutrons from the reactor by a vertical tube (Steyerl 1969, 1972c), or reflection of the neutrons from moving surfaces or crystals (Antonov et al 1969a, Buras and Giebultowicz 1972, Steyerl 1975) cannot produce a higher UCN density than that which existed in the initial source. This is because the neutrons move through such systems (for which the Hamiltonian formalism is valid) in such a way as to keep the phase space density constant. This is a statement of Liouville's theorem (Tolman 1938, Goldstein 1950), which says that the volume of phase space occupied by a group of particles is constant if the particles move under the action of forces which are derivable from a potential, which holds in the case of reflections from moving matter where the potential is described by equation (2.46) (Shapiro 1972a). There is a relationship between the phase space volume occupied by a system and its

entropy. In an isolated system both quantities can only increase. In order
to decrease the phase space volume occupied by a system (and the entropy)
the original system (A) must be allowed to interact with another system (B)
so that the entropy and phase volume of (A) decrease while that of (B) as
well as the entire system (A+B) increases. This is what happens when the
neutrons are diffusing through a colder moderator. In Chapter 4 we discuss
the applications of Liouville's theorem to the case where particle number is
not conserved.

We note that a UCN gas with density distribution, given by (3.1) for $v \leq v_c$
and zero otherwise, is not in equilibrium. Left to itself in a vacuum the gas
would relax through neutron–neutron interactions to a temperature given
by

$$k_B T = \tfrac{2}{5} V \qquad (3.4)$$

although, in practice, it will never reach equilibrium because neutrons in the
high-energy tail (with $E > V$) of the Maxwell distribution corresponding to
the temperature (3.4) will escape from the container.

Needless to say, the collisional relaxation time is enormous, being given
by

$$\tau \sim 10^{21}/\rho\sigma_b \sim 10^{19} \text{ s}$$

where ρ is the UCN density in n cm^{-3} and σ_b is the neutron–neutron cross
section in barns (1 b $= 10^{-24}$ cm^2) and may be about 34 b (Haddock *et al*
1965).

The strongest interaction of the UCN gas will be with the walls of the
container. The result of this is that the gas will tend to equilibrium at the
temperature of the walls ($k_B T \gg V_W$) so that this interaction will contribute
to the loss of UCN as discussed in section 2.4.6. In any event the neutron
β-decay limits the times of interest to a few thousand seconds ($\tau_\beta \sim 10^3$ s).

We now consider some of the effects which occur in the moderator and
reduce the density of the available UCN below that given by our calculations
so far. If we wish to trap some UCN inside a closed container we have the
problem that the only neutrons which can enter the cavity by penetrating the
walls will have too large an energy to be contained by the walls ($E > V_W$).
To overcome this we must (a) fit the container with a window made of
a material whose potential energy is considerably lower than that of the
material forming the walls of the container; (b) place some material, called
a 'converter', which is capable of converting the neutrons from $E > V_W$ to
$E < V_W$ inside the container; or (c) provide a movable valve made of the
same material as the walls. In cases (a) and (b) the window or converter
surfaces provide a path for the UCN to leave as well as enter the container, as
does the valve (case (c)) when it is open. At equilibrium the number of UCN
entering the container per second will equal the number leaving per second
when the UCN density just inside the container equals that just outside the

entrance (a) or in the converter (b). In case (c) equation (3.2) gives the equilibrium density.

In order to calculate the UCN density available from a moderator which may be either the converter itself or the moderator adjacent to the window, we note that the total cross section for UCN, which is the sum of capture and inelastic scattering cross sections, can be very large since both these cross sections vary as $1/v$. For hydrogenous substances the total mean free path, λ_{UCN} is of the order of a millimetre. The only UCN which are able to emerge from the material are those which reach the UCN energy range within a distance of the order of λ_{UCN} from the surface, so we only have to consider this thin surface layer in calculating the production of UCN. However, the remainder of the moderator material may affect the energy distribution of the neutrons incident on this 'active' layer.

UCN are produced by the inelastic down-scattering of higher energy neutrons, i.e. neutrons which lose energy by creating excitations in the moderator material. The number of UCN produced in an energy range between E_{UCN} and $E_{\text{UCN}} + \mathrm{d}E_{\text{UCN}}$ per unit volume of moderator per second is found by multiplying the incident neutron flux $\varphi_0(E)$ by the macroscopic differential energy transfer cross section for down-scattering

$$\Sigma_{\text{s}}(E \longrightarrow E_{\text{UCN}}) = N \frac{\mathrm{d}\sigma_{\text{s}}(E \longrightarrow E_{\text{UCN}})}{\mathrm{d}E_{\text{UCN}}} \tag{3.5}$$

where N is the density of atoms in the moderator material and $\mathrm{d}\sigma_{\text{s}}(E \longrightarrow E_{\text{UCN}})$ is the (microscopic) differential cross section for scattering neutrons of energy E into the energy range E_{UCN} to $E_{\text{UCN}} + \mathrm{d}E_{\text{UCN}}$ (see appendices A2 and A3). (Note that Σ_{s} for a given process is the inverse mean free path for that process.) This gives a UCN source strength (UCN produced per unit volume per second) of

$$S(E_{\text{UCN}})\,\mathrm{d}E_{\text{UCN}} = \int_0^\infty \mathrm{d}E\, \Sigma_{\text{s}}(E \longrightarrow E_{\text{UCN}})\, \varphi_0(E)\,\mathrm{d}E_{\text{UCN}}. \tag{3.6}$$

We assume $\varphi_0(E)$ to be homogeneous and isotropic. The UCN produced according to (3.6) travel through the moderator until they are either captured, scattered inelastically (in which case they leave the UCN energy range) or elastically scattered. In the absence of elastic scattering we obtain the flux of UCN leaving the surface of the moderator by multiplying $S(E_{\text{UCN}})\,\mathrm{d}E_{\text{UCN}}$ by the effective thickness of the active layer previously described, $(\Sigma_{\text{in}} + \Sigma_{\text{a}})^{-1}$. Calculations have been carried out by Golub (1972) and Golikov et al (1973). Using the principle of detailed balance (A2.29) we can write $\Sigma_{\text{s}}(E \longrightarrow E_{\text{UCN}})$ in terms of $\Sigma_{\text{s}}(E_{\text{UCN}} \longrightarrow E)$. While the former is in general difficult to calculate as it involves multi-phonon processes, the latter is predominantly one-phonon up-scattering for the conditions important here and is proportional to the excitation spectrum $g(\omega)$ of the moderator (Appendix A3). Thus we can rewrite (3.6) as

$$S(E_{\text{UCN}}) = \varphi_0 f_T(E_{\text{UCN}}) \left(\frac{T}{T_0}\right)^2 \int_{E_{\text{UCN}}}^{\infty} dE \, \Sigma_s(E_{\text{UCN}} \longrightarrow E) \, e^{E/k_B T^*} \quad (3.7)$$

where we have taken the incident flux $\varphi_0(E)$ as a Maxwellian at temperature T_0 with φ_0 the total incident flux,

$$1/T^* = 1/T - 1/T_0 \quad (3.8)$$

and

$$f_T(E) = \frac{E}{(k_B T^2)} e^{-E/k_B T} \quad (3.9)$$

is the normalized Maxwell distribution at the moderator temperature T. We use the incoherent one-phonon cross section in equation (3.7).

UCN produced at the rate (3.7) per unit volume travel through the moderator making elastic collisions, until they (a) leave the moderator at a boundary, (b) are captured or inelastically up-scattered out of the UCN energy region. The elastic scattering is non-zero only because of inhomogeneities in the moderator material. Inhomogeneity scattering has been studied by Steyerl and Vonach (1972) and we will discuss it in Chapter 8. If we assume the moderator to be a plane infinite slab of thickness $2d$ the corresponding transport problem has been solved by Marshak (1947). The current emerging from the slab is given by

$$J(E_{\text{UCN}}) = \frac{S(E_{\text{UCN}})}{\Sigma_s} \left(\frac{\tanh[\nu_0(d+z_0)\Sigma_t]}{\nu_0} - 1 - F(\nu_0)\right)$$
$$\equiv \tfrac{1}{4}\varphi_0 f_T(E_{\text{UCN}}) \, \eta_m(T, T_0, E_{\text{UCN}}) \quad (3.10)$$

where the last expression defines the moderation efficiency η_m and ν_0 is the solution of

$$(1/\nu_0)\tanh^{-1}\nu_0 = \Sigma_t/\Sigma_s = \sigma \quad (3.11)$$

and $F(\nu_0)$ and z_0 are functions of ν_0 (and hence of σ) which must be calculated numerically. z_0 is always positive so that we can set the tanh in (3.10) equal to 1 for the case of a thick slab, $\nu_0 \Sigma_t d \gg 1$. The Marshak calculation is only valid for $\Sigma_s d \gg 1$ but this condition can be relaxed for strong absorption $\sigma \gg 1$ ($\Sigma_t \longrightarrow \Sigma_{\text{abs}} + \Sigma_{\text{inel}}$) in which case $F(\nu_0)$ is

$$F(\nu_0) \approx -\tfrac{1}{4}\Sigma_s/\Sigma_t \quad (3.12)$$

which only gives an error of 15% at $\sigma = 1$.

Using (3.7), (3.10) and (3.12) we have for $T = T_0$

$$J(E_{\text{UCN}}) = \tfrac{1}{4}S(E_{\text{UCN}})/\Sigma_t(E_{\text{UCN}})$$
$$= \tfrac{1}{4}\varphi_0 f_T(E_{\text{UCN}}) \left\{\frac{\Sigma_{\text{in}}(E_{\text{UCN}})}{\Sigma_{\text{abs}} + \Sigma_{\text{in}}}\right\} \quad (3.13)$$

in the strong absorption limit. The factor in brackets gives the reduction in output current due to absorption. For $T \neq T_0$ we assume a Debye spectrum for the excitations in the moderator

$$g(\epsilon) = \begin{cases} 3\epsilon^2 / (k_B \theta)^3 & \epsilon \leq \theta \\ \\ 0 & \text{otherwise} \end{cases} \tag{3.14}$$

and a hydrogenous moderator (bound atom cross section for hydrogen $=$ 81.5 b, $\sigma_{abs} = 0.33$ b for thermal neutrons) and consider the one-phonon incoherent inelastic scattering cross section (A3.36). The results are shown in figure 3.1 (Golub 1972) for a moderator temperature of 20 K. We see that, in general, there is an optimum temperature for the incident spectrum, T_0, although the peak is quite broad. As η is essentially independent of E_{UCN} the spectrum of UCN emerging from the moderator will be very nearly Maxwellian. For temperature independent $g(\epsilon)$ the UCN current will increase as the moderator temperature T is decreased until $\Sigma_{in}(E_{UCN})$ becomes less than Σ_{abs}. Calculations based on equation (3.13) have been performed for hydrogenous materials with a Debye phonon spectrum (Golub 1972), for polythene, beryllium, zirconium hydride, aluminium and magnesium using realistic phonon spectra by Golikov *et al* (1973) and for graphite using a realistic spectrum by Steyerl (1972c). Some results of the calculations of Golikov are shown in figure 3.2.

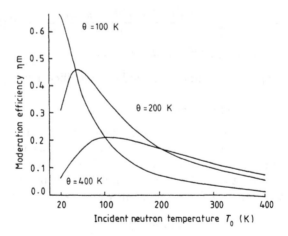

Figure 3.1   Moderation efficiency for neutrons with $\lambda = 640$ Å in a cold ($T = 20$ K) hydrogenous moderator with a Debye spectrum against incident neutron temperature (Golub 1972).

It is important to consider the effects of the potential energy in the moderator material on the emerging flux (Golikov *et al* 1971). On emerging

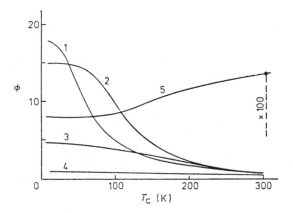

Figure 3.2 Calculated UCN flux versus converter temperature T_c, relative to a polyethylene converter at $T_c = 300$ K. The incident neutrons are assumed to have a Maxwellian spectrum at 300 K (from Golikov *et al* 1973). 1, polyethylene; 2, beryllium; 3, zirconium hydride; 4, magnesium; 5, aluminium.

from the moderator into a vacuum the UCN will gain an amount of kinetic energy equal to the magnitude of the moderator's potential energy and this increase of energy will be in a direction normal to the moderator surface. As a result of collisions with the container walls the neutron gas will have its direction of motion randomized. Equating the number of UCN entering the vacuum with the number going from the vacuum to the moderator we find the spectrum given by equation (3.1) is unchanged except for the loss of UCN with $E < V_{mod}$. Figure 3.3 shows the spectrum in the moderator (broken line) and the vacuum (full line).

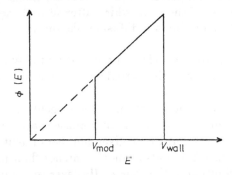

Figure 3.3 Energy spectrum of UCN flux in the moderator (broken curve) and in vacuum (full curve).

The potential energy in the medium has a further effect on the UCN density in the vacuum. The potential barrier represented by the moderator surface will cause a wave-mechanical reflection to take place for neutrons entering and leaving the moderator (see Chapter 2). The effects of this have been calculated by Golikov *et al* (1971). It is clear from figure 3.3 that one of the criteria for selecting a moderator material in the case of straightforward horizontal extraction is that $V_{mod} \ll V_{wall}$. If this can be achieved then the effects of V_{mod} will be small.

There are several methods which have been used for extracting UCN from steady sources. These are extraction through vertical and horizontal guide tubes with or without spectral transformation of the neutrons leaving the guide tube. With a vertical guide tube the neutrons are slowed down by gravity. Vertical extraction, pioneered by Steyerl (1969), has the following advantages:

(i) vertical installations seem to be able to be built with less window material in the beam than is the case with horizontal guide tubes and the windows transmit the higher energy neutrons more efficiently;

(ii) the effects of the potential barrier at the surface of the moderator are reduced; and

(iii) because the neutrons enter the guide at an energy large compared with the UCN energy (the gravitational energy loss being 10^{-7} eV m^{-1}) they enter with a smaller divergence and make fewer collisions with the guide wall than would be the case for a horizontal UCN guide.

As discussed in Chapter 2, surface roughness causes some non-specular reflection which can result in the UCN approaching the next collison with the guide at a glancing angle greater than the critical angle and hence in the loss of the neutron. In horizontal extraction it would appear that this problem would not be so serious as the UCN remain in the guide for any angle of incidence. However the result of several collisions can be a reversal of the direction of travel of the UCN, which after several more collisions can be reversed again. This produces a diffusion-like motion (Chapter 4) which results in increased losses.

It seems that the quality of surfaces produced by today's technology is so good that the losses are not such a problem for vertical extraction (Steyerl and Malik 1989).

As previously discussed, deceleration of faster neutrons into the UCN region by a potential cannot produce a higher phase space density (or change the phase space volume occupied by a given number of neutrons) so the UCN density produced by such methods cannot be higher than that in the tail of the Maxwellian distribution in the source. However the shape of the volume can be transformed into a more useful form by these methods.

Those neutrons which enter the decelerating device with a velocity v_1 in a time interval δt_1, are located in a volume of linear extent $v_1 \delta t_1$ and when they leave at a velocity v_{UCN} they will occupy a volume $v_{UCN} \delta t_{UCN}$. If the input and output cross-sectional areas are equal and, since the velocity space volumes are also equal for input and output, we have

$$v_1 \delta t_1 = v_{\text{UCN}} \delta t_{\text{UCN}} \qquad \delta t_{\text{UCN}} = \frac{v_1}{v_{\text{UCN}}} \delta t_1 \qquad (3.15)$$

that is the pulse length of the UCN is increased with respect to the pulse length of the incoming v_1 neutrons. This can be useful in the case of a pulsed source (see later). In the case of a 'neutron turbine' (Steyerl 1975) the UCN leave the device at almost 90° with respect to the incoming direction so this increase shows up as an increase in UCN source area with respect to the area of the incoming guide tube (see later).

3.2.2 Use of pulsed sources for UCN production

As mentioned in section 3.1 it is possible to obtain higher peak thermal fluxes with a pulsed source than with a steady-state source. By covering the end of the UCN gas collecting vessel with a fast acting shutter which is only open during the time the flux pulse is on, the UCN can build up to a density which corresponds to the flux in the peak of the pulse. In practice, the increase in density will be limited by the loss processes in the walls of the vessel, any spread in the length of the pulse as the neutrons undergo moderation and any excess of the window opening time over the pulsewidth (Antonov *et al* 1969b, Shapiro 1972b).

If the thermal neutron pulse lasts a time τ_{p} and the shutter of area S is open for a time $\tau_1 > \tau_{\text{p}}$, ρ is the average UCN density in the cavity, ρ_{p} is the UCN density during the pulse, T is the pulse period and μ is the loss probability per bounce on the cavity walls whose total area is A, then by equating the number of neutrons entering and leaving the cavity per cycle

$$S \rho_{\text{p}} v \tau_{\text{p}} = \rho v \left(\tau_1 S + A \mu T \right) \qquad (3.16)$$

we obtain

$$\rho = \frac{\rho_{\text{p}} \tau_{\text{p}}}{\tau_1 \left(1 + A \mu T / \tau_1 S \right)}. \qquad (3.17)$$

For $\tau_{\text{p}} = \tau_1 = T/(200)$ and $A/S = 10^2$ we obtain

$$\rho = \rho_{\text{p}}/1.2 \qquad \text{for } \mu = 10^{-5}$$

which is possible theoretically. Thus we might hope for a density

$$\rho = 10^{-13} \Phi_{\text{p}} = 10^3 \text{ n cm}^{-3}$$

using the Dubna source, with a possible further increase on cooling the moderator.

Golikov and Taran (1975) suggested the use of switched magnetic fields as a means of providing a fast-acting UCN valve. This would rely on the magnetic interaction (2.2) to provide a valve for neutrons with the proper spin direction and would provide a source of polarized UCN. Note that the use of a cold source to further cool the incident spectrum will produce an additional increase in τ_p (about a factor of 2), and that τ_p for the UCN will be increased by the kinematic factor explained at the end of the last section (in the case of deceleration of the neutrons), so that, in principle, it is possible for the UCN pulse to last as long as the time between incoming pulses and thus provide a steady-state source of UCN at a density corresponding to the peak flux of the pulsed source.

3.3 WORKING SOURCES OF UCN

The two strongest UCN sources currently available are at the Institut Laue–Langevin (ILL) in Grenoble, France, and the Leningrad Institute of Nuclear Physics (LINP) in Gatchina, USSR. Both use vertical extraction from a cold ($\sim$ 20 K) source of neutrons, and both involved rebuilding the cold source to allow installation of the vertical guides.

3.3.1   The Leningrad (LINP) ultra-cold and cold neutron source

This source (Altarev et al 1986a) consists of a 1 litre vessel capable of being cooled below 20 K and of being filled with liquid hydrogen or deuterium or mixtures of the two. The vessel is placed in the centre of the core of the reactor at Gatchina where the thermal neutron flux is given as 1.5–2×10^{14} n cm^2 s^{-1}, see figure 3.4. The UCN are separated from faster neutrons by means of the bent guide (radius 1 m) shown in the figure. The spectrum of the neutrons emerging from this guide peaks at a velocity of 18.5 m s^{-1}, large compared with usual UCN velocities so that the intensity of the UCN source depends strongly on the cut-off velocity chosen to define the UCN energy range. The authors used a thin aluminium foil with a coating having a critical velocity of 7.8 m s^{-1}, the same as the walls of the guide tube. This was made of a non-magnetic alloy of ^{58}Ni, the isotope with the highest known neutron potential (see table 1), and molybdenum (Steyerl and Malik 1989). By comparing measurements with and without the foil and correcting for neutrons above the critical velocity which are absorbed or scattered by the foil the Leningrad group concluded that the flux density for UCN with $v_x, v_y, v_z < 7.8$ m s^{-1} is 6×10^3 n cm^{-2} s^{-1} or a total current of 2.5×10^5 n s^{-1} for each of the two 40 cm^2 UCN outlets, and an expected UCN density of 16 cm^{-3}.

The source has been used for studies of the dependence of UCN yield on source temperature. Figure 3.5 shows the temperature dependence of UCN

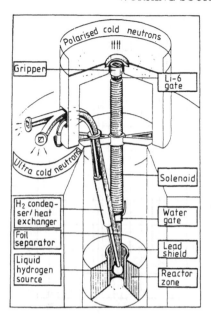

Figure 3.4 Schematic diagram of the liquid hydrogen source of polarized cold and UCN at Leningrad (Altarev *et al* 1986a).

yield as a function of temperature, the dramatic increase at 23.5 K corresponds to the source filling with liquid hydrogen; further cooling produces a continuing increase in UCN yield. The curve is normalized to the yield obtained at $T = 285$ K with gaseous hydrogen. Comparisons of H_2, D_2 and various mixtures were made. The highest gains (66) were obtained with D_2 at 17 K. At higher temperatures H_2 is more efficient. A mixture of 40% H_2 and 60% D_2 was used as the source for other experiments. The UCN source had been used for the search for the neutron electric dipole moment (EDM), (Altarev *et al* 1986b) and for studies of neutron β-decay (Kharatinov *et al* 1989)

It should be emphasized that the bold decision to place the cold source in the centre of the core leads to a very efficient use of the neutrons produced by the reactor.

3.3.2   The ILL neutron turbine UCN source

3.3.2.1   *The ILL source of UCN*
In this source the benefits of vertical extraction are further enhanced by a 'neutron turbine' which transforms incident neutrons with a velocity of 50 m s^{-1} into the UCN region. Thus throughout their transit through the vertical guide the neutrons have a relatively large velocity and are thus

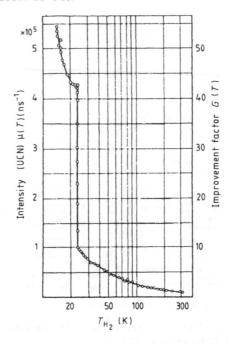

Figure 3.5   UCN yield as a function of source temperature for the Leningrad source shown in figure 3.4. The dramatic increase is due to the source filling with liquid hydrogen (Altarev *et al* 1986a).

confined to a narrow cone of solid angle making fewer wall collisions than would be the case if they slowed down to the UCN region in the guide. It is felt that the system will be less sensitive to slight deterioration in guide surface with time because of this feature.

These guides were produced by a new technology specially developed for this application (Steyerl *et al* 1986). Plane or cylindrical glass surfaces which can be produced to higher standards of micro-roughness than other materials, are coated by evaporating about 1000 Å of nickel and copper. The coating is further thickened by electroplating with nickel or copper (or cadmium to absorb the unreflected neutrons) until the total thickness is between 0.15 and 0.30 mm. The resulting metal sheet can be gently removed resulting in a metal surface which retains the smoothness of the original glass. Measurements with neutron reflection and an electron microscope show that surfaces produced in this way have better micro-roughness than surfaces produced by directly evaporating nickel onto glass. The evaporation seems to result in a slightly roughened external surface, whereas the replication technique produces a more exact replica of the original glass surface.

In the ILL source (figure 3.6) a round guide 7 cm in diameter and 5 m long was produced by this technique on the outside of a glass tube. The guide was produced in 70 cm sections which were cut along their length before

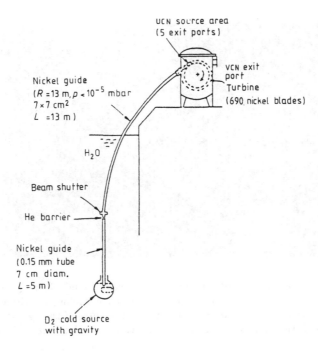

UCN source area
(5 exit ports)

Nickel guide
(R =13 m, p < 10^{-5} mbar
7 × 7 cm²
L =13 m)

VCN exit port
Turbine
(690 nickel blades)

H_2O

Beam shutter

He barrier

Nickel guide
(0.15 mm tube
7 cm diam.
L =5 m)

D_2 cold source
with gravity

Figure 3.6   (*a*) A diagram of the UCN/VCN turbine source at ILL, Grenoble. A straight and a curved guide tube conduct slow neutrons (λ > 20 Å) from the 'vertical' cold source to a Garching-type turbine where one-half of the beam (3.4 × 7 cm²) is decelerated to the UCN regime by multiple reflection on the receding nickel blades, thereby producing a 16 × 8 cm² wide UCN beam with a current density of 2.6 × 10⁴ UCN cm⁻² s⁻¹ and a total UCN current of 3.3 × 10⁶ s⁻¹ (for v_z < 6.2 m s⁻¹). The second half of the beam will bypass the turbine wheel and will be used for cold-neutron experiments (Steyerl *et al* 1986).

removal from the glass tube, and then spot welded together. This guide is mounted in an aluminium vacuum tube which has a hemispherical nose 1 mm thick dipping into the liquid D_2 filled cold source. The cold source was located in thermal flux of 4.5 × 10¹⁴ n cm⁻² s⁻¹. This material which the neutrons must traverse to leave the cold source compares favourably with the 1.5 mm wall thickness of the cold source itself which would have to be traversed (in addition to a thick vacuum vessel) in the case of horizontal extraction. Another advantage of the vertical extraction is that the neutrons rise to a height of 17 m so that a neutron with UCN energy $\sim$ 170 neV at the experimental platform would have a ten times greater energy in the cold source, i.e. a three times greater velocity, resulting in a similar decrease in loss cross section for neutrons traversing the window. The 5 m long round guide is followed by 13 m of bent square guide (radius of curvature 13 m) with a 7 cm × 7 cm cross section. The guide was bent in order to separate

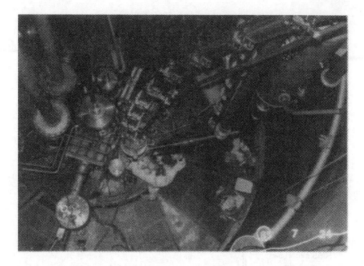

Figure 3.6   (*b*) Installation of the curved neutron guide feeding the ILL UCN turbine source. The bent pipes on the right provide cooling for the cold source. The reactor core is located under the flange at the centre of the cylindrical reactor vessel on which the man is standing. The control rod enters the core through the flange.

the very cold neutrons (VCN) from the faster neutrons and γ-rays.

3.3.2.2   The neutron turbine (Steyerl 1975)

Neutrons which leave the vertical guide system with velocities of approximately 40 m s^{-1} are brought into the UCN range by reflection from a set of curved turbine blades moving with a velocity 20 m s^{-1} in the same direction as the neutrons. Considering a neutron with velocity v striking a wall moving in the same direction with a velocity V, we see that with respect to the wall the neutron has a velocity $v - V$ before the collision and $V - v$ after the collision. Thus, after the collision, the neutron is moving at a velocity $2V - v$ in the laboratory system. This relation is shown in the vector diagram in figure 3.7(*b*), while figure 3.7(*a*) shows the geometry of the turbine blades. The turbine has a diameter of 1.7 m and consists of 690 thin nickel curved blades, manufactured by the replica technique and mounted with a spacing of 7.7 mm. The turbine rotates at 230 rev min^{-1} giving a peripheral velocity $V = 20$ m s^{-1}. During manufacture of the blades care was taken with the reverse (convex) surface because there are a small number of neutron reflections on this surface. The increase in area A_2/A_1 corresponding to equation (3.15) is also shown in figure 3.7. Each turbine blade functions as a curved guide, the neutrons making about ten reflections before leaving with their direction essentially reversed with respect to the moving blades.

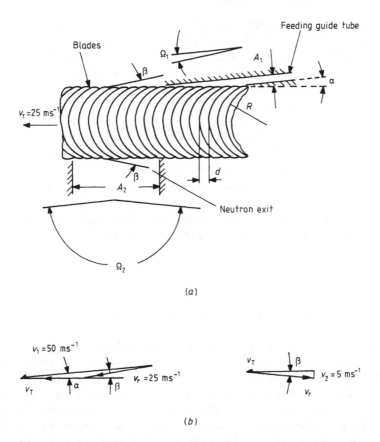

Figure 3.7 Principle of the neutron turbine. Neutrons provided by a guide tube are decelerated by several total reflections from the moving curved blades (*a*). In this process the beam cross section (A_1 and A_2) and divergence (Ω_1 and Ω_2) increase. The velocity triangles at the entrance and exit (*b*) illustrate the deceleration from an original velocity v_1 to a final velocity v_2 (v_T, blade velocity, v_r, velocity relative to the blades) (Steyerl 1975).

3.3.2.3 *UCN intensities at the ILL source*

Measurements of UCN density were carried out using a storage bottle in the form of a stainless steel tube 70 mm in diameter. The critical velocity for the bottle walls was 6.2 m s^{-1} and the storage time was about 70 s. By measuring the decay of UCN and extrapolating back to zero time the authors (Steyerl *et al* 1986) obtained 19 UCN cm^{-3} as the UCN density produced by the vertical guide without using the turbine. This increased to 41 UCN cm^{-3} when a 0.2 mm thick aluminium window separating two vacuum systems was removed. Similar measurements with the same bottle at the turbine exit gave a UCN density of 36 UCN cm^{-3}.

Time-of-flight measurements of the emerging neutron spectrum gave a value of 0.25 cm^{-3} (m s^{-1})$^{-3}$ for the phase space density leaving the vertical guide at a velocity of 50 m s^{-1} and 0.084 cm^{-3} (m s^{-1})$^{-3}$ for UCN leaving the turbine. The latter corresponds to UCN densities of 87 UCN cm^{-3} for $|v| <$ 6.2 m s^{-1} and 110 UCN cm^{-3} for $|v| < 7$ m s^{-1} so the storage experiments seemed to be about 40% efficient. UCN current densities of 2.6×10^4 cm^{-2} s^{-1} for $v_z < 6.2$ m s^{-1} and 3.3×10^4 cm^{-2} s^{-1} for $v_z < 7$ m s^{-1} were deduced from measurements. The overall extraction efficiency for UCN was estimated as about 8%. Figure 3.8 shows the spectra measured at the output of the turbine and its angular and position dependence.

Later measurements with the UCN storage bottle used in the search for a neutron electric dipole moment (see Chapter 7) gave UCN densities compatible with the time-of-flight measurements when allowing for the losses in the extra guides used in this installation (Pendlebury 1990).

3.3.3   A UCN source using horizontal extraction

As an example of horizontal extraction from a high flux reactor we describe the UCN installation at the Research Reactor Institute in Dimitrovgrad, USSR (Kosvintsev *et al* 1977). In this installation UCN are produced in a H$_2$O-cooled convertor of zirconium hydride located inside a stainless steel UCN pipe in a thermal flux $2-4 \times 10^{14}$ n cm^{-2} s^{-1} (reactor power 100 MW). The UCN pipe leading the neutrons out of the reactor had several bends (figure 3.9) in order to separate the UCN from fast neutrons and γ-rays. The design also allowed the straight through beam to be used for other work. There was a 100 μm Al foil which separated the in-pile vacuum system from the rest of the installation. The UCN pipe outside of the reactor was made of copper tubing (critical velocity 5.7 m s^{-1}) with a 100 mm diameter. The UCN current was measured to be 1200–1300 s^{-1} correcting for the 50% detector efficiency. Energy spectrum measurements using a gravitational spectrometer showed a peak at an energy considerably lower than the copper cut-off energy, i.e. at a velocity of about 4.8 m s^{-1}. Storage measurements in a 340 litre stainless steel bottle gave 5×10^3 UCN i.e. 5/340=1.5 $\times 10^{-2}$ UCN cm^{-3}. As the storage time was comparable with the filling time one can deduce the filling was about 50% efficient so the source was providing a density of about 3×10^{-2} UCN cm^{-3}. When comparing this with the ILL source (located in approximately the same thermal flux) one should consider that the cold source gain for cold neutrons at ILL is a factor of 50. The remaining factor of 20 is perhaps a measure of the advantages of vertical extraction coupled with the neutron turbine. The energy spectrum of this source is shown in section 6.3.2.

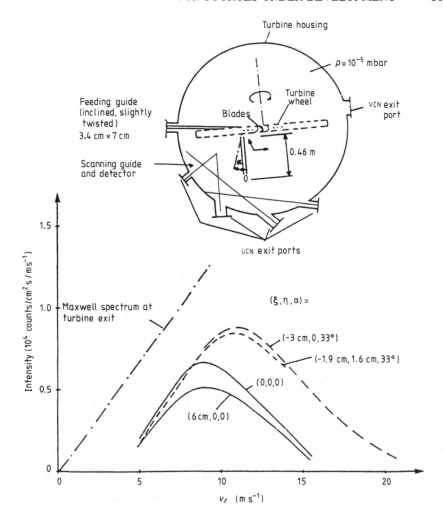

Figure 3.8 Time-of-flight data for the turbine UCN source plotted as a function of axial velocity v_z, using the set-up sketched in the inset diagram. The weak dependence of UCN output on position and angle α is typical of the turbine. The spectral cut-off depends clearly on α as expected from the turbine theory. The chain curve refers to the Maxwell spectrum at the turbine blade ends (up to a lateral velocity of 6.2 m s^{-1}) (Steyerl *et al* 1986).

3.4   UCN SOURCES UNDER DEVELOPMENT

Having described some UCN sources which are currently being used for experiments with UCN, in this and the following section we turn to a description of work directed to studying the possibilities of more intense sources in the future.

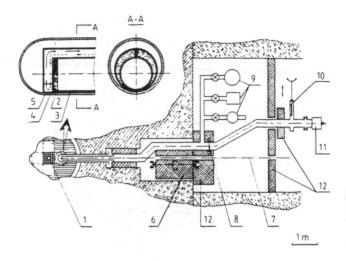

Figure 3.9   The neutron guide lay-out at Dmitrovgrad, USSR: 1, The reactor active zone; 2, the converter (zirconium hydride); 3, the UCN neutron guide; 4, water cooling system of the convertor; 5, zirconium envelope of the channel; 6, the direct beam shutter; 7, the direct beam; 8, the aluminium foil; 9, vacuum pumps; 10, the vacuum shutter; 11, the UCN detector; 12, additional protection system (Kosvintsev *et al* 1977).

3.4.1   Possible improvements to the neutron turbine

Several groups are attempting to produce a modified design for a neutron turbine so that it can accept faster neutrons at the input and/or work with a high-intensity pulsed source of cold neutrons. (a) *Radial turbine*. Working in Bulgaria, Kashoukeev *et al* (1975) have developed a neutron turbine consisting of 12 flat mirrors arranged like the fins of a paddle wheel. When the turbine rotates at 100 rad s^{-1} ($\sim 10^3$ rev min^{-1}) neutrons of 50 m s^{-1} can be slowed down to the UCN region after making about twenty collisions with the moving mirrors. As they decelerate the neutrons move radially to the centre of the wheel, leaving as UCN through the axle (figure 3.10).

(b) *Super-mirror turbine*. The group at Kyoto (Utsuro *et al* 1988) has constructed and tested a turbine whose blades are coated with super-mirrors to increase the critical velocity of the blade surface. Super-mirrors (Mezei 1976, Mezei and Dagleish 1977) are produced by coating a surface with a large number of alternating layers of two materials with different neutron potentials. If the layers are all equivalent the system acts as an artificial crystal providing a large reflection when

$$k_\perp a = \pi n$$

where $\hbar k_\perp$ is the component of neutron momentum perpendicular to the layers and a is the repetition period of the structure. Mezei showed that by

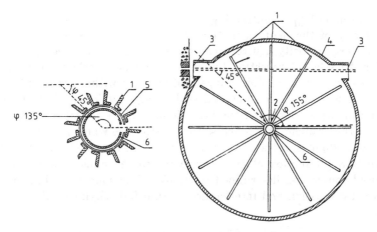

Figure 3.10 Diagram of the mechanical generator for UCN (horizontal cross section): 1, mobile neutron reflectors (rotating paddles); 2, hub; 3, Al mirror; 4, steel cover; 5, Cd protection; 6, neutron channel (fixed) (Kashoukeev *et al* 1975).

continuously varying a with depth this condition can be satisfied for a range of $k_\perp$ and one can produce a surface with significant reflection up to perpendicular velocities reaching as much as three times the usual critical velocities. In this velocity range reflectivities on the order of 80–90% are achieved. The main result of using super-mirrors as turbine blades seems to be to increase the yield at velocities slightly above the UCN range ($\gtrsim 20$ m s^{-1}). Further advantages are that the turbine can be built with only 32 turbine blades compared with the 690 blades used in the ILL turbine and that the blades are built up out of flat mirrors.

(c) *'Crystal' turbine.* In the production of UCN by reflection from a moving wall as in the neutron turbine only a narrow band of incident velocities is effective in producing UCN. Thus if one wants to work with larger incident velocities it is only necessary to achieve good reflection properties for the narrow band of incident velocities in question, and this can be achieved using Bragg reflection from a moving crystal. This idea was suggested by Buras under whose guidance a series of works investigating this question has been carried out (Steenstrup and Buras 1978 and references therein). A group at Argonne has developed this idea further to work with a pulsed spallation source. Their apparatus is shown in figure 3.11. Neutrons with velocity 400 m s^{-1} produced in a hydrogenous cold source (20 K) arrive in a pulse (width 200 μs) at the crystal which is moving with a velocity of 200 m s^{-1} away from the source. Figure 3.12 shows the principle of the technique (Dombeck *et al* 1979). Neutrons travelling with a velocity V_{nL} in the laboratory, have a velocity V_{nR} with respect to the crystal, which is moving with a velocity V_R. These neutrons are then diffracted by the crystal, G being a reciprocal lattice vector, so that their velocity in the moving frame is V'_{nR}. Subtracting the motion of the crystal we obtain the final velocity of the neutron in the laboratory frame, V'_{nL}. We see that relatively large velocity neutrons can be reflected, determined by the value of G for the selected reflection but it is necessary to select a low value of G in order to keep the rotor's tangential velocity within reasonable limits. This means we must select a crystal with a large lattice spacing. Thermica (artificial mica) is considered the best choice. There is also a complication caused by the gradient of V_R across the crystal face which favours a smaller G. Time-of-flight spectra of UCN have been measured (Brun *et al* 1980) and the UCN production efficiency, relative to the phase space density in the pulsed source was determined to be 6%. The source has been used to measure transmission and interference of thin foils (Lynn *et al* 1983). See section 6.1.

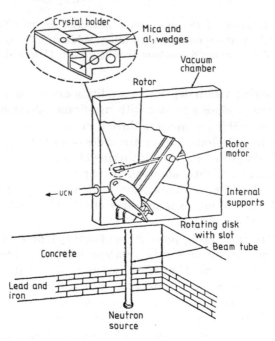

Figure 3.11   Doppler shifter UCN source using a rotating crystal as realized at Argonne (Brun *et al* 1980, Dombeck *et al* 1979).

3.5   SUPERTHERMAL SOURCES OF UCN

All the UCN production methods previously discussed have one thing in common. They are all limited by the application of Liouville's theorem to the neutron system alone. This means that any energy transformations carried out on a beam of neutrons, e.g. deceleration by gravity or by reflection

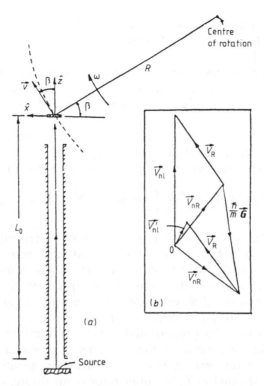

Figure 3.12   Physics of the 'crystal turbine' Doppler-shifter UCN source: $\boldsymbol{V}_R$, velocity of rotor; $\boldsymbol{V}_{nL}, \boldsymbol{V}'_{nL}$, velocities of neutron before and after reflection in laboratory; $\boldsymbol{V}_{nR}, \boldsymbol{V}'_{nR}$, the same measured with respect to the rotor. $\boldsymbol{G}$ is a vector of the crystal inverse lattice (Dombeck *et al* 1979).

from a moving surface, cannot increase the phase space density above its value in the primary source moderator. In an ideal cold source consisting of say, liquid deuterium in which the neutrons reach thermal equilibrium, the phase space density is limited according to the following conditions (Shapiro 1972a):

(i) If the thermal neutron diffusion length is small compared with the slowing down length and with the size of the moderator, the phase space density is limited to that which would be produced by cooling the incident neutrons at constant density to the temperature of the moderator.

(ii) In the opposite case the limiting phase space density is reached by cooling the incident neutrons at constant flux.

In case (i) the thermal flux is, so to say, held in the moderator as in a container so the cooling is at constant density, while in case (ii) the moderator is essentially transparent to thermal neutrons and the cooling is determined by the rate of neutrons incident on the moderator, i.e. the incident flux.

In case (i) the gain due to the cooling is $(T_0/T)^{3/2}$ while in case (ii) it is $(T_0/T)^2$ (T_0 is the temperature of incident neutron flux, T the temperature of the moderator).

However, this is not the complete story. In certain cases, where the neutrons do not reach thermal equilibrium with the moderator, it is indeed possible to achieve higher phase space densities than those predicted by the previous discussion.

We begin by considering a moderator with only two energy levels, a ground state and an excited state separated by an energy Δ from the ground state. Applying the principle of detailed balance (A2.29) to this case we find

$$\sigma\left(E_{\text{UCN}} \longrightarrow E_{\text{UCN}} + \Delta\right) = \frac{(E_{\text{UCN}} + \Delta)}{E_{\text{UCN}}} e^{-\Delta/k_B T} \sigma\left(E_{\text{UCN}} + \Delta \longrightarrow E_{\text{UCN}}\right).$$

(3.18)

In general $\sigma\left(E_{\text{UCN}} + \Delta \longrightarrow E_{\text{UCN}}\right)$ is practically independent of temperature so that for $\Delta \gg k_B T \gg E_{\text{UCN}}$ the up-scattering cross section for UCN can be made arbitrarily small by decreasing the temperature. If the moderator is now placed in a neutron flux at a temperature $T_n \geq \Delta$ there will be a significant number of down-scattering events and a negligible number of up-scattering events and if there were no further losses, the UCN density would build up to a value proportional to $(E_{\text{UCN}}/\Delta)e^{\Delta/k_B T}$ increasing exponentially as the temperature decreases rather than with $1/T^2$ as for case (ii). The situation is shown in figure 3.13. If the temperature of the incident neutrons was equal to T, the moderator temperature, the intensity of incident neutrons at the energy $E_{\text{UCN}} + \Delta$ would also decrease as $e^{-\Delta/k_B T}$ and there would be no 'superthermal' enhancement of the UCN density (see discussion following (A2.29)). Thus we see that the system is somewhat like a gas refrigerator, requiring a high-temperature input to produce significant cooling.

If the moderator is contained in a vessel whose walls are good UCN reflectors with potential $V \gg V_m$, the UCN potential of the moderator, and the walls are transparent to neutrons of energy Δ, we see that the UCN density will build up in the moderator to a density such that the rate of loss will be equal to the rate of UCN production. The loss rate will be due to up-scattering (equation (3.18)) and absorption in the moderator, β-decay, wall losses or losses through cracks. In the following we will use the macroscopic cross section Σ where $\Sigma = N\sigma$, N is the density of scatterers in the moderator and σ is the usual cross section.

The number of UCN produced per second in the volume V is

$$\begin{aligned} Q\left(E_{\text{UCN}}\right) \mathrm{d}E_{\text{UCN}} &= \int_V \mathrm{d}V \int \mathrm{d}E_0\, \Phi_0\left(E_0, \boldsymbol{r}\right) \Sigma\left(E_0 \longrightarrow E_{\text{UCN}}\right) \mathrm{d}E_{\text{UCN}} \\ &= V \int \mathrm{d}E_0\, \Phi_0\left(E_0\right) \Sigma\left(E_0 \longrightarrow E_{\text{UCN}}\right) \mathrm{d}E_{\text{UCN}} \qquad (3.19) \\ &\equiv V P\left(E_{\text{UCN}}\right) \mathrm{d}E_{\text{UCN}} \end{aligned}$$

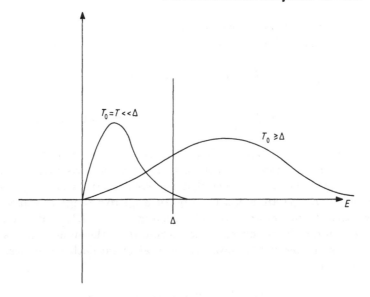

Figure 3.13 Maxwell–Boltzmann energy distributions for neutrons in equilibrium at $T_0 \ll \Delta$, and $T_0 \geq \Delta$, where Δ is the excitation energy of a two-level system. For $T_0 \ll \Delta$, the probability of the neutron losing energy by exciting the system is very small.

where the last relation assumes the incident flux $\Phi_0(E_0)$ is independent of position and we have defined $P(E_{\text{UCN}})$ as the UCN production rate per unit volume. Since the loss rate of UCN is given by $V\rho_{\text{UCN}}/\tau$ with ρ_{UCN} the UCN density and τ the mean lifetime of the UCN in the moderator we have

$$\frac{V\rho_{\text{UCN}}(E_{\text{UCN}})}{\tau} = Q(E_{\text{UCN}}) = VP(E_{\text{UCN}})$$

or

$$\rho_{\text{UCN}} = \tau P(E_{\text{UCN}}) \tag{3.20}$$

where τ is in general a function of E_{UCN}. Considering for the moment only wall losses and up-scattering (further details are given by Golub and Pendlebury (1975)) we have

$$1/\tau = 1/\tau_{\text{wall}} + 1/\tau_{\text{up}} \tag{3.21}$$

with

$$\frac{1}{\tau_{\text{wall}}} = \frac{\bar{\mu}(E_{\text{UCN}})\,v_{\text{UCN}}S}{4V} = \bar{\mu}(E_{\text{UCN}})\,\frac{v_{\text{UCN}}}{l} \tag{3.22}$$

where V is the volume and S is the surface area of the storage vessel while $\bar{\mu}(E_{\text{UCN}})$ is the average loss probability per bounce (2.70), and $l = 4V/S$ is the average distance between wall collisions. See Chapter 4 for a derivation of (3.22) and a discussion of wall losses. The up-scattering rate is

$$1/\tau_{\text{up}} = v_{\text{UCN}} \int \Sigma \left(E_{\text{UCN}} \longrightarrow E_2 \right) dE_2$$

$$= v_{\text{UCN}} \int \frac{E_2}{E_{\text{UCN}}} e^{-E_2/k_{\text{B}}T} \Sigma \left(E_2 \longrightarrow E_{\text{UCN}} \right) dE_2 \qquad (3.23)$$

using the principle of detailed balance (3.18) ($E_{\text{UCN}} \ll E_2$). Now we limit our attention to moderators with the behaviour of the two-level system previously discussed, i.e. we consider that the cross sections $\Sigma \left(E_{\text{UCN}} \longrightarrow E_2 \right)$ and $\Sigma \left(E_2 \longrightarrow E_{\text{UCN}} \right)$ are non-zero only for a narrow range of values of E_2 which we take to be centred around an energy $E_2 = E^*$; that is UCN can only scatter into or be produced by scattering from the small region around E^*. Neutrons of energy E^*, however, can scatter into other energies. With this assumption

$$1/\tau_{\text{up}} = \frac{v_{\text{UCN}} E^*}{E_{\text{UCN}}} e^{-E^*/k_{\text{B}}T} \Sigma \left(E^* \longrightarrow E_{\text{UCN}} \right) \qquad (3.24)$$

and we can neglect the up-scattering if

$$\frac{\tau_{\text{wall}}}{\tau_{\text{up}}} \ll 1 \Longrightarrow \frac{E^*}{E_{\text{UCN}}} \frac{e^{-E^*/k_{\text{B}}T}}{\bar{\mu}} \Sigma \left(E^* \longrightarrow E_{\text{UCN}} \right) l \ll 1. \qquad (3.25)$$

Now $l \Sigma \left(E^* \longrightarrow E_{\text{UCN}} \right) \lesssim l \int \Sigma \left(E^* \longrightarrow E_2 \right) dE_2$ and it does not make sense to make the latter quantity greater than one as the incident flux would begin to be attenuated inside the moderator of linear dimension l. Thus we can neglect the up-scattering if

$$\frac{E^*}{E_{\text{UCN}}} e^{-E^*/k_{\text{B}}T} \ll \bar{\mu} \qquad (3.26)$$

and in this case the steady-state UCN density will be determined by the wall loss rate and (3.20).

3.5.1 Spin-polarized matter as a two-level superthermal UCN source

Namiot (1973) suggested the use of a moderator whose nuclei (with positive magnetic moments) were polarized along a strong magnetic field. Then, remembering that the neutron magnetic moment is negative (μ directed opposite to the spin σ) we see that a neutron in the high-energy magnetic state (equation (2.3)) (σ parallel, μ anti-parallel to the magnetic field) cannot flip its spin to the anti-parallel direction by collision with one of the polarized nuclei since conservation of angular momentum would require a compensating change (into the direction parallel to the field) in the nuclear spin and the nuclei are initially all polarized parallel to the field.

The application of an AC field of frequency resonant with the neutron magnetic energy ($\hbar\omega = 2\mu_n \cdot B$) will induce transitions of the neutron from this state to the low-energy magnetic state (μ parallel, σ anti-parallel to the field). This latter state can undergo a spin-flip scattering, resulting in the increase of the interaction energies of both the neutron and the nucleus with the magnetic field. This increase in potential energy causes the kinetic energy of the neutron to decrease by an amount $\Delta = 2B \cdot (\mu_n + \mu_N)$ where μ_n is the neutron and μ_N the nuclear magnetic moments. Thus this system is a realization of the two-level system previously discussed, where the UCN upscattering is suppressed by the conservation of angular momentum and the assumed perfect polarization of the moderator, and could serve as the basis of a superthermal source of UCN. However, it requires further consideration of the ultimate cooling that can be achieved and the effects of such things as imperfect polarization of the nuclei.

From equation (2.3) we see that even in fields of 3×10^5 G (30 T) the energy change per collision is about 5 μeV (taking the nuclear magnetic moment of the deuteron) so that many collisions will be necessary to cool a neutron from 12 K (1 meV). Note that this system satisfies only weakly $\Delta \gg E_{\text{UCN}}$.

3.5.2   Coherent neutron scattering and superfluid helium as the basis of a superthermal UCN source

Returning to the discussion of coherent inelastic scattering in section 2.4.6 and Appendix A3 we see that this can provide the equivalent of a two-level system for UCN. Thus a neutron at rest can only absorb a phonon of energy $\hbar\omega$ and momentum $\hbar q$ given by the solution of

$$\omega(q) = \frac{\hbar q^2}{2m} \tag{3.27}$$

or

$$q^* = \frac{2mc}{\hbar} \tag{3.28}$$

in the case of a linear dispersion relation $\omega = cq$.

However neutrons with a momentum $\hbar q^*$ can come to rest by emission of a single phonon. This argument, applicable to neutrons at rest, can be applied to UCN with a non-zero velocity by means of figure 2.8. Thus neutrons in a narrow band around q^* can scatter to k_{UCN} and *vice versa*. By considering figure 2.8 we see that the limits of the region of k which can scatter to k_{UCN} are given by the solution of

$$c(k \pm k_{\text{UCN}}) = \frac{\hbar}{2m}(k^2 - k_{\text{UCN}}^2) \tag{3.29}$$

or

$$k_{1,2} = k^* \pm k_{\mathrm{UCN}} + k_{\mathrm{UCN}}^2/k^* \qquad (3.30)$$

for a linear dispersion relationship.

Thus the width, δk, of the region of k that can scatter to (or from) k_{UCN} is $\delta k \approx 2k_{\mathrm{UCN}}$. We have already derived this result in the discussion following (2.104).

All the neutrons in this region of k space do not scatter into the UCN region; the majority of scattering events result in neutrons outside this region. Nevertheless it is interesting to note the enormous compression in the momentum space occupied by the neutrons as a result of the scattering (Steyerl and Malik 1989). The incident neutrons occupy a volume in k space of $8\pi k^{*2} k_{\mathrm{UCN}}$ and the UCN occupy a volume of $\frac{4}{3}\pi k_{\mathrm{UCN}}^3$ a reduction by a factor $\frac{1}{6}(k_{\mathrm{UCN}}/k^*)^2$. This is only possible because the phase space occupied by the emitted phonons is as large.

Turning now to a search for possible materials we note from (3.20) that the achievable UCN density is limited by the storage time of the UCN in the vessel containing the moderator. Therefore a fundamental limit is the absorption time in the moderator material

$$\tau_{\mathrm{abs}} = [N\sigma_{\mathrm{abs}}(v)v]^{-1}. \qquad (3.31)$$

Even for a relatively weak absorber like pure deuterium ($\sigma_{\mathrm{abs}} = 4.6 \times 10^{-4}$ b for thermal neutrons) this time is only 200 ms. Since obtainable storage times are now limited by the β-decay time (see Chapter 7) it would be desirable to have a material with a smaller absorption.

Helium 4 is the most tightly bound nucleus and as a result of this can be shown to have a rigorously zero absorption cross section. The result of the tight binding is that the lowest energy state for two protons and three neutrons is a ^{4}He nucleus and a free neutron, so neutron absorption by ^{4}He is forbidden on energy grounds.

In addition the UCN potential of liquid ^{4}He is about ten times smaller than for some common wall materials (Chapter 2, table 2.1) so another condition for a superthermal source is met. Note that ^{16}O, which has an absorption cross section five times smaller than ^{2}D, has a large UCN potential.

Figure 3.14 shows the dispersion curve of liquid helium, the famous phonon–roton dispersion curve, whose form was first deduced by Landau and which was studied by Feynman and many others, as measured by neutron scattering. See Wilks (1967) for a detailed discussion of the physics of liquid helium. Also shown in the figure is the 'free neutron dispersion curve', i.e. the right-hand side of (3.27) and the intersection q^*. To simplify the calculation of the UCN production rate (Golub and Pendlebury 1977) we follow Jewell (1983), by first calculating the up-scattering rate for a neutron at rest:

$$\lim_{k_{\mathrm{u}} \longrightarrow 0} \int k_{\mathrm{u}}\sigma\left(E_{\mathrm{u}} \longrightarrow E_{\mathrm{f}}\right) \mathrm{d}E_{\mathrm{f}} = \frac{\sigma_{\mathrm{coh}}}{2k_{\mathrm{u}}} \int_{Q_1}^{Q_2} \mathrm{d}Q \int \mathrm{d}\omega \, Q S(Q,\omega) \qquad (3.32)$$

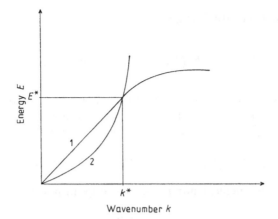

Figure 3.14   Conservation of energy and momentum in coherent scattering of UCN from phonons: 1, phonon dispersion curve; 2, free neutron energy, $E = \hbar k^2/2m$.

using (2.109) and (A2.11) with $\sigma_{\text{coh}} = 4\pi a_{\text{coh}}^2$ and $Q_{1,2} = k_{\text{f}} \pm k_{\text{u}}$. We take

$$S(Q,\omega) = S(Q)\frac{\delta\left(\omega - \omega_0(Q)\right)e^{-\hbar\omega/2T} + \delta\left(\omega + \omega_0(Q)\right)e^{\hbar\omega/2T}}{e^{-\hbar\omega_0/T} + e^{\hbar\omega_0/T}} \qquad (3.33)$$

as the simplest form satisfying detailed balance (A2.28) and the first sum rule (A2.26). Then for $\hbar\omega_0 \gg T$ and for up-scattering we have

$$S(Q,\omega) = S(Q)\delta\left(\omega - \omega_0(Q)\right)e^{-\hbar\omega_0/T}. \qquad (3.34)$$

Using

$$\omega = \frac{\hbar}{2m}k_{\text{f}}^2 \qquad \mathrm{d}\omega = \frac{\hbar}{m}k_{\text{f}}\,\mathrm{d}k_{\text{f}} = v_{\text{f}}\,\mathrm{d}k_{\text{f}} \qquad Q = k_{\text{f}}$$

where v_{f} is the velocity of a neutron with k_{f}, we evaluate the integral (3.32) by expanding the argument of the δ-function as we did in calculating (2.107). Using $\mathrm{d}Q = 2k_{\text{u}}$ we find

$$\int \sigma\left(E_{\text{u}} \longrightarrow E_{\text{f}}\right)\,\mathrm{d}E_{\text{f}} = \sigma_{\text{coh}}\left(\frac{k^*}{k_{\text{u}}}\right)\alpha S\left(k^*\right)e^{-\hbar\omega(k^*)/T} \qquad (3.35)$$

where $\alpha = |v_{\text{n}}/(v_{\text{n}} - v_{\text{g}})|$ and $v_{\text{g}} = \partial\omega\left(Q^*\right)/\partial Q$ is the phonon group velocity. Using detailed balance (A2.29)

$$\int \mathrm{d}E_{\text{f}}\,\sigma\left(E_{\text{f}} \longrightarrow E_{\text{u}}\right) = \int \mathrm{d}E_{\text{f}}\,\frac{E_{\text{u}}}{E_{\text{f}}}e^{E_{\text{f}}/T}\sigma\left(E_{\text{u}} \longrightarrow E_{\text{f}}\right)$$

$$= \sigma_{\text{coh}}\frac{k_{\text{u}}}{k^*}\alpha S\left(k^*\right). \qquad (3.36)$$

Thus the production rate P is given by (3.19)

$$P\left(E_{\text{UCN}}\right) \, dE_{\text{UCN}} = \Phi_0\left(E_0^*\right) N\sigma_{\text{coh}} \sqrt{\frac{E_u}{E^*}} \alpha S\left(k^*\right) \, dE_{\text{UCN}} \qquad (3.37)$$

a result first obtained by Pendlebury (1982). For a linear dispersion relation $\alpha = 2$ while for the measured dispersion curve of helium we find $\alpha = 1.45$. Taking $k_u = 10^{-2}$ Å^{-1}, $k^* = 0.7$ Å^{-1}, $\sigma_{\text{coh}} = 1.3$ b, $N = 2.18 \times 10^{22}$ atom/cm^3

$$P\left(E_{\text{UCN}}\right) = \Phi_0\left(E^*\right) \times 10^{-7} \text{ UCN cm}^{-3} \text{ s}^{-1} \qquad (3.38)$$

where $\Phi_0(E)$ is the incident flux per unit energy in n s^{-1} cm^{-2} K^{-1}. Using

$$\Phi_0\left(E^*\right) = \Phi_{\text{th}} \frac{E^*}{T^2} e^{-E^*/T} \qquad (3.39)$$

we have $P = \Phi_{\text{th}} \times 1.3 \times 10^{-11}$ cm^{-3} s^{-1} for a total thermal flux Φ_{th} assumed to be Maxwellian at $T = 300$ K. We expect a steady-state UCN density

$$\rho_{\text{UCN}} = 1.3 \times 10^3 \tau \qquad (3.40)$$

for $\Phi_{\text{th}} = 10^{14}$ cm^{-2} s^{-1}.

For wall loss times of several hundred seconds this represents a formidable UCN density but its achievement would require the installation of several litres of liquid helium at a temperature below 1 K inside a high-flux reactor. Preliminary cooling of the incident flux by means of a conventional cold source would provide still higher UCN densities. These densities are so large that it is interesting to think of placing such a source outside the reactor on a guide tube. This would result in the reduction of UCN density by the small solid angle transmitted by the guide (as well as guide losses). The solid angle factor is

$$\frac{\Delta\Omega_{\text{guide}}}{4\pi} = \frac{\theta_c^2}{4} = 0.7 \times 10^{-4} \qquad (3.41)$$

where θ_c, the critical angle, is taken for a nickel guide, $\theta_c = 1.7 \times 10^{-2}$ rad for $k_0^* = 0.7$ Å^{-1}. The question of which processes contribute to the upscattering , neglected in this calculation, and the related question of how cold the system must be so that up-scattering can be neglected, will be discussed in Chapter 8.

As the production rate P (3.19) is independent of time for a steady flux, the UCN density will behave according to

$$\frac{d\rho_u}{dt} = P - \rho_u/\tau \qquad (3.42)$$

with the solution

$$\rho_u = P\tau \left(1 - e^{-t/\tau}\right) \qquad (3.43)$$

for the case when $\rho = 0$ at $t = 0$. Thus the source is best adapted to operate in a cyclical mode. The UCN density will build up according to (3.43) for a given time after which the UCN will be admitted into a neighbouring apparatus. While measurements are made in this apparatus the helium source can be filled for the next cycle.

However it is also possible to operate in a steady-state mode if the exit port is kept small enough so that its effect on the overall storage time τ is not too large. We return to this point in a later section.

3.5.3 Studies of a superthermal source based on superfluid ^{4}He

In order to test the ideas of the last section some preliminary measurements were made on a small ($\frac{1}{2}$ litre) cryostat (Ageron *et al* 1978) and the production of UCN was found to be in reasonable agreement with the calculations of the last section.

A large cryostat holding 10 litres of liquid helium in a guide tube 3 m long (in order to increase the volume of helium in contact with the beam) and cooled to 0.5 K by a closed circuit ^{3}He refrigeration system was then constructed by the Oxford Instruments Company and placed in a cold neutron beam at the ILL. Figure 3.15 shows a sketch of the apparatus. UCN collected in the helium are released by opening the UCN valve B, and exit through the windows at 0.6 and 77 K. The maximum number of UCN detected (~ 1 UCN cm^{-3}) was less than expected but correcting for measured losses in the aluminium windows and UCN valve arrangement resulted in a value of 20 UCN cm^{-3} in the helium in reasonable agreement with the value of 30 UCN cm^{-3} calculated on the basis of the corrected production rate of $P \sim 0.5$ UCN cm^{-3} (Golub *et al* 1983) with the measured storage time of 60 s.

Later measurements (Kilvington *et al* 1987) were made of the UCN up-scattered by the helium as a function of temperature. The storage time is given by

$$1/\tau(T) = 1/\tau_{\text{wall}} + 1/\tau_{\text{up}}(T) \qquad (3.44)$$

where the wall loss rate is independent of temperature at such low temperatures. The rate of up-scattering of neutrons is given by

$$I_{\text{up}} = \frac{\rho_{\text{UCN}}}{\tau_{\text{up}}(T)} = \frac{P\tau}{\tau_{\text{up}}(T)}$$
$$= \left(1 - \frac{\tau(T)}{\tau_{\text{wall}}(T)}\right) P. \qquad (3.45)$$

Figure 3.16 shows the measured up-scattering rate as a function of the measured storage time $\tau(T)$. Extrapolation to $\tau(T) = 0$ gives the value of P

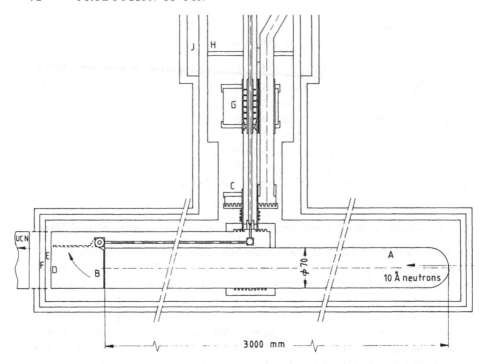

Figure 3.15 Superthermal source UCN storage cryostat: A, stainless steel UCN storage vessel; B, UCN flap valve; C, ^{3}He evaporation pot; D, 0.6 K window; E, 4 K window; F, 77 K window; G, 1 K ^{4}He pot; H, 4 K ^{4}He reservoir; J, 77 K liquid nitrogen reservoir (Golub *et al* 1983).

which, after corrections for various losses in the measurement system, was found to be 1 UCN cm^{-3} s^{-1}. The calculated value, based on more careful measurements of the incident neutron spectrum and gold foil measurements of the total incident intensity was about 2.2 times greater.

The superthermal source based on liquid helium seems ideal for installation on a small reactor where the reduced flux can be partially compensated by placing the source closer to the reactor core (Golub 1984). In any consideration of the installation of a low-temperature UCN source in a high-flux environment the distance from the source will be determined by the amount of radiation (neutron and γ-ray) heating of the low-temperature installation (Yoshiki *et al* 1987) so that the question of shielding becomes crucial (Golub and Böning 1981).

3.5.4   A thin film superthermal source of UCN

In section 3.5.2 we mentioned that absorption would limit the storage time to 1 s or less for UCN moving in those materials with the smallest known

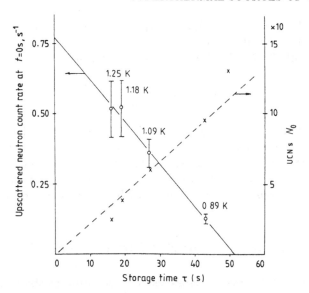

Figure 3.16 Temperature dependence of the counting rate of the up-scattered neutrons (full line, left axis) and number of UCN remaining in the vessel (broken line, right axis), both obtained by extrapolating the data to $t = 0$, plotted as a function of the measured UCN storage time $\tau(T)$ (Kilvington *et al* 1987).

absorption cross section. We now ask what would happen if the material is located in a thin film on the surface of an otherwise empty UCN storage vessel, thus reducing the time the UCN spend inside the material and reducing the average absorption rate (Golub and Böning 1983).

Considering a UCN storage vessel with area A, volume V, whose area is covered with a layer of thickness t of material, we have for the production of UCN s^{-1} (equation (3.19)) where the volume is the volume of the thin film, tA

$$Q\left(E_{\text{UCN}}\right) dE_{\text{UCN}} = tAP\left(E_{\text{UCN}}\right) dE_{\text{UCN}}. \qquad (3.46)$$

The density of UCN follows from (3.20)

$$\rho\left(E_{\text{UCN}}\right) dE_{\text{UCN}} = \frac{Q\left(E_{\text{UCN}}\right) \tau_0\left(E_{\text{UCN}}\right) dE_{\text{UCN}}}{V}. \qquad (3.47).$$

We now assume that the temperature is low enough so that the absorption in the film is greater than the up-scattering rate and the film is thick enough so that absorption in the film dominates the losses on the underlying substrate. Then

$$\tau_0\left(E_{\text{UCN}}\right) = \tau_{\text{abs}}\left(E_{\text{UCN}}\right) = \frac{4V}{v_{\text{UCN}}A\bar{\mu}} \qquad (3.48)$$

where $\bar{\mu}$ is the average loss probability per bounce

$$\bar{\mu} = 4N\sigma_{\text{abs}}\left(v_{\text{UCN}}\right)t \qquad (3.49)$$

in the case where $\bar{\mu} \ll 1$, the film absorption dominates, and we neglect the
UCN potential of the film material.

The result of (3.47)–(3.49) is that the UCN density

$$\rho\left(E_{\text{UCN}}\right) \mathrm{d}E_{\text{UCN}} = \frac{\mathcal{P}\left(E_{\text{UCN}}\right) \mathrm{d}E_{\text{UCN}}}{v_{\text{UCN}}\, \sigma_{\text{abs}}\left(v_{\text{UCN}}\right)} \tag{3.50}$$

where $\mathcal{P} = P/N$ (P given by (3.19)), is independent of the geometry of
the vessel, and the thickness and density of the film material. Increasing
the density or the thickness decreases the storage time but increases the
production rate so that the UCN density is unchanged. Thus we are able to
adjust the time constant of the source with no effect on the UCN density.
From (3.19)

$$\mathcal{P}\left(E_{\text{UCN}}\right) = \int \mathrm{d}E_0 \Phi_0\left(E_0\right) \sigma\left(E_0 \longrightarrow E_{\text{UCN}}\right) \tag{3.51}$$

and we will now calculate the cross section according to the incoherent approximation (Appendix A3) starting from (A3.36) for down-scattering

$$\left(\frac{\mathrm{d}^2\sigma}{\mathrm{d}\Omega\,\mathrm{d}\omega}\right)_{\text{inc}} = a_c^2 \frac{k_u}{k_0} e^{-\alpha Q^2} \frac{\hbar}{2M}\frac{1}{\omega} \frac{g(\omega)Q^2}{\left(1 - e^{-\hbar\omega/k_B T}\right)} \tag{3.52}$$

where we put

$$\left|\boldsymbol{Q}\cdot\boldsymbol{\gamma}\left(s,q\right)\right|^2 = \tfrac{1}{3}Q^2 \tag{3.53}$$

valid for cubic symmetry and substances containing only one type of atom,
and α follows from (A3.23) under the same assumptions.

Using the Debye model for $g(w)$

$$g(w) = 3\frac{\omega^2}{\left(k_B\Theta/\hbar\right)^3} \qquad \omega \leq \Theta$$

$$= 0 \qquad \omega > \Theta \tag{3.54}$$

and equation (2.109) and noting that $\omega \gg T$ where $g(\omega)$ is significant ($T \ll \Theta$) we have

$$\frac{\mathrm{d}\sigma\left(E_0 \longrightarrow E_u\right)}{\mathrm{d}E_u} = a^2 \frac{\hbar^2}{2M}\frac{1}{k_0^2}\frac{g(\omega)}{\omega} 2\pi \int_{k_0-k_u}^{k_0+k_u} \mathrm{d}Q\, Q^3 e^{-\alpha Q^2}$$

$$= 4\pi a^2 \frac{\hbar^2}{2M} k_u k_0 e^{-\alpha k_0^2} \frac{3}{\left(k_b\Theta\right)^3} E_0 \tag{3.55}$$

($\hbar\omega = E_0$, $E_u \ll E_0$, $k_u \ll k_0$) so that taking a Maxwell distribution for the
incident flux

$$\Phi_0\left(E_0\right) = \frac{\Phi_0 E_0}{T_n^2} e^{-E_0/T_n} \tag{3.56}$$

we have

$$
\mathcal{P} = \int \Phi_0 (E_0)\, \sigma\, (E_0 \longrightarrow E_{UCN})\, \mathrm{d}E_0
$$

$$
= \Phi_0 \frac{4\pi a^2}{T_n^2} \frac{3}{(k_B \theta)^3} \frac{m_n}{M} (E_u)^{1/2} \int_0^\theta E_0^{5/2} \mathrm{e}^{-\beta E_0}\, \mathrm{d}E_0 \qquad (3.57)
$$

where

$$
\beta = \left(\frac{\alpha 2m}{\hbar^2} + \frac{1}{T_n} \right)
$$

putting $y^2 = \beta E_0$ we transform the integral in (3.57) to

$$
\frac{2}{\beta^{7/2}} \int_0^{(\beta\theta)^{1/2}} y^6 \mathrm{e}^{-y^2} \mathrm{d}y = 2\theta^{7/2} \sum_{n=0}^{\infty} \frac{(-\beta\theta)^n}{n!(2n+7)} \qquad (3.58)
$$

where the integral is evaluated by expanding the exponential in a power series. Yu $et\ al$ (1986) have carried out similar calculations for measured $g(\omega)$ for solid hydrogen and deuterium. Figure 3.17 shows the results for the UCN density as a function of the temperature of the incident neutron Maxwellian spectrum. The results have been corrected for the effect of the deuterium potential on the production of UCN.

As with the helium source we see that if the temperature of the incoming neutrons becomes too low the UCN density goes down. In this case the neutron spectrum should overlap the excitation spectrum $g(\omega)$ for maximum gain. The cross section for up-scattering of UCN has been calculated in the same way; the results are shown in figure 3.18 for D_2 and H_2. The absorption cross section is also shown and we see that by 4 K the absorption in D_2 dominates.

We now turn to the remark we made at the end of section 3.5.2 concerning the use of a superthermal source as a continuous source of UCN. Since the number of UCN hitting a surface per cm^2 per second is $\frac{1}{4}\rho_u v_u$ the loss rate of UCN through the exit hole is

$$
\frac{1}{\tau_h} = \frac{v_u A_h}{4V} \qquad (3.59)
$$

where A_h is the area of the hole and V the volume of the storage vessel. If the storage time in the absence of the hole is τ_0, the storage time in the presence of the hole will be given by

$$
\frac{1}{\tau_{tot}} = \frac{1}{\tau_h} + \frac{1}{\tau_0} \qquad (3.60)
$$

so that if $\tau_h \geq \tau_0$ the UCN density with the hole will be greater than half of its value without the hole.

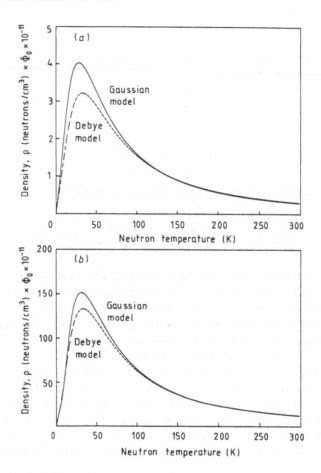

Figure 3.17 (a) The variation of the UCN density ρ as a function of neutron temperature, T_n, for a film of para-hydrogen. The full curve is for ρ obtained using a Gaussian model while the broken curve is obtained using a Debye model for the excitation spectrum of the film. ρ has a maximum for $T_n = 31$ K. (b) Same as (a) except that the film material is ortho-deuterium. Maximum ρ occurs for $T_n = 29$ K (Yu et al 1986).

For a 30 litre helium-filled UCN source with $\tau_0 = 10^2$ s this condition implies an exit area $A_h \leq 2$ cm^2, and this will be reduced as the storage time increases. For those applications of UCN scattering (e.g. to biology), where samples are only available in small sizes, this will not be a problem and the current densities will be determined by the extraordinarily high UCN densities in the helium ($j_{UCN} = \rho_u v_u / 4$).

While the thin film source produces lower UCN densities than the helium-based source, the ability of this source to operate with shorter storage times without further loss of UCN density will allow either a larger beam area or

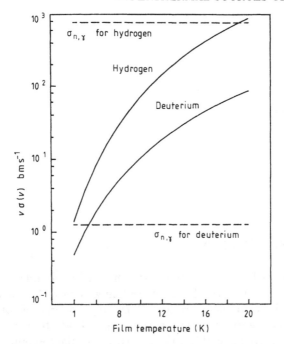

Figure 3.18   The dependence of $v_{\mathring{u}}\sigma(v_{\mathring{u}})$ i.e. the product of initial neutron speed, $v_{\mathring{u}}$, and the up-scattering cross section (phonon absorption), $\sigma(v_{\mathring{u}})$ as a function of the film temperature, T. At a given temperature, the product $v_{\mathring{u}}\sigma(v_{\mathring{u}})$ is independent of the neutron speed for UCN. The horizontal lines show the values of $v_{\mathring{u}}\sigma_{n,\gamma}(v_{\mathring{u}})$ which is independent of the film temperature. A comparison of $v_{\mathring{u}}\sigma_{n,\gamma}(v_{\mathring{u}})$ and $v_{\mathring{u}}\sigma(v_{\mathring{u}})$ shows that $\sigma(v_{\mathring{u}})$ is too small to effect the neutron storage lifetime, τ_0, for a film temperature $\lesssim 10$ K for hydrogen and $\lesssim 4$ K for deuterium (Yu *et al* 1986).

a smaller source volume or some combination. The possibility of a smaller source volume combined with the higher operating temperature of the thin film source offers significant technical simplification when we consider the possibility of installing a superthermal source within an intense neutron source.

4

UCN gas

4.1 INTRODUCTION

In Chapter 2 we have shown that UCN should be capable of making many reflections from material surfaces before being absorbed or inelastically scattered out of the UCN energy range. It is this property of total internal reflection, coupled with their low speeds, that leads to their most striking property: confinement within material traps of macroscopic dimensions for time periods approaching those of the neutron β-decay lifetime. Since UCN have wavelengths 10^{-7} times smaller than the dimensions of the guides and vessels in which they are trapped they can be considered to follow classical trajectories between collisions with the walls of any confining vessel (see section 2.5). In many ways they are very similar to the classical particles of kinetic theory and not too surprisingly many results from kinetic theory can be invoked to describe the behaviour of UCN in bottles and neutron guides. It is this similarity that leads to the idea of UCN gas. In this chapter we consider the concept of UCN gas and derive the necessary theory to describe the various features of neutron storage and guide transmission. Many of these results will be required in later chapters in which results from UCN storage measurements are presented and discussed.

4.2 THE PROPERTIES OF UCN GAS

UCN gas, although very similar to an ideal gas, has several distinct properties quite peculiar to itself. The most important and unique of these characteristics are listed here;

  (i) UCN/wall collisions are for the most part elastic and specular. Any inelastic scattering will, almost invariably, result in the neutron being heated and lost from the UCN energy range. Since such processes always occur UCN gas can never be in true statistical thermal equilibrium. However, as the relaxation rate is often comparatively slow, the UCN gas does achieve a sort of quasi-statistical equilibrium which we will call mechanical equilibrium. Mechanical equilibrium is characterized by isotropy of the neutron gas in velocity space at all points within the storage volume. The degree of specularity of the neutron reflection can be important when trying to decide whether we can truly apply those kinetic theory results which are valid only in conditions of mechanical equilibrium.

(ii) The low UCN densities obtainable outside of reactors (at most 80 n cm^{-3}) and the small size of the neutron mean that neutron–neutron collisions have a negligible probability of occurrence. Neutron–neutron energy exchange therefore does not occur and individual neutron trajectories are completely independent of each other. The UCN mean free path between wall collisions is totally characterized by the geometry of the confinement system and the details of the UCN–wall collision mechanism. It is also important to note that randomization of the velocity direction in a UCN gas results purely from non-specular reflection at the bottle walls.

(iii) Bottled UCN gas densities decay during storage. This decay is due both to the effects of wall collisions and β-decay. Since both the rate of wall collisions and the loss probability per bounce for a UCN is energy-dependent this density decay rate will also be energy-dependent for a given system.

(iv) Since UCN have very low speeds ($v < 6$ m s^{-1}) their motion is greatly affected by their gravitational interaction. A neutron rising against gravity loses 1.02 neV of kinetic energy per cm. As well as affecting UCN velocities gravity also has a significant effect on neutron density distributions within bottles.

Although the microscopic behaviour of UCN is somewhat different from that of a classical gas particle many kinetic theory and statistical mechanical results relevant to the bulk properties of typical gases are still relevant to UCN. It is important, however, to always bear in mind their different nature when deciding which results apply. In the following section we demonstrate the use of kinetic theory in describing the storage of UCN in material bottles in the presence of a gravitational field.

4.3   STORAGE OF UCN

4.3.1   Kinetic theory treatment of UCN storage

In this section we derive expressions for density distributions and decay rates as a function of time within a neutron vessel. Similar formulations have been presented by Ignatovich and Terekhov (1976), Burnett (1982) and Pendlebury and Richardson (1990). We follow the approach of the latter two authors. The results obtained are only valid when the system is in a state of mechanical equilibrium during the entire period of neutron storage. Exactly what criteria are required to ensure that a system will attain and remain in a state of mechanical equilibrium will be discussed in the next section, it suffices to say that the theory is applicable to most systems. We will derive completely general results. The results are applied in several of the following chapters. Storage experiments performed with UCN are described in Chapter 5.

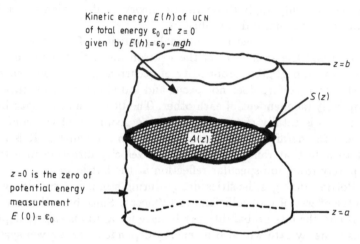

Figure 4.1   Generalized bottle geometry for neutron storage calculations.

We consider neutron storage in the vessel shown in figure 4.1. We denote the cross-sectional area at height h of the bottle by $A(h)$ and the surface area at height h by $S(h)\,dh$. The earth's gravitational field will be taken to be in the z-direction. The presence of the earth's gravitational field means that the kinetic energy of a given neutron will vary depending on its height within the bottle. We will label neutrons by their total energy at height $z = 0$. We use $z = 0$ as our datum level for the potential energy, i.e. we simply define the potential energy to be zero at this height. Our measure of total energy for a given neutron is thus by definition numerically equal to the kinetic energy of that neutron at $h = 0$. A group of UCN with total energy in the range $\epsilon_0, \epsilon_0 + d\epsilon_0$ will thus have kinetic energy $E(h)$ in the range $(\epsilon_0 - mgh), (\epsilon_0 - mgh) + d\epsilon_0$ at height $z = h$. It is important to note that a given neutron group has the same value of $d\epsilon_0$ at all h. The magnitude of the momentum $p(h)$ at height h of a neutron defined to have energy ϵ_0 is simply $p(h) = \sqrt{2mE(h)} = \sqrt{2m(\epsilon_0 - mgh)}$. An equilibrium state is defined in statistical mechanics to be the state in which the system is equally likely to be found in any one of its accessible microstates (Reif 1965). Mechanical equilibrium for an individual UCN with total energy in the infinitesimal range $\epsilon_0, \epsilon_0 + d\epsilon_0$ is thus represented by a uniform density over the phase space volume $d^3r\,d^3p$ accessible to that particular neutron. If one now considers an ensemble of such neutrons one can establish a phase space density for the total system

$$\rho = \frac{\mathrm{d}^6 N}{\mathrm{d}^3 r \, \mathrm{d}^3 p} \tag{4.1}$$

which gives the total number of neutrons to be found in a given region of the new system's phase space. Since the particles are mutually non-interacting this density is also uniform over the accessible phase space for an equilibrium state. This statement is true despite the presence of a gravitational field acting collectively on the particles and is known as Liouville's theorem (for a proof and discussion of this classical statistical mechanical result see Tolman (1938)). Liouville's theorem is usually invoked in instances where equilibrium also includes conservation of particle number. Bottled UCN gas has the additional feature that neutrons are continuously being lost from the system via wall interaction and β-decay and so we need to extend Liouville's theorem to cover the case of non-conservation of particles. We can proceed by making the assumption that scattering of the UCN at the walls is sufficiently diffuse, due to the surface roughness, and the total neutron loss rate sufficiently slow, to allow the gas always to be in a state of mechanical equilibrium (see the following section for justification). In this case we can extend Liouville's theorem to cover our decaying gas of mono-energetic UCN subject to a gravitational field. The theorem becomes:

The instantaneous phase space density $\rho(\epsilon_0, t)$ of neutrons with total energy in the range ϵ_0, $\epsilon_0 + \mathrm{d}\epsilon_0$ is at all times t uniform over the entire phase space volume accessible to those neutrons.

The phase space volume accessible to a given neutron energy group ϵ_0, $\epsilon_0 + \mathrm{d}\epsilon_0$ is determined by the real space volume of the neutron bottle and also the magnitude and range of the momentum $p(h)$ and $\mathrm{d}p(h)$ accessible to the group at each point in real space within the bottle. It is important to remember that the magnitude of $p(h)$ and $\mathrm{d}p(h)$ will be different at different points in the system due to the influence of the gravitational field on total kinetic energy. The restriction to mono-energetic neutron groups is also necessary in allowing for the energy dependence of the neutron loss rate. During storage the phase space density corresponding to a particular mono-energetic group decays exponentially in time with a decay constant which depends on the total energy ϵ_0 of the group and the geometry of the bottle. The initial value of the density is determined by the UCN source. The real space UCN density at each point in real space $n(\epsilon_0, t, h)\mathrm{d}\epsilon_0$ (the density of neutrons characterized by having total energy in the range $\epsilon_0, \epsilon_0 + \mathrm{d}\epsilon_0$ at height $z = h$ and at time t) is directly related to the phase space density relevant to the group and the available velocity space associated with the UCN momentum at that position

$$n(\epsilon_0, t, h) \, \mathrm{d}\epsilon_0 = \frac{\mathrm{d}^3 N(r, p(r))}{\mathrm{d}^3 r} = \rho(\epsilon_0, t) \, \mathrm{d}^3 p(r). \tag{4.2}$$

If we consider UCN with total momentum in the range $p(h)$, $p(h) + dp(h)$ at height $z = h$, (4.2) becomes

$$n(\epsilon_0, h, t)\, d\epsilon_0 = 4\pi \rho(\epsilon_0, t) p(h)^2 dp(h). \tag{4.3}$$

We now relate the momentum space volume $4\pi p(0)^2 dp(0)$ at height $z = 0$ accessible to a given neutron group of total energy in the range ϵ_0, $\epsilon_0 + d\epsilon_0$ to the phase space volume accessible to the same group at height h. The momentum space volume accessible to the group becomes smaller due to the slowing down of the neutrons as they rise the distance h in the gravitational field. We can relate $p(0)$ to $p(h)$ by conservation of energy

$$\frac{p(0)^2}{2m} = \frac{p(h)^2}{2m} + mgh. \tag{4.4}$$

Differentiation for fixed h gives

$$p(0)dp(0) = p(h)dp(h). \tag{4.5}$$

The momentum space volume available to the neutron at $z = h$ is given by

$$\begin{aligned}
d^3 p(h) &= 4\pi p(h)^2 dp(h) \\
&= 4\pi p(0) p(h)\, dp(0) \\
&= \frac{p(h)}{p(0)} d^3 p(0).
\end{aligned} \tag{4.6}$$

This contraction in phase space is illustrated in figure 4.2. Substituting for the momentum at the different heights and using results (4.3) and (4.6), we obtain the following relationship between the real space densities of neutrons, defined by energy ϵ_0, at the two heights and at all times

$$\begin{aligned}
n(\epsilon_0, t, h) &= \sqrt{\frac{E(h)}{E(0)}} n(\epsilon_0, t, 0) \\
&= \sqrt{\frac{(\epsilon_0 - mgh)}{\epsilon_0}} n(\epsilon_0, t, 0).
\end{aligned} \tag{4.7}$$

To find the total number of neutrons confined within the vessel within the total energy range ϵ_0, $\epsilon_0 + d\epsilon_0$ at any time, we integrate the real space density of neutrons over the real space volume of the bottle:

$$N(\epsilon_0, t)\, d\epsilon_0 = \int_a^{b*} n(\epsilon_0, t, h) A(h)\, dh\, d\epsilon_0 \tag{4.8}$$

$$= n(\epsilon_0, t, 0)\, d\epsilon_0 \int_a^{b*} A(h) \sqrt{\frac{(\epsilon_0 - mgh)}{\epsilon_0}}\, dh \tag{4.9}$$

$$= \gamma(\epsilon_0) n(\epsilon_0, t, 0)\, d\epsilon_0 \tag{4.10}$$

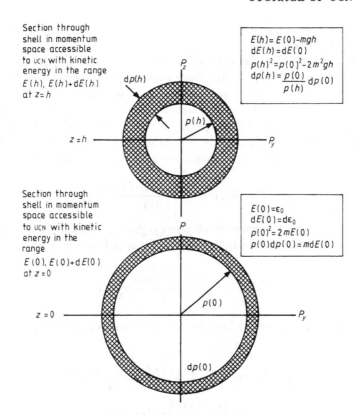

Figure 4.2 Diagram to illustrate the contraction in accessible phase space associated with the decrease in kinetic energy as UCN, within a well-defined range of total energy, rise in a gravitational field.

where

$$\gamma(\epsilon_0) = \int_a^{b*} A(h) \sqrt{\frac{(\epsilon_0 - mgh)}{\epsilon_0}} \, \mathrm{d}h \qquad (4.11)$$

and

$$b* = \begin{cases} b & \text{if } \epsilon_0/mg \geq b \\ \epsilon_0/mg & \text{if } \epsilon_0/mg < b. \end{cases} \qquad (4.12)$$

We define $\gamma(\epsilon_0)$ to be the effective real space volume of the bottle for the neutron group of energy ϵ_0. It is the true measure of the neutron holding capacity of the system and incorporates all the effects of gravity on the neutron density distribution. To obtain the total number of UCN within the bottle at any time and within a given energy range, it is necessary to integrate (4.10) over the required neutron energy range. In order to evaluate the rate at which the phase space density of a given energy group of neutrons decay in time we need to evaluate the rate at which neutrons are removed from the system via collisions on the bottle walls. This requires a knowledge of the wall collision rate $J(\epsilon_0, t, h)$ and the energy-dependent UCN loss per bounce function $\bar{\mu}(E)$. The rate of collision per unit area of wall for a gas in equilibrium is given by the standard kinetic theory result $J = \frac{1}{4}n\bar{v}$ where $\bar{v}$ and n are the local gas average particle velocity and density respectively. Since the derivation of this result relies solely on an assumption of isotropy in velocity space we are able to carry the result over to the case of a gas of UCN in mechanical equilibrium. If we include the effects of gravity, we need to rewrite the expression for the collision rate $J(\epsilon_0, h, t)$ at $z = h$ as

$$J(\epsilon_0, h, t) = \frac{n(\epsilon_0, t, h)v(\epsilon_0, h)}{4}. \tag{4.13}$$

Once again we use the concept of energy groups. Using result (4.7) and relating the UCN speed at height $z = h$ to its speed at $z = 0$ via

$$v(\epsilon_0, h) = \sqrt{\frac{(\epsilon_0 - mgh)}{\epsilon_0}} v(\epsilon_0, 0) \tag{4.14}$$

we can express the current at height $z = h$, in terms of the current at $h = 0$ by

$$J(\epsilon_0, h, t) = \frac{(\epsilon_0 - mgh)}{\epsilon_0} \frac{n(\epsilon_0, t, 0)v(\epsilon_0, 0)}{4}$$

$$= \frac{(\epsilon_0 - mgh)}{\epsilon_0} J(\epsilon_0, t, 0). \tag{4.15}$$

We can now write a loss rate equation for neutrons of energy $\epsilon_0, \epsilon_0 + d\epsilon_0$ within our vessel. The expression incorporates both the effect of wall collision and β-decay.

$$\frac{d}{dt}N(\epsilon_0, t) = -\left(\int_a^{b*} S(h)\bar{\mu}(\epsilon_0, h)J(\epsilon_0, h)\, dh\right) - \frac{N(\epsilon_0, t)}{\tau_\beta}$$

$$= -\left(J(\epsilon_0, t, 0)\int_a^{b*} \frac{(\epsilon_0 - mgh)}{\epsilon_0} S(h)\bar{\mu}(\epsilon_0, h)\, dh\right) - \frac{N(\epsilon_0, t)}{\tau_\beta}$$

$$= -N(\epsilon_0, t)\left(\frac{v(\epsilon_0, 0)}{4\gamma(\epsilon_0)}\int_a^{b*} \frac{(\epsilon_0 - mgh)}{\epsilon_0} S(h)\bar{\mu}(\epsilon_0, h)\, dh + \frac{1}{\tau_\beta}\right)$$

$$= -\frac{N(\epsilon_0, t)}{\tau_{tot}(\epsilon_0)} \tag{4.16}$$

using (4.10), where

$$\frac{1}{\tau_{tot}(\epsilon_0)} = \left(\frac{v(\epsilon_0, 0)}{4\gamma(\epsilon_0)}\int_a^{b*} \frac{(\epsilon_0 - mgh)}{\epsilon_0} S(h)\bar{\mu}(\epsilon_0, h)\, dh\right) + \frac{1}{\tau_\beta} \tag{4.17}$$

where $\bar{\mu}(\epsilon_0, h) = \bar{\mu}(E(h))$ is the averaged loss probability per bounce for the neutron of energy ϵ_0 (at $h = 0$) at height h on the particular bottle surface. If we were to assume the bottle surface to be represented by a complex step change in Fermi potential then $\bar{\mu}(E)$ would be given by equation (2.70). Any holes or gaps within the bottle could also have their loss rate contribution incorporated within the expression for surface loss. UCN admitted into the bottle with an energy greater than the Fermi potential of the bottle confining them at some point within the vessel will normally be lost within a period of a few seconds. The exact decay time required for them to disappear from the vessel depends on the area of bottle surface from which they can leak, their energy above the cut-off and finally the roughness of the bottle surface. The decay time is usually very short, except for UCN very close to the cut-off but can be calculated from the previous theory if required. The final expression for the total number of neutrons residing in the bottle in the total energy range $E_{\min} < \epsilon_0 < E_{\max}$, at time t, in terms of the initial UCN spectrum and $\tau_{tot}(\epsilon_0)$, is given by the integral

$$N(t) = \int_{E_{\min}}^{E_{\max}} N(\epsilon_0, 0) \exp(-t/\tau_{tot}(\epsilon_0)) \, d\epsilon_0 \qquad (4.18)$$

$$N(t) = \int_{E_{\min}}^{E_{\max}} \gamma(\epsilon_0) n(\epsilon_0, 0, 0) \exp(-t/\tau_{tot}(\epsilon_0)) \, d\epsilon_0. \qquad (4.19)$$

In order to use these equations we will need information on the initial starting spectra $n(\epsilon_0, 0, 0)$. The initial spectrum will depend on the characteristics of the UCN source and the source loading effect of neutron losses occurring during the filling process. For convenience we shall assume that the neutrons enter the vessel at $z = 0$. The area of the input pipe is A_{pipe} and $n_{\text{source}}(\epsilon_0, 0) \, d\epsilon_0$ the maximum real space density of UCN in the range ϵ_0, $\epsilon_0 + d\epsilon_0$ obtainable from the UCN source at the bottle input port. We expect this density to be approximately of the form

$$n(\epsilon_0, 0) \, d\epsilon_0 = \rho_0 \epsilon_0^k \, d\epsilon_0 \qquad (4.20)$$

with a k value of around $\frac{1}{2}$ ($k = \frac{1}{2}$ corresponds to a Maxwellian spectrum). Assuming (4.13) to hold in the vicinity of the input port we can write

$$\frac{d}{dt} N(\epsilon_0, t_{\text{fill}}) = \frac{n_{\text{source}}(\epsilon_0, 0) v(\epsilon_0, 0) A_{\text{pipe}}}{4} - \frac{N(\epsilon_0, t) v(\epsilon_0, 0) A_{\text{pipe}}}{4\gamma(\epsilon_0)} - \frac{N(\epsilon_0, t)}{\tau_{tot}(\epsilon_0)}. \qquad (4.21)$$

This is easily solved explicitly but if we take the case where we fill the bottle for sufficient time to fill up the entire accessible phase space to the dynamic equilibrium density ($\dot{N}(\epsilon_0, t_{\text{fill}}) = 0$) we obtain

$$n_{\text{eqm}}(\epsilon_0, \infty) = \frac{n_{\text{source}}(\epsilon_0, 0)}{[1 + (\tau_{\text{fill}}(\epsilon_0)/\tau_{tot}(\epsilon_0))]} \qquad (4.22)$$

where we define $\tau_{\text{fill}}(\epsilon_0)$ to be the fill time constant given by

$$\tau_{\text{fill}}(\epsilon_0) = \frac{4\gamma(\epsilon_0)}{v(\epsilon_0)A_{\text{pipe}}}. \tag{4.23}$$

If $\tau_{\text{tot}}(\epsilon_0) \gg \tau_{\text{fill}}(\epsilon_0)$ the equilibrium density within the bottle at $z = 0$ is essentially that of the UCN source at the source side of the input port. A similar result to this can be obtained when considering neutron loss during emptying of a neutron vessel. A rate equation for the emptying of confined UCN through the output port can be written and solved and the following expression for $D(\epsilon_0, t_{\text{empty}})$, the probability that a given neutron of energy ϵ_0 initially confined within the bottle actually escapes from the vessel (instead of decaying whilst trying), is obtained

$$D(\epsilon_0, t_{\text{empty}}) = \frac{1 - \exp\{-t_{\text{empty}}[1/\tau_{\text{tot}}(\epsilon_0) + 1/\tau_{\text{empty}}(\epsilon_0)]\}}{\tau_{\text{empty}}(\epsilon_0)[1/\tau_{\text{empty}}(\epsilon_0) + 1/\tau_{\text{tot}}(\epsilon_0)]} \tag{4.24}$$

where t_{empty} is the time that the neutron door is left open and τ_{empty} is the empty time constant. If the same port is used for both emptying and filling $\tau_{\text{empty}} = \tau_{\text{fill}}$. We now have all the general results required to describe neutron storage in material vessels. It can be seen that the expressions are complicated and require computer calculation.

In many practical instances, we are not that interested in such a precise description of a system's losses and the theory can be simplified by neglecting the effect of gravity over the bottle volume, and treating the spectral effects by working in terms of ensemble average values of $\bar{\mu}(\epsilon_0)$ and $v(\epsilon_0)$. The corresponding equation describing the decay of the neutron population then reduces to a simple single exponential form

$$N(t) = N(0)\exp(t/\bar{\tau}) \tag{4.25}$$

where

$$\frac{1}{\bar{\tau}} \approx \frac{A}{4V}\bar{v}\bar{\mu}_{\text{eff}} + \frac{1}{\tau_\beta} \tag{4.26}$$

with

$$N(0) = \int_0^{V_s} N(\epsilon_0, 0)\,d\epsilon_0 \tag{4.27}$$

A and V are the total surface area and volume of the bottle respectively, $\bar{v}^2 = 2m\bar{\epsilon}_0$ and $\bar{\mu}_{\text{eff}}$ is the effective average loss per bounce. $4V/A$ is immediately identifiable as the neutron mean free path λ within the bottle in the absence of gravity. (4.26) is sometimes rewritten in the following form

$$\frac{1}{\bar{\tau}} = \frac{\bar{v}}{\lambda\bar{K}} + \frac{1}{\tau_\beta} \tag{4.28}$$

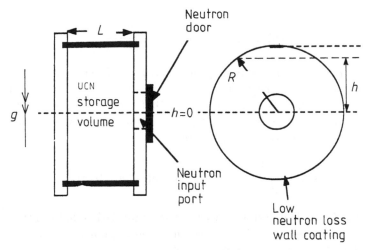

Figure 4.3   A cylindrical storage vessel.

where $\bar{K} = 1/\bar{\mu}_{\text{eff}}$ is the mean number of collisions before UCN loss due to a wall interaction for the given system.

As an example of the application of the more exact theory (up to equation (4.24)), we consider the case of neutron storage in a cylindrical vessel of radius R and length L oriented such that its symmetry axis lies orthogonal to the direction of the earth's gravitational field. Our datum level for potential energy measurement is $h = 0$ (see figure 4.3). Let us once again consider monochromatic neutrons with total energy in the range $\epsilon_0, \epsilon_0 + d\epsilon_0$. Using (4.10) and (4.11) we can relate the total number of neutrons within this range, confined within the bottle at time t, to the density at $h = 0$ by the relationship

$$N(\epsilon_0, t)\,d\epsilon_0 = \gamma(\epsilon_0)n(\epsilon_0, t)\,d\epsilon_0 \tag{4.29}$$

where

$$\gamma(\epsilon_0) = 2l \int_{-r}^{b} \sqrt{(1 - mgh/\epsilon_0)(r^2 - h^2)}\,dh \tag{4.30}$$

with

$$b = \begin{cases} +r & \text{if } \epsilon_0/mg \geq r \\ \epsilon_0/mg & \text{if } \epsilon_0/mg < r. \end{cases}$$

The decay of the neutron population during storage is given by (4.16) with

$$\frac{1}{\tau_{\text{tot}}(\epsilon_0)} = \frac{1}{\tau_{\text{walls}}(\epsilon_0)} + \frac{1}{\tau_\beta} \tag{4.31}$$

where

$$\frac{1}{\tau_{\text{walls}}(\epsilon_0)} = \frac{1}{4}\frac{v(\epsilon_0)}{\gamma(\epsilon_0)}\int_{-r}^{b}\frac{(\epsilon_0 - mgh)}{\epsilon_0}\bar{\mu}(\epsilon_0, h)S(h)\,dh \tag{4.32}$$

and

$$S(h) = 4\sqrt{(r^2 - h^2)} + \frac{2lr}{\sqrt{(r^2 - h^2)}}. \qquad (4.33)$$

The number of neutrons remaining in the vessel (of Fermi potential V_F) at time t can be calculated by integrating (4.19) over the range of confinable neutron energies. In doing this we assume that the only neutrons we need to consider are those whose kinetic energy is less than the mean Fermi potential at every point in the container yielding

$$N(t) = \int_0^{V_F - mgr} N(\epsilon_0, 0) \exp(-t/\tau_{\text{tot}}(\epsilon_0)) \, d\epsilon_0 \qquad (4.34)$$

where $N(\epsilon_0, 0) \, d\epsilon_0$ is determined by the spectrum derivable from the source. If gravity and the spectral shape of the neutron population are ignored then the population decay can be modelled by

$$N(t) \approx N_0 \exp(-t/\bar{\tau}) \qquad (4.35)$$

where

$$\frac{1}{\bar{\tau}} = \frac{(R + L)}{2LR} \bar{v} \bar{\mu}_{\text{eff}} + \frac{1}{\tau_\beta}. \qquad (4.36)$$

It has been pointed out by Pendlebury (1984) that (4.35) is a good approximation to (4.31) for neutrons with $\epsilon_0 > mgr$ since $mghS(h)$ is an odd function of h and integrates to zero over the height of the bottle. This is true for neutrons with sufficient energy to hit the roof of any vessel with a symmetry axis perpendicular to the earth's gravitational field and is an important feature of the liquid walled bottle neutron lifetime measurement described in Chapter 7. The case of UCN trapped on a plane with the effective bottle roof defined by gravity is considered in section 6.3.3.

4.3.2   Validity of the kinetic theory approach

As stated earlier the validity of the kinetic theory approach is limited to the case in which the UCN distribution can be considered to be in a state of mechanical equilibrium throughout the neutron storage volume and throughout the storage period. The ability of a system to meet this criteria is determined primarily by the relative probabilities of diffuse reflection to loss on collision with the bottle surfaces, although the time taken by the system to attain an initial state of mechanical equilibrium is also an important factor. Ignatovich (1975b) has considered storage of UCN in spherical and cylindrical walled bottles, in which the scattering on the walls was assumed to be predominantly specular but to contain a small component of diffuse reflection. His results show that if the probability of diffuse reflection is very much greater than the probability of loss, as it always is for neutron bottle surfaces, this theory remains valid; only when one considers extreme cases of specularly reflecting surfaces do significant deviations occur. Richardson (1985) has written a Monte Carlo program to examine the rate of attainment of mechanical equilibrium in neutron gases, confined within material bottles filled from a UCN source and subject to a variety of scattering laws at the bottle walls. The results show that, for a rough bottle surface, the characteristic time taken to attain equilibrium will be of the order of 0.5 s. Polished (micro-rough) surfaces, in particular those to which the scattering theory of Steyerl (1972b) (see section 2.4.5) can be applied, result in settle times of the order of 3 s. In both cases this time is very much shorter than typical neutron bottle lifetimes and so providing we do not require the theory to describe neutron losses over such a small time scale the adoption of this theory is valid. For the case of mirror-like, or liquid bottle surfaces, the settle times can become much longer due to the specularity of the reflection and some care has to be taken in deciding exactly when application of the theory becomes valid. Although the effects of non-specular reflection on the measured neutron decay times have been noted for short storage times by Mampe et al (1989b) during storage experiments on liquid surface bottles (see section 7.2.3), the theory derived in the preceding sections is applicable to essentially all cases of practical interest.

4.4   TRANSPORTATION OF UCN

4.4.1   Introduction

The pioneering work showing the advantages to be gained by the use of neutron guides was that of Maier-Leibnitz and Springer (1963). The possibility of collimating and guiding neutrons away from a reactor through the required biological shielding to experimental apparatus has been central to the expansion of the field of neutron physics. The ability to guide neutrons relies upon the fact that neutrons with a normal velocity component less than that defined by the mean Fermi potential of the surface on which they are incident, are specularly reflected away from the surface with high probability. Subsequently neutrons of any energy can be guided away from a reactor if the angular divergence of the beam is such that the neutron's transverse velocity component relative to the guide axis, is less the guide's cut-off velocity. An effective way of expressing this is to state that the guide places restrictions in phase space on the transverse velocities it can pass. In this respect neutron transmission in guides is almost identical to the transportation of light down a highly multi-mode optical fibre. Loss of neutrons in the guides is highly sensitive to any process that can alter the angle a neutron makes with the guiding surface normal between successive wall collisions, be the effect due to non-specularity of the neutron reflection process due to surface roughness (Steyerl 1972b, see also section 2.4.5) or physical curvature of the neutron guide. In fact giving neutron guides curvature enables one to remove faster moving VCN or UCN from the input beam and has been put to good use at several UCN/VCN installations (Steyerl 1977, Steyerl *et al* 1986). The effects of gravity can also be used to select and transform neutron velocities (Steyerl 1972c, 1977). When working with UCN the neutron's total kinetic energy is less than the guide material's Fermi potential, and the UCN are confined to the space within the guide, no matter what the direction of their velocity. UCN are only lost from the guide by absorption or up-scattering on the guide walls, or less likely by β-decay, and behave in a gas-like manner. UCN transmission in guides is thus quite different from the general case of neutron guides. In the following sections we concern ourselves solely with the case of UCN transmission in guides. We first give some relevant general results concerning rarified gas flow, before presenting the various approaches that have been adopted to describe UCN flow in guides. We close the chapter by reviewing the experimental situation.

4.4.2   UCN and the theory of rarified gas flow

The flow of UCN in guides has many features in common with the problem of rarified, or molecular gas flow in tubes, a subject reviewed by Steckelmacher (1966). The principal differences between the two processes are:

(i) that UCN undergo specular reflection with a much larger probability than gas molecules;

(ii) individual UCN reflections are predominantly elastic; and

(iii) UCN have the additional possibility of being absorbed or scattered out of the system.

It should also be pointed out that UCN have relatively low speeds and subsequently gravity can play a large role in neutron flow within vertical guides (Steyerl 1972c).

The ultimate objective in developing a theory to describe UCN guide transmission is to arrive at results that enable us to predict and optimize the transmission probabilities and corresponding velocity distributions associated with particular guide configurations. We begin by mentioning a few relevant results from molecular flow theory which relate to the case of gas flow through long pipes, of length Z and circular cross section A, in which the surface scattering is completely diffuse and described by the cosine law, i.e. at each reflection on the guide walls the probability that a particle leaves the surface at an angle θ into the solid angle $d\Omega$, is given by $P(\theta)\,d\Omega = \cos\theta\,d\Omega$.

If the entrance to the guide is illuminated by an isotropic angular distribution of velocities, having a density $n(0)$, the flow into the guide is obtained from (4.13)

$$J_+(0) = \tfrac{1}{4}n(0)\bar{v}A \qquad (4.37)$$

where A is the area of the input aperture. The aperture conductance U_0 or flow rate into the aperture in terms of volume units per second is given by

$$U_0 = \tfrac{1}{4}\bar{v}A. \tag{4.38}$$

For a long circular tube $Z/R \gg 1$ the transmission probability $J(Z)/J(0)$ can be shown to be given by

$$W = \frac{8R}{3Z} \tag{4.39}$$

which is known as Knusden's long tube formula (Knusden 1909, Kennard 1938, De Marcus 1961). The corresponding tube conductance is given by

$$U_1 = \frac{8RU_0}{3Z}. \tag{4.40}$$

For short tubes Knusden's law breaks down as can be seen by examination of (4.39) in the limit $R/Z > 1$ where $W \to \infty$. A correction to (4.39) was suggested by Dushman (1922) who suggested that the tube should be considered as having a flow impedance which was the sum of an element due to its length and a part due to an aperture with an area corresponding to that of the guide. The conductance of the tube is then given by

$$\frac{1}{U_t} = \frac{1}{U_0} + \frac{1}{U_1} \tag{4.41}$$

and the corresponding transmission probability by

$$W = \left(1 + \frac{3Z}{8R}\right)^{-1}. \tag{4.42}$$

The justification for this approach has been discussed in more detail by Clausing (1932), De Marcus (1961) and Berman (1965).

These results are of limited use when considering the flow of UCN as their derivation is based on an assumption of completely diffuse wall reflection. However the model has been extended by Von Smoluchowski (1910) and subsequently by De Marcus (1961) to cover the case in which scattering at the tube walls is no longer completely diffuse but contains an element of specular reflection. If the scattering law is represented by a probability $(1 - f)$ of a molecule undergoing specular reflection and a probability f of diffuse scattering, it is found that

$$W = \frac{(2 - f)}{f}\frac{8R}{3Z} \tag{4.43}$$

i.e. the transmission probability is enhanced by a factor $(2 - f)/f$. For short tubes one can employ the Dushman result as before to obtain the result

$$W = \left(1 + \frac{3Zf}{8R(2 - f)}\right)^{-1}. \tag{4.44}$$

In the limit $f = 1$ the expression returns to the conventional form (4.42). For UCN with typical f values in the range 0.2 to 0.05 within the standard electropolished stainless steel guides, this theory predicts large increases in UCN transmission coefficients of guides relative to the gas case. This theory is closely related to results obtained in the diffusion theory approximation discussed below.

4.4.3   Diffusion theory

The theory of gas diffusion has been applied to UCN transportation through guides (Luschikov *et al* 1969, Groshev *et al* 1971). The diffusion equation can be written as

$$\frac{\partial^2 n}{\partial z^2} = \frac{n}{L^2} + \frac{1}{D}\frac{\partial n}{\partial t}. \tag{4.45}$$

This can be obtained by altering the derivation in Appendix A2.2.1 equations (A2.32)–(A2.34), to include the effects of a constant loss rate (τ^{-1}). In (4.45) the diffusion length L is given by

$$L^2 = D\tau \tag{4.46}$$

with τ the mean lifetime of the neutrons within the guide and D the diffusion coefficient, defined by

$$D = \tfrac{1}{3}\bar{v}\bar{l} \tag{4.47}$$

where $\bar{v}$ is the average neutron velocity and $\bar{l}$ is the mean free path between wall collisions. In the case of a guide of circular cross section and radius R

$$D = \tfrac{2}{3}\bar{v}R. \tag{4.48}$$

The net current $J(Z)$ crossing a plane at a given value of z is given by

$$J(z) = -D\frac{\mathrm{d}n(z)}{\mathrm{d}z}A \tag{4.49}$$

where A is the cross-sectional area of the guide. In the instance of steady flow conditions n depends solely on z and satisfies the equation

$$\frac{\mathrm{d}^2 n}{\mathrm{d}z^2} = \frac{n}{L^2}. \tag{4.50}$$

As an example of the application of such a theory we calculate the steady-state flow rate of neutrons into a device of albedo R, through a guide of length Z and cross-sectional area A. We consider there to be an isotropic distribution of UCN of density n_0, at the input to the guide and, that the velocity distribution is isotropic at the device output i.e. $J(Z) = A(1 - R)n(Z)\bar{v}/4$. The boundary conditions derived from these constraints are thus

$$n(0) = n_0 \qquad (4.51)$$

and

$$-DA\frac{dn(z = Z)}{dz} = \tfrac{1}{4}A(1 - R)n(Z)\bar{v}. \qquad (4.52)$$

The general solution to (4.50) is given by

$$n(z) = B\sinh(z/L) + C\cosh(z/L). \qquad (4.53)$$

The boundary conditions (4.51) and (4.52) thus imply

$$C = n_0 \qquad (4.54)$$

and

$$B = -n_0\frac{\sinh(Z/L) + \gamma\cosh(Z/L)}{\cosh(Z/L) + \gamma\sinh(Z/L)} \qquad (4.55)$$

with

$$\gamma = \frac{L}{4D}(1 - R)\bar{v}. \qquad (4.56)$$

The flow rate into the device is thus given by

$$J(Z) = \frac{n_0\bar{v}A}{4[\cosh(Z/L) + \gamma\sinh(Z/L)]}. \qquad (4.57)$$

In the limit $Z/L \gg 1$ this expression simplifies to

$$J(Z) = \frac{n_0\bar{v}A\exp(-Z/L)}{2(1 + \gamma)} \qquad (4.58)$$

and we see that the UCN current falls off exponentially with an attenuation length given by the diffusion length L. It is also worth noting that, in the limit of a lossless guide ($L \to \infty$) with no albedo ($R = 0$), (4.58) reduces to an expression consistent with the Dushman result (4.42).

In order to include the effects of a component f of specular reflection on UCN diffusion we need to consider the result of an increased average path length, and time interval between successive diffuse reflections on the diffusion constant. The most illuminating way is to treat the process of diffusion within the framework of one-dimensional random walk theory.

Consider the random walk problem illustrated in figure 4.4. We denote the probability that a given UCN undergoes a displacement in the z direction in the range $\alpha, \alpha + d\alpha$ in a time interval t, by $P(\alpha, t)\,d\alpha$. For now we only assume that $P(-\alpha, t) = P(\alpha, t)$. We focus our attention on UCN within the

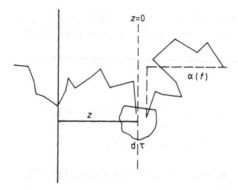

Figure 4.4   Diffusion as a random walk.

volume $d\tau$ at $z = 0$ at $t = 0$. The initial UCN density within this volume is n. After a time t the total number of these UCN having crossed the plane $z = -z$ with a z coordinate in the range $-\infty, -z$ is given by

$$n\,d\tau \int_{-\infty}^{-z} P(\alpha, t)\,d\alpha. \tag{4.59}$$

The total number of neutrons crossing unit area of guide, moving in the negative direction in time t, is thus given by

$$N_-(0) = J_-(0)t = \int_0^\infty n(z)\,dz \int_{-\infty}^{-z} P(\alpha, t)\,d\alpha \tag{4.60}$$

$$= \int_0^\infty n(z)\,dz \int_z^\infty P(\alpha, t)\,d\alpha. \tag{4.61}$$

Similarly the total flow in the positive direction is given by

$$N_+(0) = J_+(0)t = \int_{-\infty}^0 n(z)\,dz \int_{-z}^\infty P(\alpha, t)\,d\alpha \tag{4.62}$$

$$= \int_0^\infty n(-z)\,dz \int_z^\infty P(\alpha, t)\,d\alpha. \tag{4.63}$$

If we replace $n(z)$ by $n(z) + z\,dn/dz$ we find for the net current through the plane

$$J(0) = J_+(0) - J_-(0) = -\frac{2}{t}\frac{dn}{dz} \int_0^\infty z\,dz \int_z^\infty P(\alpha, t)\,d\alpha \tag{4.64}$$

which after integrating by parts and noting that

$$\frac{d}{dz} \int_z^\infty P(\alpha, t)\,d\alpha = -P(z, t) \tag{4.65}$$

leads to the result

$$J = -\frac{1}{2t}\frac{dn}{dz}\int_{-\infty}^{\infty}\alpha^2 P(\alpha,t)\,d\alpha \qquad (4.66)$$

$$= -\frac{1}{2t}\overline{\alpha(t)^2}\frac{dn}{dz}. \qquad (4.67)$$

Comparing (4.49) with (4.67) we find that

$$D = \frac{\overline{\alpha(t)^2}}{2t}. \qquad (4.68)$$

If we assume initially that the wall scattering is completely diffuse then we can write

$$\overline{\alpha(t)^2} = k\bar{l}_0^2 t \qquad (4.69)$$

where k is the average number of diffuse wall collisions per second ($k = \bar{v}/\bar{\lambda}$, where $\bar{\lambda}$ is the UCN mean free path, and $\bar{v}$ is the UCN velocity) and $\bar{l}_0$ is the average separation of the wall collisions along z. Inserting (4.69) in (4.68) we obtain the following expression for D_{diffuse}

$$D_{\text{diffuse}} = \frac{k\bar{l}_0^2}{2} \qquad (4.70)$$

$$= \bar{v}\frac{\bar{l}_0^2}{2\bar{\lambda}}. \qquad (4.71)$$

If we now incorporate an element f of specular reflection into the reflection process (assuming the probability of specular reflection $P_s = 1 - f$) then we will need to find a new expression for $\alpha^2(t)$. All that is required is to evaluate a new value for the mean z displacement between diffuse collisions and the corresponding mean diffuse reflection rate. Using elementary probability theory we can write the following expression

$$\overline{l_s^2} = \sum_{i=0}^{\infty}(1-f)^i(i+1)^2 f\bar{l}_0^2 \qquad (4.72)$$

where the sum is over the number of specular reflections between successive diffuse wall reflections. (The probability of i consecutive specular reflections is $(1-f)^i$. The probability of a single diffuse reflection is f. After i specular and one diffuse reflection the mean distance travelled along the direction of the guide is $(i+1)\bar{l}_0$.) We can rewrite (4.72) as

$$\overline{l_s^2} = \bar{l}_0^2 f\sum_{i=0}^{\infty}\Big(i(i+1)+(i+1)\Big)(1-f)^i \qquad (4.73)$$

$$= \bar{l}_0^2 f\Big((1-f)\sum_{i=0}^{\infty}i(i+1)(1-f)^{i-1}+\sum_{i=0}^{\infty}(i+1)(1-f)^i\Big). \qquad (4.74)$$

The summations can be performed by differentiation of the relevant expressions for the sums of infinite geometric progressions and the following result is obtained

$$\overline{l_s^2} = \frac{(2-f)}{f^2}\overline{l_0}^2. \tag{4.75}$$

The mean number of collisions between successive diffuse reflections can similarly be written

$$\bar{N} = \sum_{i=0}^{\infty}(1-f)^i f(i+1) \tag{4.76}$$

$$= \frac{1}{f}. \tag{4.77}$$

If we now allow for the new length step and diffuse reflection rate in (4.70) we obtain the final result

$$D_{\text{specular}} = \frac{(2-f)}{f}D_{\text{diffuse}}. \tag{4.78}$$

We thus see that the existence of specular reflection manifests itself in an enhancement of the diffusion coefficient. This factor is the same as that derived by Von Smoluchowski (1910) to account for the effects of specular reflection on gas pipe transmittance (4.43). The enhancement in D leads to an enhanced UCN diffusion length given by

$$L_s = \sqrt{\frac{(2-f)}{f}}L_d \tag{4.79}$$

and subsequently from (4.56) and (4.58) to an improvement in guide transmission properties, although it should be noted that the lengths of guide to which the diffusion approach remains a reasonable approximation are also increased, i.e. what might be considered a long guide to gas molecules may exhibit considerable end effects with specularly reflected UCN.

A theory of this type was tested experimentally by Groshev et al (1971). A detector of area A_d and an absorber of area A_a were placed at the end of a guide of cross-sectional area A and the rest of the output plane was made from a good UCN reflector. The effective albedo of the detection system that the guide was feeding was thus $R = (1 - (A_d + A_a)/A)$. The count rate at the detector using this theory is given by (see 4.58, 4.56)

$$J(Z, A_a) = \frac{n_0\bar{v}\exp(-Z/L)}{2(1+\gamma)} \tag{4.80}$$

with $\gamma = L\bar{v}(A_a+A_d)/4DA$. The authors verified the dependence of the UCN count rate on both detector area and guide length and found the respective dependencies to be in agreement with (4.80), despite the fact that they were using UCN in a wide range of energies. They did, however, find that the value of the average diffusion coefficient required to describe the data (D = 1.7(.2) m^2 s^{-1}, τ =13 s, L =4.7 m for $\bar{v}$ = 4.8 m s^{-1} for cylindrical guides of radius 47 mm) was 10 to 13 times larger than the value calculated on the assumption that UCN undergo completely diffuse reflection at the guide walls. Using (4.79) we see that this implies a probability of specular reflection of $\approx$ 0.85.

The full time-dependent diffusion theory has been applied by several authors (Sumner 1977, Ignatovich and Terekhov 1977), to describe the emptying of bottles full of UCN to detectors via initially empty connecting guides. Although reasonable fits to the data can be obtained, the presence of a wide neutron spectrum and the large degrees of specular reflection within the guides used, make the approach somewhat questionable. Diffusion theory has also been extended by Ignatovich (1987) to cover the case of UCN diffusion in the presence of a gravitational field and applied to predict the transmission properties of both inclined and vertical guides.

4.4.4   The two-current model

In conventional diffusion theory as previously described, the neutron velocity distribution is usually considered to be close to isotropic at all points within the guide. For UCN in short highly polished guides this is never going to be the case; the backward travelling current J_- may never constitute more than a small fraction of the total current J, implying a large difference between the two half spaces of the velocity distribution, the diffusion approach is thus liable to give errors in this limit. For this reason a new model for UCN transportation along guides was developed by Frank (1976). The neutron flow at any point within the guide (assumed horizontal and straight) is assumed to consist of two components, one travelling in the positive z-direction and the other in the negative z-direction, denoted $J_+(z)$ and $J_-(z)$ respectively. The two currents are coupled in the sense that there is an associated probability per unit length α that UCN are scattered from one current to the other. One can immediately write a set of coupled equations to describe the current evolution along a length of guide

$$\frac{\mathrm{d}J_+(z)}{\mathrm{d}z} = \alpha(J_-(z) - J_+(z)) \qquad (4.81)$$

$$\frac{\mathrm{d}J_-(z)}{\mathrm{d}z} = -\alpha(J_+(z) - J_-(z)). \qquad (4.82)$$

The net flow along the guide is given by

$$J = J_+(z) - J_-(z) \qquad (4.83)$$

and is seen to be a constant along the length of the guide. Equations (4.81) and (4.82) are easily solved for a given set of boundary conditions. As an example consider a current of UCN $J_+(0)$ entering a guide at $z = 0$ and a perfect absorber placed at $z = Z$, or the UCN enter free space at $z = Z$, such that $J_-(Z) = 0$. The solutions to (4.81) and (4.82) are given by

$$J_+(z) = J_+(0)\left(\frac{1 + \alpha(Z - z)}{1 + \alpha Z}\right) \tag{4.84}$$

$$J_-(z) = J_+(0)\left(\frac{\alpha(Z - z)}{1 + \alpha Z}\right) \tag{4.85}$$

respectively. The transmission of the pipe is thus seen to be

$$T = \frac{J_+(Z)}{J_+(0)} = \frac{1}{(1 + \alpha Z)}. \tag{4.86}$$

For the more general case when a guide is feeding a device with an albedo R, i.e.

$$R = \frac{J_-(Z)}{J_+(Z)} \tag{4.87}$$

one obtains

$$J_+(z) = J_+(0)\left(\frac{1 + \alpha(Z - z)(1 - R)}{1 + (1 - R)\alpha Z}\right) \tag{4.88}$$

$$J_-(z) = J_+(0)\left(\frac{R + \alpha(Z - z)(1 - R)}{1 + \alpha Z(1 - R)}\right). \tag{4.89}$$

The value of α for a given guide system and velocity range of neutrons is determined by the average number of collisions the neutrons make per unit length of guide and the probability that a reflection is such that the neutron velocity component along the guide is reversed. α will thus depend on the ratio of average longitudinal to transverse velocity within the guide, which will be expected to change along the length of the guide as the beam gets more collimated and will lead to departures from the simple theory for long guides. Typical values for α in polished stainless steel guide of diameter 0.067 m are found to be in the range 0.13 to 0.10 m^{-1}.

The two-current model has been extended (Pendlebury 1978) to include loss of neutrons from the guide system by absorption and up-scattering at the guide walls. In this instance (4.81) and (4.82) need modifying to incorporate this loss. If the induced loss per unit length is given by β, we can write

$$\frac{dJ_+(z)}{dz} = \alpha J_-(z) - (\alpha + \beta)J_+(z) \tag{4.90}$$

$$\frac{dJ_-(z)}{dz} = (\alpha + \beta)J_-(z) - \alpha J_+(z). \tag{4.91}$$

By differentiation and substitution these can be rewritten as

$$\frac{\mathrm{d}^2 J_+(z)}{\mathrm{d}z^2} = (2\alpha + \beta)\beta J_+(z) \tag{4.92}$$

$$\frac{\mathrm{d}^2 J_-(z)}{\mathrm{d}z^2} = (2\alpha + \beta)\beta J_-(z). \tag{4.93}$$

These have general solutions of the form

$$J_+(z) = A\cosh(qz) + B\sinh(qz) \tag{4.94}$$

$$J_-(z) = \left(\frac{qB + (\alpha + \beta)A}{\alpha}\right)\cosh(qz) + \left(\frac{qA + (\alpha + \beta)B}{\alpha}\right)\sinh(qz) \tag{4.95}$$

with

$$q^2 = (2\alpha + \beta)\beta \tag{4.96}$$

Typical values for β for UCN in a stainless steel guide of diameter 0.066 m turn out to be 0.01 to 0.02 m^{-1}. Although the two-current model has its limitations it provides a convenient method for estimation of the properties of simple guiding systems.

4.4.5   Monte Carlo simulations

The most satisfactory and complete way to analyse neutron flow in guides is by way of computer simulation (Monte Carlo) methods. In principle the UCN transport properties of any guide system can be calculated as a function of any physically reasonable wall-scattering distribution. The Monte Carlo method works on a probabilistic approach to modelling the system and relies upon tracing the trajectories of many individual UCN through the given guide configuration and building up an average picture of the system's bulk guiding properties.

The conventional procedure for modelling a guide system is as follows. The initial trajectory parameters of each UCN (velocity components and initial coordinates at the entrance to the system) are generated subject to initial distributions, by the inverse transform method (De Groot 1975), which enables one to generate random variables subject to a given probability distribution from pseudo random numbers generated uniformly in the range [0,1]. The position and time of the first collision point with the guide walls is evaluated using classical mechanics from these initial parameters and the specification of the guide geometry. The angle of incidence of the UCN to the surface at the point of impact θ_i is evaluated using the trajectory parameters prior to collision. The result of the collision in terms of the UCN's new velocity components can then be generated by the inverse transform method for a given scattering function, from the relevant scattering parameters (resulting angle to surface normal θ_f, surface normal vector $\hat{n}$ and azimuthal scattering angle ϕ_f (the angle between the plane containing the incident neutron velocity vector and the plane containing the final velocity vector), and finally the probability of absorption or up-scattering). Commonly used scattering distributions are $P(\theta_f, \phi)\, d\Omega = \cos\theta_f d\Omega$ in the case of completely diffuse scattering and $\theta_f = \theta_i, \phi_f = 0$ for specular reflection. From these new trajectory parameters the next wall collision point can be calculated, and the resulting trajectory parameters generated. This process is continued until the UCN are found to leave the guide, either by exit or entrance apertures or by being absorbed or up-scattered. On leaving the system the parameters of interest, e.g. exit coordinate, angle to guide axis, time of exit, mean number of collisions or axial velocity direction reversals etc, can be evaluated and recorded. By following many neutrons through the system probability distributions for these parameters can be determined and the system behaviour categorized.

Although simple in principle, Monte Carlo calculations can rapidly become extremely involved when complicated geometries and scattering laws are introduced. The required CPU times, even on a large mainframe computer can become forbiddingly long. For simple geometries, e.g. circular guides and scattering laws, considerable reduction in computing time and algebraic complexity can be avoided by making use of the symmetry of the system and carefully choosing coordinate systems with which to describe the trajectories. The other major problem is knowing exactly which scattering law to invoke for a given guide material and surface quality.

Many Monte Carlo programs have been written to analyse rarified gas flow through capillaries where the wall reflections are diffuse with a small component of specular reflection (e.g. Davis 1960). Scattering of UCN in polished guides is, as previously discussed, at the other extreme; i.e. specular scattering is by far the predominant event. In addition the assumption that the non-specular component of reflection is diffuse is only correct for a macroscopically rough surface. UCN guides are usually only microscopically rough and, as shown by Steyerl (1972b) and Ignatovich (1973) (see section 2.4.5), the non-specular reflection is dependent on the neutron's angle of incidence to the surface normal. For microscopically rough surfaces the probability of non-specular reflection of a UCN of wavevector k, incident at an angle θ_i to the normal of a surface of RMS height b and correlation length w (assumed Gaussian) is given by

$$I_{ns}(\theta_f, \phi_f) = \frac{3}{2\pi} G \cos\theta_i \cos^2\theta_f$$
$$\times \exp[-w^2 k^2(\sin^2\theta_i + \sin^2\theta_f - 2\sin\theta_i \sin\theta_f \cos\phi_2)/2] \tag{4.97}$$

where

$$G = \tfrac{4}{3} k^4 w^2 b^2. \tag{4.98}$$

The total probability of non-specular reflection is given by

$$P_{ns}(\theta_f) = \int I_{ns}(\theta_f, \phi_f) \, d\Omega_f. \tag{4.99}$$

The results are valid only when $2bk < 1$ which restricts the theory to highly polished surfaces with $b < 50$ Å, such as electrolytically polished copper ($b = 35$ Å, $w = 250$ Å) and polished glass coated with nickel (Steyerl 1972b). This expression is too complex to integrate analytically, although it can be reduced to a single numerical integration by integrating over ϕ (Brown et al 1975). For long correlation lengths $c = w^2 k^2 \gg 1$ examination of (4.97) shows that the non-specular scattering component is sharply peaked about the direction of specular reflection. However, in the limit of short correlation lengths $c = w^2 k^2 \ll 1$, (4.97) simplifies to

$$I_{ns}(\theta_f, \phi_f) = \frac{3}{2\pi} G \cos \theta_i \cos^2 \theta_f \tag{4.100}$$

which implies a somewhat broader angular distribution, although still considerably different from a diffuse distribution. Within the short range limit we are able to use (4.100) within (4.99), and obtain a closed expression for the probability of non-specular reflection:

$$P_{ns}(\theta_i) = G \cos \theta_i. \tag{4.101}$$

Monte Carlo programs have been written to examine the transmission characteristics and angular distribution of velocities of the transmitted UCN, for horizontal, circular cross-sectional guides, using both the exact scattering laws (4.97) (Brown et al 1975), and its approximation (4.100) (Berceanu and Ignatovich 1973) which is only valid in the short correlation length regime. A theoretical model for the guide transmission based on the theory of rarified gas flow developed by Von Smoluchowski (1910) was further developed by Bercenau and Ignatovich who incorporated an element of scattering described by (4.100) instead of diffuse scattering. The model gave results consistent with their Monte Carlo data. The two sets of Monte Carlo experiments gave consistent results over the c value ranges in which the simplified results (4.100) and (4.101) are valid as is shown in figure 4.5, although significantly different results were obtained outside of the short correlation length regime. A polar plot of the angular distribution functions obtained by various authors for an electropolished guide with a length-to-radius ratio of 64 is shown in figure 4.6. The results clearly show the pronounced effect of forward peaking of the beam with increased specularity of reflection. Winfield and Robson (1975) have also performed Monte Carlo calculations for a proportion of completely diffuse reflections including a small probability ($\bar{\mu}$) of loss at each reflection.

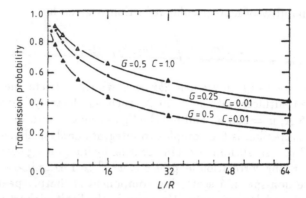

Figure 4.5 Transmission probabilities for cylindrical tubes against length for different roughness parameters as determined by Monte Carlo calculation. Triangles refer to the results of Brown *et al* (1975) and full points refer to the results of Bercenau and Ignatovitch (1973) (Brown *et al* (1975)).

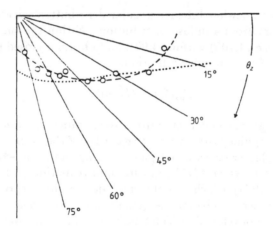

Figure 4.6 Polar diagrams for the distribution function $f(\theta_z)$ of neutron velocities with respect to the angle θ_z with the z-axis showing the asymptotic form appropriate to long round tubes with diffuse reflection as given by Bercenau and Ignatovich (1973) and Steckelmacher (1978) (dotted line), and the Monte Carlo results of Brown *et al* (1975) for a length-to-radius ratio $Z/R = 64$ with the reflection relation and the parameters for electropolished copper given by Steyerl (1972b) (—0—0—). For an isotropic distribution, $f(\theta_z)$ is independent of θ_z (from Golub and Pendlebury (1979)).

Monte Carlo programs have also been written to calculate the transmission properties of sharp bends for UCN with energies both above and below the guide cut-off (Maysenholder 1976)—UCN reflections are taken to be completely specular in these calculations. Robson (1976) has also evaluated the

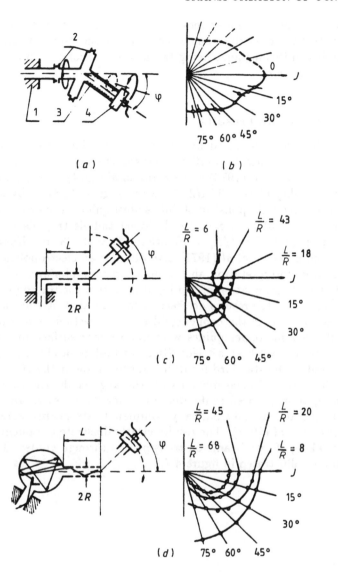

Figure 4.7   Experimental results showing the angular distribution $f(\theta_z)$ of UCN emerging from guides, for a variety of guide configurations. (a) The scheme for measurements of the UCN angular distribution into the forward direction: 1, the neutron guide; 2, the rotating vacuum junctions; 3, the segment of the tube covered from the inner side with the UCN absorbing material (polythene); 4, the UCN detector. (b) The UCN angular distribution at the neutron guide exit. (c) The angular distribution of UCN measured at different distances from the rectangular bend of the neutron guide. (d) A similarly measured distribution with a hollow sphere (Kosvinstev *et al* 1977).

transmission properties of a system comprising three straight guide sections coupled with two short right-angle bends using the Monte Carlo method, the scattering in this instance being taken to be specular with a component f of diffuse reflection.

4.4.6   Experimental results

Several authors have reported measurements of guide transmission probabilities for UCN in guides with different cross-sectional shapes, materials and lengths. Lobashov *et al* (1973) and Egorov *et al* (1974) both report a transmission probability of $W = 0.71(3)$ for a rectangular (1.82 m by 60 mm by 70 mm) electrolytically polished stainless steel guide. Taylor (1975) quotes a value $W = 0.89(5)$ for a cylindrical, honed and electropolished type 304 stainless steel guide for $Z/R = 16$, where $R = 0.033$ m. Kosvinstev *et al* (1977) and Groshev *et al* (1971) give data for an electropolished copper guide. A direct interpretation and comparison of results are made difficult by incomplete determination of the input spectrum for the majority of the measurements, but the results do seem to indicate that attenuation in steel guides was half that of the copper guides. The question as to whether the additional losses in copper guides were due to poor surface preparation of the guides has not yet been settled. Stainless steel guides have subsequently been adopted as the standard for UCN transportation at the ILL.

A careful series of measurements on the angular distribution of velocities issuing from guides, configured in a variety of ways and fed with both isotropic and already partially collimated UCN beams were reported by Kosvinstev *et al* (1977). The results clearly illustrate the phenomenon of forward peaking of UCN beams on transmission through guides. The principal results are illustrated in figure 4.7.

5

UCN storage measurements

5.1 INTRODUCTION

In the preceding chapter we discussed the gas-like behaviour of UCN and derived the results required to describe the decay of a neutron population undergoing storage in a material bottle. Our earlier calculations based on the assumption that the neutron interaction with the walls could be represented by reflection from an abrupt step in mean Fermi potential showed that, for reasonable geometry and choice of material, UCN should be confinable for time periods determined almost solely by β-decay. In this chapter we concern ourselves with the experimental situation with respect to UCN storage. For many years reconciling experiment with theory proved a problem for UCN experimenters. We briefly review the early work on UCN storage before moving on to describe the more recent developments in the field.

5.2 EARLY NEUTRON STORAGE MEASUREMENTS

The first direct measurements of UCN storage properties were made in 1971 by Groshev *et al.* Experiments were performed to determine the mean rate of neutron loss within bottles constructed from a variety of materials: copper, beryllium, pyrolytic graphite and Teflon. Some of the results of these first measurements are shown in figure 5.1; note the low count rates obtained from the early sources. The measurements showed that the average UCN loss rates were very much greater than those predicted on the basis of the theory presented in Chapter 4, assuming the form of $\bar{\mu}(E)$ in (2.70) corresponding to an abrupt step change in mean Fermi potential and evaluated using the appropriate values of loss cross sections, and scattering lengths. The average number of bounces before loss on each surface tested ($\bar{K} = 1/\bar{\mu}$) was always found to lie in the range $500 < K < 1500$. The predicted values were always greater, sometimes a hundred times larger. The findings of these initial experiments were soon confirmed by further measurements (Robson and Winfield 1972, Groshev *et al* 1973, Lobashov *et al* 1973, Egorov *et al* 1974 amongst others; a complete compilation of the early experimental results is to be found in Steyerl (1977); see also Golub and Pendlebury (1979)). Several of these authors did find that slight improvements in storage characteristics (by up to a factor of two) could be obtained by careful cleaning of the bottle surfaces.

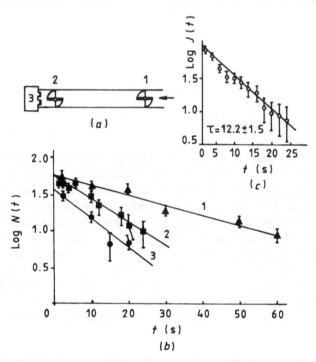

Figure 5.1   Apparatus (*a*) and early UCN storage results (*b*) and (*c*). (*a*) The vessel was of length 174 cm and diameter 14 cm. 1 and 2 represent in-and-out valves; 3, the UCN detector. (*b*) Number of UCN retained in tank plotted against containment time (*t*) for: 1, the chemically polished copper, mean lifetime $T = 33$ s; 2, the untreated copper foil $T = 14$ s; 3, the pyrolytic graphite, $T = 11$ s. (*c*) UCN counting rate against time after opening of valve 2 (time analyser channel width 2 s) (Groshev *et al* 1971).

The temperature dependence of the loss was investigated by Groshev *et al* (1971): a copper bottle was heated to 520 K and a graphite bottle to 370 K, but no temperature dependence of the bottle lifetime was detected. Additional investigations were performed by Groshev *et al* (1973) and Luschikov *et al* (1976, 1977) but no evidence of a thermal effect was ever found. Results from an experiment by Groshev *et al* (1975) in which a bottle of boron-free aluminosilicate was both heated and cooled over a temperature range that spanned 670 °C are shown in figure 5.2. Although these experiments did really seem to indicate that UCN storage times were independent of temperature, it was pointed out by several authors (Steyerl and Trustedt 1974, Ignatovich and Satarov 1977, Golub and Pendlebury 1974) that any increase in loss cross section with temperature could have been partially masked by a decrease in impurity concentration with temperature.

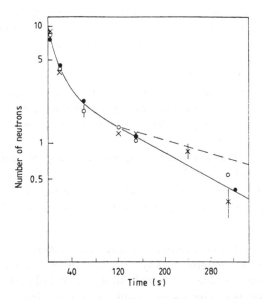

Figure 5.2   Time dependence of the number of UCN remaining in a glass bottle for different bottle wall temperatures. Open circles, temperature varying along the length of the bottle between 200 500 °C; full circles, room temperature; crosses, temperature varying along the length of the bottle between −40 and −170 °C. The broken curve corresponds to twice the storage time of the full curve. (Data from Groshev *et al* (1976), see also Luschikov (1976).)

The only conclusion that could be derived from these early experiments was that as far as UCN loss rates were concerned, all material surfaces seemed more or less identical, $\bar{\mu} \approx 10^{-3}$. Either all substances possessed some common feature that was responsible for the high loss per bounce values or some unexpected fundamental principle was involved. Whatever the cause might be, it seemed that its effects were temperature independent.

5.3 THE SEARCH FOR THE CAUSE OF THE ANOMALOUSLY LARGE NEUTRON LOSS RATES

A wide range of ideas as to the cause of the excess losses was proposed and investigated (for a more complete review see Golub and Pendlebury 1979 or Steyerl 1977). Amongst the ideas investigated were:

(i) the effects of pores and surface roughness on the loss per bounce function (Ignatovich 1974, see section 2.4.5);

(ii) the effects of specular reflection and non-attainment of isotropy in velocity space on the average loss per bounce function (Ignatovich 1975b see section 4.3.2);

(iii) scattering of neutrons out of the UCN range by inelastic scattering ($\Delta E \approx 1$ meV) (two processes—scattering from atoms undergoing diffusive motion in the surface and magnetic scattering from atoms possessing a permanent electron magnetic moment—were considered (Blokhinstev and Plakida 1977, see section 2.4.6));

(iv) gradual warming of UCN to energies above the mean Fermi potential by acoustic vibration of the bottle walls (Gerasimov et al 1973, see section 5.5.2), by thermally excited Rayleigh-type surface waves (Frank 1975) and ultrasonic vibrations (Ignatovich 1975a);

(v) leakage of UCN between the atoms constituting the bottle surface (Ignatovich and Luschikov 1975);

(vi) scattering from loosely bound clusters of atoms (60–70 Å in diameter), vibrating independently of each other at frequencies of up to 10^{10} Hz (Ignatovich 1975a).

All these postulated loss mechanisms gave loss rates far below those required to explain the experimentally observed values for physically reasonable model parameters consistent with all direct or indirect observations.

The possibility that the losses could be due to impurities or surface films had been considered almost immediately (Groshev et al 1971, Shapiro 1972b). However the idea was rejected by most UCN workers on two counts: first the amount of surface contamination required to describe the losses would need to be considerable even for materials with a large neutron loss cross section; and second the loss rates seemed to be temperature independent (as shown in Chapter 2 any cross section for inelastic scattering should be temperature-dependent; the exact dependence depending on the details of the impurity binding and whether the scattering is coherent or incoherent).

As an example of the degrees of surface contamination required to describe the experimental results consider an impurity species of total loss cross section 50 b at $\lambda = 10$ Å (a reasonable figure for hydrogen) within a surface material of $a_{coh} = 10^{-12}$ cm and negligible loss cross section relative to hydrogen. If we assume a uniform distribution for the impurity within the surface, then using $\bar{\mu} \approx 2f = 2W/V_f$, we find that we require a 40% (atomic) contamination of the surface over the first 100 Å of the surface in order to obtain a value of $\bar{\mu} = 10^{-4}$. Alternatively if one assumes the contaminant forms a pure surface layer it needs to be 50 Å thick to obtain the same loss rate. Surface contamination to such an extent and with such a uniformity over the entire range of materials tested was not thought to be at all likely; it was, however, not completely out of the question.

In 1977 Blokhinstev and Plakhida performed a detailed calculation on the effects of phonon scattering on UCN loss probabilities (section 2.4.6) and showed that hydrogen concentrations of 15 to 30% were required to explain the experimental observations with respect to the crude estimate. Calculations were also performed by Ignatovich and Satarov (1977). The temperature dependence of the loss rate depended on the form of binding of the hydrogen within the surface, but fairly weak dependencies, e.g. $\sqrt{T}$, could be obtained if the hydrogen was assumed to behave as a weakly diffusing gas (see section 2.4.6). An exponential dependence was predicted for strongly bound hydrogen. In 1974 Golub and Pendlebury emphasized the point that the experimental evidence at that stage was not conclusive enough to rule out the impurity hypothesis and suggested searches for the presence of impurities independent of UCN experiments. Improved measurements of the temperature dependence were also required.

Attempts were soon made to investigate the energy dependence of the loss rate in the hope that this might give information regarding the shape of the Fermi potential distribution at the material surfaces. Steyerl and Trustedt (1974) used a system in which the UCN entered a bottle through a horizontal port and left through a secondary port across which various foils could be positioned (see figure 5.3). These foils of differing mean Fermi potentials acted as high pass velocity filters—only UCN with an energy greater than the foil's Fermi potential could reach the detector. The resolution was crude but the authors found that there was a more rapid increase in $\bar{\mu}(E)$ as one approached the cut-off than predicted by (2.70) for the copper and graphite bottles they investigated. Crude energy dependence measurements were also made in a large 340 litre stainless steel vessel which could be physically raised relative to the UCN source (Kosvinstev et al 1977). This movement effected a change in the initial UCN spectrum; the authors observed an energy dependence for the average decay time but were unable to arrive at any firm conclusions concerning the loss function. The most careful measurements were made by Groshev et al (1976). In this experiment the storage bottle constituted the horizontal guide section of a crank spectrometer (section 6.3.2). The height of the bottle could be raised relative to the source. By the appropriate placement of foils in the downward exit guide leading to the detector (see figure 5.4) and subsequent vertical displacement of the detector relative to the neutron bottle, measurements on UCN within reasonably well-defined energy bands $\Delta E = 25$ neV, could be performed.

The average $\bar{\mu}(E)$ function determined for copper is shown in figure 5.5. As can be seen these experiments seem to provide evidence that wall losses are well described by the standard loss probability function, albeit with an enhanced $(2f)$ parameter. It should be noted that several of the other model surfaces provide a reasonable fit to the data. Burnett (1982) tried fitting decay curves obtained with a copper bottle using a full UCN spectrum, with the techniques presented in Chapter 4. He was unable to describe his data

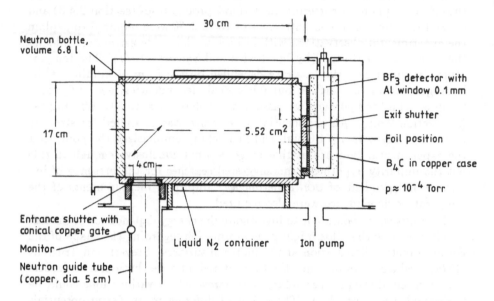

Figure 5.3 An early scheme for determining the energy dependence of the UCN loss rates within material walled traps. The foils could be positioned over the detector entrance enabling the decay of UCN within well-defined energy bands to be studied. The bottle interior is coated uniformly with pyrolytic carbon. Oil-free pumping was ensured by the use of zeolite adsorption pumps and an ion pump. The graphite exit (entrance) shutter moves in (perpendicular to) the drawing plane (Steyerl and Trustedt 1974).

by assuming the standard form of the loss function. His data, like Steyerl's, also indicated too strong an energy dependence for $\bar{\mu}(E)$ as the UCN energy approached the cut-off. The contradictory conclusions arising from these investigations provided little help in finding the cause of the large UCN loss rates.

The first experiment that really shed any light on the cause of the anomalous loss rates was performed by Stoika et al in 1978. They performed UCN storage experiments in a thin-walled neutron vessel surrounded by a system of ^{3}He proportional counters (see figure 5.6). The ^{3}He concentration could be adjusted in these counters as could the detector windows in order to vary their spectral sensitivity. The object of their experiment was to determine whether the UCN lost from the bottle were absorbed at the bottle walls or were inelastically scattered out of the bottle. The results of the experiments showed conclusively that the anomalous losses were due to inelastic heating of the UCN. This statement was true even for materials such as copper in which the absorption cross section far exceeds the inelastic scattering cross

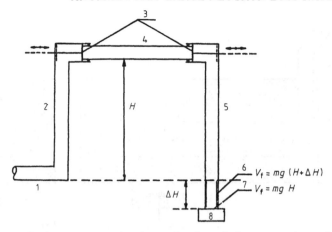

Figure 5.4   Early apparatus for determination of the energy dependence of the UCN loss rate. 1, input UCN guide; 2, vertical guide; 3, shutters; 4, copper neutron bottle; 5, output guide (copper); 6, section of UCN guide with mean Fermi potential greater than that of copper; 7, copper shutter (Groshev *et al* 1976).

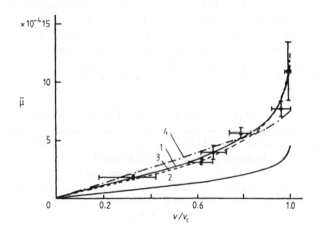

Figure 5.5   Average probability of loss per wall collision $\bar{\mu}$ against relative UCN velocity v/v_c for a copper surface. Measured values shown by points. Curve 1, calculated for loss parameter 2.6 times larger than expected; curve 2, calculated for an ideal copper surface; curve 3, calculated for a rough surface, RMS roughness 100 Å, roughness correlation length $w \approx 0$; curve 4, showing the effect of a 55 Å layer of water (Groshev *et al* 1976).

section. Further still, it was demonstrated that the majority of UCN scattered from the bottle had energies in the thermal range 1–10 meV (see figure 5.7). The results of these experiments were consistent with the hypothesis of inelastic neutron scattering by a light impurity such as hydrogen on the bottle walls.

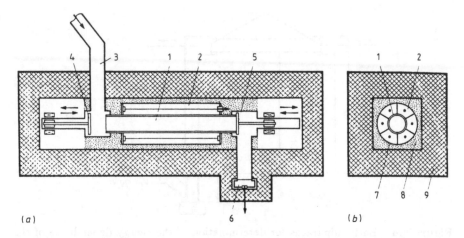

(a)   6   (b)

Figure 5.6 (a) Experimental set-up for detecting up-scattered UCN leaving a neutron bottle. 1, Neutron storage vessel. Two bottles were tested; bottle one was made from thin (10 μm) electrolytic copper sheet, cleaned in HNO_3 and rinsed with distilled water (length of bottle = 96 cm and 8.2 cm diameter); bottle two was made of electropolished solid copper tubing (length = 70 cm, diameter = 9 cm, wall thickness = 1.5 mm); 2, six-chamber ^{3}He counter; 3, UCN-channel; 4, entrance valve; 5, exit valve; 6, UCN detector; 7, Cd 1.5 mm; 8, B_4C; 9, $(CH_2)_n$ + B. (b) Details of the detector systems used to determine the energy of the up-scattered neutrons are given in table 5.1 (Stoika *et al* 1978).

Table 5.1   Technical details regarding the two detector systems

Detector system number	Outer diameter	Inner diameter	Active length	Material and thickness of the inner wall	Gas filling	
I	180 mm	85 mm	650 mm	Duralumin 0.5 mm	He^3 $+CO_2$ $+Ar$	53 Torr 20 Torr 1140 Torr
II	230 mm	100 mm	380 mm	Duralumin 3 mm	He^3 $+CO_2$ $+Ar$	400 Torr 5 Torr 400 Torr

These experiments were closely followed by direct measurements of surface hydrogen contamination on typical neutron bottle surfaces (Lanford and Golub 1977, see also La Marche *et al* 1981). The experiments involved bombardment of neutron bottle surfaces with ^{15}N ions and were based on the detection of 4.43 MeV γ-rays produced by the reaction ^{15}N + H = ^{12}C + ^{4}He + γ (4.43 MeV). Since this reaction is a narrow isolated resonance with a large cross section the hydrogen concentration could be determined

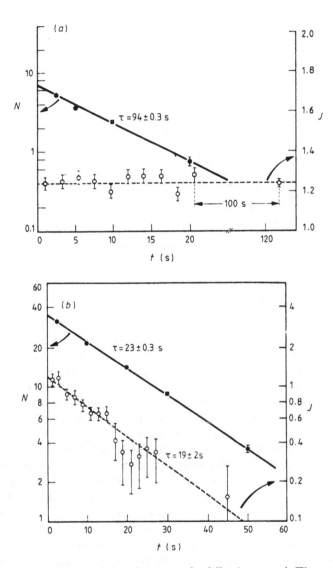

Figure 5.7 ((a) and (b), (c) is shown on the following page.) The total number N of neutrons detected by the main detector (see figure 5.6 and table 5.1) after various storage times (full curves) is compared with the count rate J of neutrons escaping from the bottle through the wall (broken curves). The data represent averages for one storage cycle. (a) Results of UCN containment in bottle no 1 enclosed by a detector system 1 (see figure 5.6 and table 5.1). (b) Results of UCN containment in bottle no 2 enclosed by a detector system 2. (c) Dependence of the counting rate in detector system 2 on ^{3}He pressure for neutrons: (1) from the walls of tube no 2; (2) from $(CH_2)_n$ inserted into the tube. (3) from a Cf-source outside the detector moderated by a water column inside the detector. All curves are normalized to the same saturation value of 1.0 (Stoika *et al* 1978).

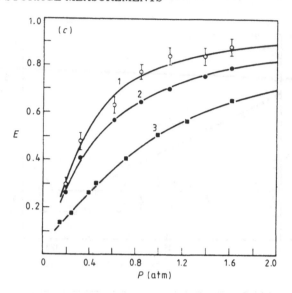

Figure 5.7   (Continued (c).)

as a function of depth into the surface by measuring the γ yield against the ^{15}N energy. The ions lose energy whilst penetrating into the surface and the reaction becomes resonant at a depth dependent on their incident energy. The depth resolution of the technique due to the width of the resonance was dependent on the material used but typically was of the order of 30 Å. Plots of typical profile measurements are shown in figure 5.8. The H impurity concentrations so determined were typically 3×10^{22} H atom/cm^3. On the basis of the theory of Chapter 3 this is sufficient to account for the anomalous UCN loss rates. In addition, the authors concluded that all common surfaces had equivalent surface hydrogen profiles to within a factor of ten, in agreement with the UCN observations. Similar measurements, this time based on the resonant nuclear reaction $H(^{11}B,\alpha)2\alpha$, were performed on typical bottle surfaces and produced the same conclusions (Bugeat and Mampe 1979). In addition, these authors succeeded in showing that surface hydrogen levels could be reduced by continued ion bombardment to about one-tenth of their original level. The recontamination as a function of time and pressure was also examined and it was found that typical recombination times were of the order of 10 h.

Further evidence concerning the hydrogen hypothesis was obtained by measurements performed with the UCN diffractometer (section 6.2.4) on UCN reflection from a glass surface (Schekenhofer and Steyerl 1977). Reflection coefficients for neutrons with a normal component of velocity slightly greater than the glass cut-off (93.6 neV) were measured (see figure 5.9) and it was found that the results could not be described on the assumption that the

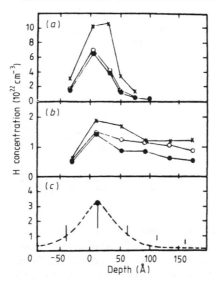

Figure 5.8 Surface hydrogen profiles for (*a*) copper (electropolished and washed in acetone), (*b*) graphite (baked under vacuum at 400 °C, then exposed to air for 6 months) and (*c*) glass (etched in HF), as determined by bombardment of the surface by ^{15}N ions. Crosses correspond to measurements made at room temperature. Empty circles correspond to measurements made after heating of the surface to 180 °C and subsequent exposure to air and full circles to measurements made after additional heating to 265 °C (Lanford and Golub 1979).

glass represented an abrupt potential step to the UCN. The results could, however, be described by postulating a smoothed wall potential

$$V(z) = V_{\text{glass}}/(1 + \exp(-z/d))$$

where z is the depth into the surface and d was found to be 73 ± 3 Å. Such a smoothed wall potential would be expected from equation (2.46) on the basis of the direct hydrogen measurements previously described (hydrogen has a negative scattering length in addition to a large inelastic cross section and would subsequently lower the real component of mean Fermi potential at the glass surface).

Although these experiments provided convincing evidence that hydrogen was responsible for the storage time anomaly the mystery still remained as to why there seemed to be no temperature dependence in the loss rate. Only when this was fully resolved could the hydrogen hypothesis be fully accepted. This did not really take place until 1985 although measurements claiming evidence of a thermal effect for low energy UCN storage in an aluminium vessel were in fact reported by Kosvinstev *et al* in 1979 and again for low energy UCN on a heavy ice surface (Kosvinstev *et al* 1982). These experiments will be discussed in more detail in section 5.4.

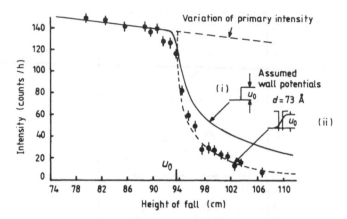

Figure 5.9   Measured UCN intensity reflected from a glass mirror (points) as measured with the gravitational UCN diffractometer, compared with theoretical curves for (i) a step function and (ii) a smoothed step function for the wall scattering potential. Dotted line, calculation for mono-energetic neutrons; full line, calculation for the instrumental resolution. Assumption (ii) may be a model for a hydrogenous surface contamination (Schekenhofer and Steyerl 1977).

As previously mentioned the early experiments with neutron bottle temperature variation could not be taken to rule out the surface contamination theory completely as care had not been taken to control the amount of potential surface contamination with temperature. This could only be rectified by paying greater attention to vacuum and cryogenic techniques. What turned out to be the definitive temperature-dependent loss-rate measurement was performed over a wide range of temperatures in the empty UCN generation chamber of the superthermal helium UCN source (described in section 3.5.2) (Ageron *et al* 1985). The experimental set-up is shown in figure 5.10. Storage experiments were performed within a beryllium-coated vessel. The major feature of the experimental arrangement was that cryo-pumping onto the vessel walls was reduced to a minimum by passing the pump line through the centre of the vertical cryostat. In addition the system was isolated from the relatively poor vacuum of the source by vacuum-tight cold windows (Miranda 1988). Measurements of the loss rate were performed at 6.5, 77, 200 and 300 K. The decay curves are shown in figure 5.11 and, as can be seen, a marked temperature dependence was found, the loss rate contribution from the walls changing by a factor of two over the temperature range covered. The authors were able to describe their data within the framework of the inelastic scattering theory developed by Blokinstev and Plakida (1977), described in Chapter 2, by assuming that the hydrogen atoms are uniformly distributed within the beryllium and tightly bound to the bulk material. Their derived value for the total amount of surface hydrogen, 1.5×10^{16} atom/cm^2, (corresponding to a surface density of 1.2×10^{22} atom/cm^3) was in good agreement with the directly observed values. The source of the anomalous neutron losses had thus been confirmed.

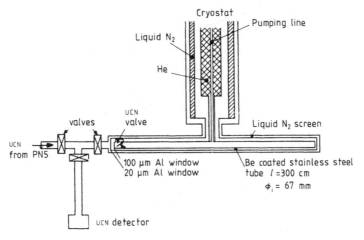

Figure 5.10   Cryostat used for determination of the temperature dependence of the UCN wall loss process (Ageron *et al* 1985).

5.4   IMPROVEMENTS IN UCN STORAGE BOTTLES AND THE SEARCH FOR LOW NEUTRON LOSS WALL MATERIALS

Once the first evidence that hydrogen contamination was likely to be the reason for the short bottle lifetimes was obtained, much effort was directed towards finding or preparing hydrogen-free surfaces. As already explained some success at hydrogen removal was obtained by Bugeat and Mampe (1979) by ion bombardment of the surface they investigated during the course of their experiments on direct surface hydrogen profiles. Effects of *in situ* gas discharge cleaning of neutron bottles were investigated by a number of experimenters (Burnett 1982, Mampe *et al* 1981). The experiments showed that considerable improvements in neutron storage properties could be obtained. Both AC and DC discharges were tried. The typical discharge gases tested included argon, helium, hydrogen and oxygen. In addition deuterium discharging was attempted as it was hoped that the chemically identical deuterium, which inelastically scatters neutrons far less than hydrogen, might replace the postulated surface hydrogen. The results of the effects of various cleaning operations are given in figure 5.12. Discharge cleaning with deuterium permitted loss rates to within a factor of two of the theoretical expectation for certain surfaces to be obtained. Extensive measurements of surface hydrogen density profiles and subsequent surface recontamination

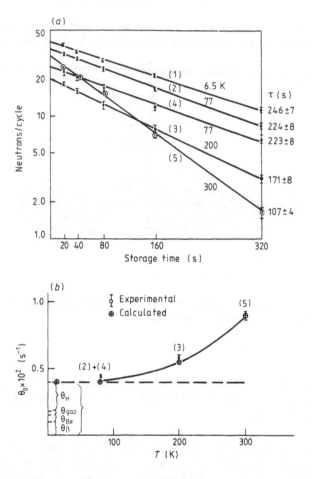

Figure 5.11 (*a*) UCN decay curves in the beryllium-coated, empty superthermal helium source at various temperatures. The numbers in brackets give the chronological order of the measurements. (*b*) Inverse UCN lifetimes in the beryllium coated, empty superthermal helium source as a function of temperature (Ageron *et al* 1985).

rates, for a variety of surfaces, after a variety of surface preparations (gas discharging, hydrogen replacement with deuterium, heating and continuous evaporation of wall coating) were performed by La Marche *et al* (1981) using the ^{15}N technique described in the previous section.

An interesting set of experiments was performed in an aluminium-walled, gravitational roofed trap by Kosvinstev *et al* (1979, 1978); these included the temperature-dependent loss-rate measurement mentioned previously. Only UCN with energies in the range 0–26 neV could get into the bottle. Their experimental arrangement is illustrated in figure 5.13.

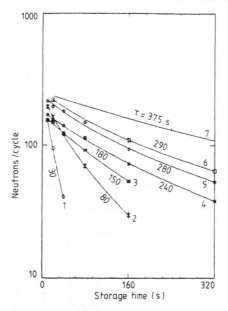

Figure 5.12 UCN storage in a horizontal quartz tube (length 1 m, diameter 6.4 cm: two steel end caps) as a function of surface cleaning. 1, alcohol cleaned; 2, ultrasonically cleaned with RBS25 detergent and washed with demineralized water; 3, discharge cleaning, 2 days at atmosphere and repeating procedure 2; 4, discharge cleaning (80% Ar, 20% D$_2$) for 30 min; 5, same discharge with admixture of D$_2$O; 6, same discharge as 3, but with admixture of deuterated methyl alcohol vapour; 7, calculated decay curve for ideal surface (Mampe *et al* 1981).

The trap was equipped with sputtering heads which enabled fresh metallic surfaces to be sputtered onto the bottle walls. Bottle walls which had never been exposed to the atmosphere could thus be fabricated. In addition the trap could also be fitted with a movable polythene roof which permitted measurement of the confined UCN spectrum. The theory of UCN storage in such a bottle is covered in section 6.3.3. Using such a surface preparation technique lifetimes as long as 645(25) s were obtained (beryllium), although the wall loss rates were still a factor of 4–100 greater than expected. Storage measurements were made in this apparatus on an aluminium surface in the temperature range 300 to 800 K. A change in $\bar{\mu}$ by a factor of 1.6 over the range was reported but the results could not be adequately explained by the assumption of inelastic scattering on hydrogen (Kosvinstev *et al* 1979). Storage experiments with a cryogenic bottle cooled to 80 K were reported by Kosvinstev *et al* in 1982 which really marked the beginning of a new era in UCN physics. Finally, after many years of experiments, bottle lifetimes in accordance with theoretical predictions were obtained.

Their experimental apparatus is shown in figure 5.14; the bottle was cylindrical, of diameter 52 cm and height 28 cm and was made from aluminium.

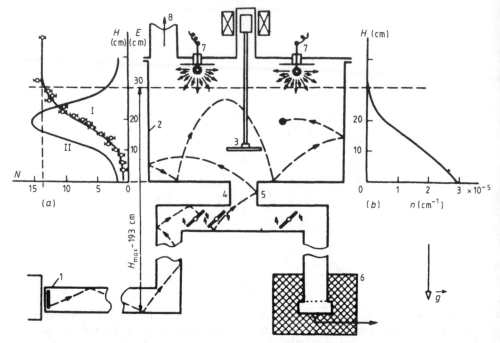

Figure 5.13 Apparatus used to investigate neutron storage in a vessel with freshly and continuously deposited metal walls: 1, converter; 2, metallic walls; 3, neutron bottle shutter; 4, UCN input valve; 5, UCN output valve; 6, shielded detector; (a) (I) total neutron count rate against bottle height of absorbing roof (not shown) and UCN spectrum (II) deduced from this data; (b) deduced UCN density (Kosvintsev et al 1978).

The bottle was also hermetically sealed and fitted with both heating coils and cooling tubes which permitted it to be heated to 750 K and cooled to 80 K under vacuum. From the appropriate cross sections the average wall loss bounce rates for aluminium and heavy ice were calculated to be $1.26(0.2) \times 10^{-3}$ and 6×10^{-5} at 300 K and 80 K respectively for the low energy UCN used. After cleaning with NaOH the aluminium bottle lifetime was 460(30) s at 300 K. On cooling this lifetime rose to 790(100) s giving a loss rate of 1.22×10^{-3} s^{-1} in agreement with the theoretical prediction. By heating the bottle to 750 K using oxygen at 0.1 Torr the authors were able to improve the bottle lifetime to 670(60) s at 300 K and 840(100) s at 80 K. By freezing 2000 Å of heavy ice onto the cleaned aluminium surface a bottle lifetime of 950(60) s was obtained. Essentially all the observed bottle loss was due to β-decay. Finally the idea of experiments such as UCN measurements of the β-decay time constant, which had been envisaged from the very start of work in the field, could be considered a realistic proposition. Subsequently two UCN storage experiments employing cryogenic bottles have been started (Kharitonov et al 1989, Morozov 1989). These experiments are

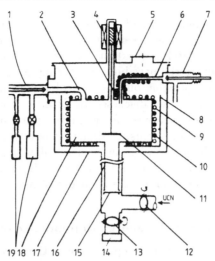

Figure 5.14   A cryogenic neutron storage system: 1 and 7, valves; 2, pipe, 3 and 9, heaters; 4, solenoid; 5, pumping pipe; 6 and 8, screen; 10, cooling tubes; 11 and 13, slide valves; 14, UCN detector; 15, aluminium foil; 16, neutron duct; 17, vacuum chamber; 18, storage vessel; 19, vacuum pumps (Kosvinstev *et al* 1982).

described and discussed in detail in Chapter 7.

In 1983 a most interesting result was obtained by Bates (1983) who attempted UCN storage measurements in a fluid walled bottle, the idea being that one could repeatedly reform the fluid walls by respraying to renew the coating. The chosen fluid was perfluoro polyether $(F_3CCF_2OCF_2CF_5)_n$ with an average molecular weight 2650. It is a chemically inert oil used in diffusion pumps and is commercially available under the name 'Fomblin'. Bates obtained a result for the mean loss per bounce of $2.2(1) \times 10^{-5}$ on the oil surface for UCN in the range 0–106.5 neV at room temperature. The result is quite remarkable especially when one notes that the measurements were made at an extremely low flux reactor (Risley) where the UCN count rates are so low that less than 1 UCN was trapped per measurement cycle. Further measurements (Ageron *et al* 1986) have shown that the Fomblin loss function is quite strongly temperature-dependent. Fomblin-based greases have also been shown to give equally low UCN loss rates (Richardson 1989). A UCN storage β-decay lifetime measurement incorporating a Fomblin oil surface has recently been completed (Mampe *et al* 1989a, see also section 7.2.3).

5.5   PRESENT STATUS OF UCN STORAGE MEASUREMENTS AND TECHNIQUES

Now that the cause of the early anomalous UCN loss rates has been determined and low UCN loss surfaces have been obtained, the major objective of

UCN experimenters has been to use the new surfaces in fundamental physics experiments; primarily in UCN measurements of the β-decay lifetime but also, albeit to a somewhat lesser extent, in searches for the neutron electric dipole moment, where the possibility of storing neutrons for longer time periods than previously possible offers a considerable improvement in sensitivity. These experiments are discussed in detail in Chapter 7. As these experiments have progressed a need for a more refined investigation of the energy exchange and loss mechanisms occurring within bottles constructed from the new materials, along with an accurate determination of their energy and temperature dependence, has developed. In this section we review and describe what measurements have been performed to this end.

5.5.1   Determination of the energy dependence of UCN loss rates

Early measurements of the energy dependence of the UCN loss rate on materials such as copper seemed to give contradictory results as to the form of the loss function (section 5.3). In light of the problems with hydrogen contamination this is not perhaps too surprising. More recent measurements of the loss function on hydrogen-free Fomblin surfaces have been performed using the UCN gravitational monochromator described in section 6.3.3 (Richardson *et al* 1991). It is important to know how well the Fomblin surface can be represented by an abrupt step change in the Fermi potential. The measurements formed part of a set of subsidiary experiments associated with the liquid bottle neutron β-decay lifetime experiment discussed in Chapter 7. The experimental set-up is shown in figure 5.15.

The monochromator operating in its storage energy selection mode was used to prepare a spectrum of UCN within a narrow energy band $0 < E < E_{max} = mg\Delta h$ where E_{max} was set to 8 neV for the Fomblin measurements. After monochromation the UCN were allowed access to a secondary cylindrical bottle ($r = 3.3$ cm and $L = 160$ cm) situated at height H below the monochromator vessel. The decay of UCN with energy in the range $mgH < E < mg(H + \Delta h)$ within the secondary vessel could then be investigated. The value of mgH could be varied over the full range of confinable UCN energies ($0 < E < 106$ neV). A plot of the lifetime data for Fomblin oil so derived is shown in figure 5.16 fitted with the theory of section 4.3.1, where a value of $\tau_\beta = 887$ s has been adopted. As can be seen the data are well fitted over almost the entire energy range by the assumption of a step potential. Only at points close to the Fermi potential cut-off (106 neV) was there any evidence of a discrepancy between theory and experiment. This is thought to be due to quasi-elastic warming of the UCN gas to energies slightly greater than the Fermi potential of Fomblin, due to interaction with low frequency vibrations of the bottle walls (see section 5.5.2).

The mean (RMS) energy exchange per bounce required to describe the effect was calculated to be $\Delta \bar{E}_{RMS} \approx 3 \times 10^{-11}$ eV/ bounce. This ties in with

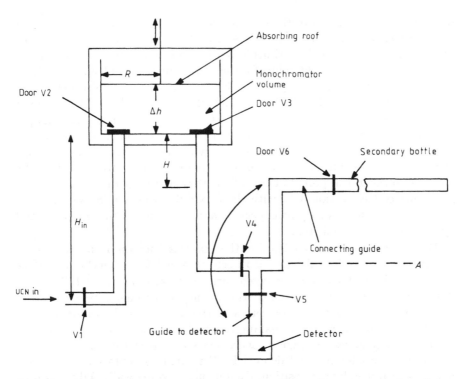

Figure 5.15   Configuration of the gravitational monochromator used to determine the energy dependence of the UCN loss rate on Fomblin liquid and grease surfaces. The experimental bottle rotates about axis A enabling the vertical separation of the mochromation and experimental bottles to be varied, while the system remained under vacuum (Richardson *et al* 1991).

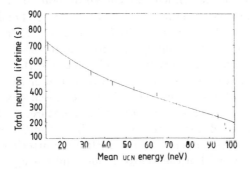

Figure 5.16   The energy dependence of the UCN lifetime in a Fomblin-oil-coated cylindrical vessel (length= 1.6 m, diameter= 6.6 cm). The full curve represents a best fit to the data using the theory of section 4.3 (Richardson *et al* 1991).

measurements looking directly for such an effect and which are discussed
later and with data obtained in the β-decay lifetime measurement (Mampe
et al 1989a and 1989b). Experiments were performed with both Fomblin oil
and grease and gave best fit values for the $2f$ parameters of 4.9 and 3.5 $\times 10^{-5}$
for oil and grease respectively at 294 K. Experiments were also performed
at temperatures of 283 and 310 K which showed that the $(2f)$ parameter
for oil varies by a factor of two over this temperature range. The magnitude and temperature dependence of the loss cross section as measured with
50 m s^{-1} neutrons is in good agreement with these observations (Mampe
1989). Measurements have also been made on the reflectivity of Fomblin
oil close to the Fermi potential cut-off (as they had been made earlier for a
glass surface (see section 5.3), with the UCN diffractometer (Steyerl 1989b);
the reflectivity was exactly as expected, assuming the oil to be represented
by an abrupt potential step.

Data on the energy dependence of the UCN loss function in the low energy
regime $0 < E < 35$ neV have also been obtained for a beryllium surface
at 10 K and also for a frozen oxygen surface (Kharitonov *et al* 1989). The
data come from a UCN storage experiment designed to determine τ_β and as
such the measurements are not very sensitive to the wall loss processes. In
fact the experiment relies on making the wall losses negligible with respect
to the β-decay losses. Storage measurements were performed in a spherical
bottle with an effective roof defined by the earth's gravitational field at a
height determined by the position of a hole in the vessel wall. Only UCN with
insufficient energy to reach the height of the hole remain confined within the
bottle. The height of the hole relative to the bottom of the bottle could be
altered by rotation of the storage vessel, enabling investigation of the decay of
UCN within well-defined energy ranges. The apparatus and principle behind
the β-decay determination are described in section 7.2.5. Energy-dependent
lifetime data from the experiment are shown in figure 5.17.

The data are well described by the theory of Chapter 4 and the assumption
that the loss probability is described by (2.70). The $(2f)$ factors deduced by
fitting the data, where $(2f)$ and τ_β are free parameters, are $2.3(0.7) \times 10^{-5}$
and $1.7(2.4) \times 10^{-6}$ for the beryllium and oxygen surfaces respectively at
10 K. The beryllium wall loss rate after a sequence of cleaning procedures,
involving heating of the surfaces to 650 K in the presence of helium and deuterium, was also examined as function of temperature. Data are presented
in figure 7.9. The fitted line corresponds to the lifetime calculated from the
known temperature dependence of the inelastic scattering cross section of
beryllium calculated within the Debye model and taking into account two-
and three-phonon processes. The data are seen to be well described by
the standard scattering theory (see section (2.4.6)). On the basis of these
experiments it seems fair to conclude that once contaminants are removed
from a material surface the average loss per bounce on that surface is well
represented by (2.70).

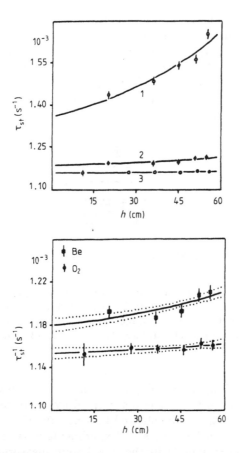

Figure 5.17 (a) Energy dependence of the UCN storage time: 1, beryllium trap (see figure 7.8, at room temperature before cleaning; 2, beryllium trap at low temperature before cleaning; 3, oxygen trap, fitted with the theory of section 4.3. (b) Energy dependence of the UCN storage time for the beryllium trap at low temperature and for the oxygen trap (Kharitonov *et al* 1989).

5.5.2   Quasi-elastic heating of UCN

The elasticity of the neutron wall reflection process is something else that needs to concern us as UCN storage measurements become more refined. The effect of vibrations of the bottle walls and the gradual effect of heating UCN to energies greater than the Fermi potential was at one stage postulated as a possible cause of the anomalously high neutron loss rates (Gerasimov *et al* 1973). The postulate that quasi-elastic heating was responsible for the anomalously large loss rates was quickly ruled out on the basis of the results from the Stoika, Hetzelt and Strelkov measurements described earlier in this chapter in which it was found that the majority of UCN leaking from bottles had energies close to thermal. However the experiments as they stood did not rule out the possibility of a smaller heating effect which might only cause a slight broadening of the UCN spectrum. In light of the current interest in UCN neutron β-decay lifetime experiments, setting a limit to the magnitude of such an effect is important. The first experiments designed to be sensitive to such an effect were performed in 1979 by Kosvinstev *et al* in the apparatus illustrated in figure 5.18.

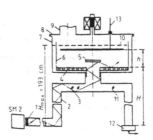

Figure 5.18   Apparatus for examining UCN storage on a plane and for detection of quasi-elastic heating: 1, converter; 2, neutron duct; 3, shutters; 4, heating elements; 5, cover plate; 6, UCN storage vessel; 7, vacuum chamber; 8, lid; 9, exhaust; 10, polythene disc; 11, shutter; 12, detector; 13, rod (Kosvinstev *et al* 1979).

A 2000 Å layer of copper was deposited on the bottle walls and the spectrum of the UCN within the bottle determined by insertion of an absorbing disc into the vessel. The results showed that no UCN with an energy greater than 28 neV had entered the vessel. Experiments were then performed in which the neutrons remaining in the vessel were measured at two storage times (50 and 190 s) as a function of the height h at which the absorbing plane was set above the base plane of the vessel. For the short storage time measurements the absorbing plane was lowered into the vessel at the start of the storage period and remained there until the vessel was emptied 50 s later. For the longer storage time, the polythene disc was inserted into the bottle 140 s into the storage period, remaining in place until the end of this counting cycle. The curves obtained are shown in figure 5.19. By comparing the shapes of the two curves (the process involved correction of the 140 s count rates for decay during storage, the standard energy dependence of $\bar{\mu}(E)$ being assumed) the authors deduced that the spectral broadening σ after the extra 140 s of storage was less than 3 neV. By adopting a one-dimensional random walk model in which the neutrons are able to either gain or lose an amount of energy ΔE per bounce at each wall reflection, by some unspecified inelastic scattering process, an upper limit on ΔE of 0.07 neV was derived. $(\sigma^2 < N\Delta E^2)$ where $N(\approx 850)$ is the mean number of wall collisions per neutron occurring during the extra storage interval.

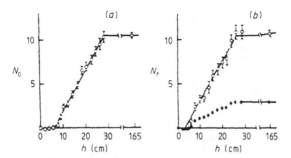

Figure 5.19   Curves used by Kosvinstev *et al* (1979) in their search for evidence of quasi-elastic heating of UCN. The plots show the number of UCN remaining in the vessel (see figure 5.18) against the height of the polyethylene disc: (*a*) the initial curve, $N_0(h_i)$, recorded immediately after filling; (*b*) the final curve, recorded 140 s after filling (full circles), and the corresponding curve, $N_t'(h_i)$, corrected for neutron decay and inelastic processes. The points for $h = 165$ cm were obtained with the polyethylene disc absent and the vessel closed with a copper lid (Kosvinstev *et al* 1979).

An experiment designed to search for the effects of quasi-elastic heating of UCN confined within a low loss Fomblin-oil-coated vessel, by low frequency surface waves or mechanical vibration of the bottle system has recently reported preliminary results (Richardson 1989). This type of effect could have important consequences in the present lifetime experiment of Mampe *et al* (1989a, 1989b), where liquid bottle surfaces are employed, and needs to be considered when calculating the gravity-induced correction to their neutron β-decay lifetime value. Surface wave heating is expected to be a more important effect on liquids because of the relatively low viscosity relative to solid and grease surfaces (Pendlebury 1990). The effects of the vibrating walls on neutron reflection leads to the UCN undergoing a random walk in velocity space, rather than a simple one-dimensional random walk in total energy. This is because the wall collision process results in discrete changes in the incident UCN's normal velocity component. The way in which this form of heating affects UCN energy distributions has been considered by Richardson (1989) and is described below.

Consider a neutron of velocity $v(\epsilon_0, 0)$ incident on a vibrating bottle wall. We assume that the velocity of the wall at any particular instant in time t is given by $v_w(t) = a\omega\sin(\omega t)$. The situation is illustrated in figure 5.20. We can treat a UCN vibrating surface collision classically providing that the wavelength of the surface vibration is greater than the wavelength of the incoming neutron ($\lambda_w > 500$ Å) and that the typical interaction time of the UCN with the moving wall ($\approx 10\times^{-7}$ s) is less than the period of wall oscillation i.e. $\omega < 10^7$ Hz. These conditions are met by low-frequency capillary waves on a liquid and low-frequency mechanical vibration of both solid and liquid bottle surfaces. (Ignatovich 1975a and Gerasimov *et al*

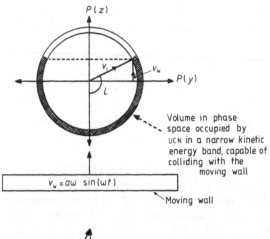

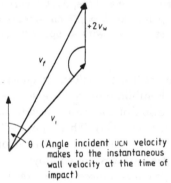

Figure 5.20   The regions in phase space capable of colliding with a vibrating wall and the effect of a wall collision on the UCN velocity.

1973, have considered the instances when both or either of these conditions are not satisfied.) In this limit we can write the following expression for the magnitude of the neutron velocity immediately after the collision $v_f(\theta, \epsilon_0)$, where θ is the angle made by the incident neutron velocity to the normal to the wall

$$v_f^2(\theta_2, \epsilon_0) = v(\epsilon_0)^2 + 4v_w(t)^2 + 4v(\epsilon_0)v_w(t)\cos\theta. \qquad (5.1)$$

A negative wall velocity means that the wall is receding from the incoming neutrons. The resulting energy change due to this collision is thus seen to be

$$\delta\epsilon_0(\epsilon_0, \theta_2) = 2m[v(\epsilon_0)v_w(t)\cos\theta + v_w^2(t)]. \qquad (5.2)$$

The average value of the energy change per collision for a neutron incident on a vibrating walled bottle is obtained by calculating the net increase in energy of the neutron gas δE due to collisions with the vibrating walls of the bottle per wall oscillation period and dividing the result for δE by the average number of wall collisions N_c occurring during this time.

In performing this calculation care has to be taken to include the effects of the wall motion and gravity on the incident neutron current distribution. Using results (4.13) and (5.2) and considering the effect of the moving walls on the equilibrium current density distribution we can write the following expression for δE where we have assumed for convenience that the whole bottle is vibrating at the frequency ω. We assume the bottle to have surface area $S(h)$ at height h and cross-sectional area $A(h)$, where $A(h)$ and $S(h)$ are even functions of h. Then

$$\int_{-R}^{R} S(h)\,dh = A_{\text{tot}}$$

is the total surface area of the bottle. As an example consider neutron storage in a cylindrical bottle lying perpendicular to the earth's gravitational field; R is seen to represent the bottle radius. It can be shown that

$$\delta E = \int_{-R}^{R} dh \int_{0}^{\tau} dt \int_{0}^{L} d\theta \, \frac{S(h)n(\epsilon_0, h, t)}{2} 2mv_{\text{w}}(t)\sin\theta[v(\epsilon_0, h)\cos\theta + v_{\text{w}}(t)]^2 \tag{5.3}$$

where

$$L = \pi/2 + \arcsin[v_{\text{w}}(t)/v(\epsilon_0, h)] \tag{5.4}$$

$$\tau = \frac{2\pi}{\omega}. \tag{5.5}$$

Performing the integration over θ we obtain

$$\delta E = \int_{-R}^{R} dh \int_{0}^{\tau} dt\, mS(h)n(\epsilon_0, h, t)v_{\text{w}}(t)\frac{[v_{\text{w}}(t) + v(\epsilon_0, h)]^3}{3v(\epsilon_0, h)}. \tag{5.6}$$

Expanding the bracket and integrating over t we find that only two non-zero terms will remain (i.e. only terms containing even powers of $v_{\text{w}}(t)$)

$$\delta E = \int_{-R}^{R} mS(h)n(\epsilon_0, h, t) \int_{0}^{\tau} \frac{[3v^2(\epsilon_0, h)v_{\text{w}}^2(t) + v_{\text{w}}^4(t)]}{3v(\epsilon_0, h)}\,dh\,dt. \tag{5.7}$$

Since we are in the limit $v(\epsilon_0, h) \gg v_{\text{w}}(t)$ we retain only the first term for the integration over h

$$\delta E = \int_{0}^{\tau} v_{\text{w}}^2(t)\,dt \int_{-R}^{R} mS(h)n(\epsilon_0, h, t)v(\epsilon_0, h)\,dh. \tag{5.8}$$

Inserting the relationships between density and height (4.7) and velocity and height (4.14) into the expression gives

$$\delta E = \int_{0}^{\tau} v_{\text{w}}^2(t)\,dt \int_{-R}^{R} mS(h)(1 - mgh/\epsilon_0)n(\epsilon_0, 0, t)v(\epsilon_0, 0)\,dh \tag{5.9}$$

which yields

$$\delta E = mA_{\text{tot}}n(\epsilon_0, 0, t)v(\epsilon_0, 0)\int_0^\tau v_{\text{w}}^2(t)\, dt. \tag{5.10}$$

The expression for the total number of collisions occurring over all surfaces in the time interval τ is obtained from

$$N_{\text{c}} = \int_{-R}^R \frac{S(h)n(\epsilon_0, h, t)}{2}\int_0^\tau\int_0^L \sin\theta[v(\epsilon_0, h)\cos\theta + v_{\text{w}}(t)]\, d\theta\, dt\, dh. \tag{5.11}$$

This can be shown on integration over θ to give

$$N_{\text{c}} = \int_{-R}^R \frac{S(h)n(\epsilon_0, h, t)}{2}\int_0^\tau v_{\text{w}}(t)(K+1) + \frac{v(\epsilon_0, h)}{2}(1 - K^2)\, dt\, dh \tag{5.12}$$

where K is given by

$$K = \frac{v_{\text{w}}(t)}{v(\epsilon_0, h)}. \tag{5.13}$$

On integration over the time we find that two terms remain

$$N_{\text{c}}(\epsilon_0, \tau) = \int_{-R}^R \frac{S(h)n(\epsilon_0, h, t)v(\epsilon_0, h)}{2}\left(\frac{\tau}{2} + \int_0^\tau \frac{K^2}{2}\, dt\right)\, dh. \tag{5.14}$$

Once again inserting the gravity factors and dropping the second-order term this simplifies to give the standard kinetic theory result

$$N_{\text{c}} = \frac{A_{\text{tot}}n(\epsilon_0, 0, t)v(\epsilon_0, 0)\tau}{4}. \tag{5.15}$$

Finally combining the two results we obtain the average energy gain per bounce as

$$\overline{\delta\epsilon_0} = 4m\frac{\int_0^\tau v_{\text{w}}(t)^2\, dt}{\tau}$$
$$= 2m(a\omega)^2. \tag{5.16}$$

If we start with a group of monochromatic neutrons of energy ϵ_0 the effect of a random walk of N_{c} steps in velocity space will be to spread the initial narrow band of neutrons centred at ϵ_0 in total energy into a Gaussian distribution with the mean centred at $\epsilon_0 + N_{\text{c}}\overline{\delta\epsilon_0}$. The dispersion $\sigma^2(\epsilon_0, N_{\text{c}})$ of this Gaussian (assuming that $N_{\text{c}}\overline{\delta\epsilon_0} \ll \epsilon_0$) is calculated from

$$\sigma^2(\epsilon_0, N_{\text{c}}) = N_{\text{c}}(\overline{\delta\epsilon_0^2}(\epsilon_0) - (\overline{\delta\epsilon_0})^2). \tag{5.17}$$

We need to calculate the average value per bounce of $\delta\epsilon_0^2$. Proceeding as before we evaluate the value of δE^2 for the whole ensemble during a wall oscillation period

$$\delta E^2 = \int_{-R}^{R} \frac{S(h)n(\epsilon_0,h,t)}{2} \int_0^T \int_0^L 4m^2 v_w^2(t) \sin\theta [v(\epsilon_0,h)\cos\theta + V_w(t)]^3 d\theta \, dt \, dh.$$

(5.18)

The angular integration gives

$$\delta E^2 = \int_{-R}^{R} 2m^2 S(h)n(\epsilon_0,h,t) \int_0^T v_w^2(t) \frac{[v_w(t) + v(\epsilon_0,h)]^4}{4v(\epsilon_0,h)} \, dt \, dh. \quad (5.19)$$

On expanding the bracket, integrating over t and only retaining the leading term we obtain

$$\delta E^2 = \int_0^T v_w^2(t) \, dt \int_{-R}^{R} 2m^2 S(h)n(\epsilon_0,h,t) \frac{v(\epsilon_0,h)^3}{4} \, dh. \quad (5.20)$$

Substituting the gravity corrections again gives

$$\delta E^2 = \int_0^T v_w^2(t) \, dt \int_{-R}^{R} m^2 S(h)(1-mgh)^2 n(\epsilon_0,0,t) \frac{v(\epsilon_0,0)^3}{2} \, dh \quad (5.21)$$

which, thanks to the symmetry and dropping the second-order terms, gives

$$\delta E^2 = \frac{m^2 A_{tot}}{2} n(\epsilon_0,0,t) v(\epsilon_0,0)^3 \int_0^T v_w^2(t) \, dt. \quad (5.22)$$

Hence on using equation (5.15) for the number of collisions per oscillation period we obtain

$$\overline{\delta\epsilon_0^2}(\epsilon_0) = 2m^2 v^2(\epsilon_0,0) \frac{\int_0^T v_w^2(t) \, dt}{\tau}$$

$$= m^2 a^2 \omega^2 v^2(\epsilon_0,0). \quad (5.23)$$

The RMS mean energy exchange per bounce $\delta\epsilon_0(\text{RMS})$ is thus given by

$$\delta\epsilon_0(\text{RMS}) = ma\omega v(\epsilon_0,0). \quad (5.23a)$$

It is to be noted that the average energy transfer per bounce (5.16) is much less than the RMS energy transfer implied by (5.23a); this is because of the near equal probability of energy loss and gain on wall collision.

The standard deviation (5.17) is obtained from (5.16) and (5.23) giving

$$\sigma^2(\epsilon_0, N_c) = N_c m^2 a^2 \omega^2 [v(\epsilon_0,0)^2 - 4a^2\omega^2] \quad (5.24)$$

which, if $v(\epsilon_0, 0) \gg 2a\omega$, becomes

$$
\begin{aligned}
\sigma^2(\epsilon_0, N_c) &\approx N_c m^2 a^2 \omega^2 v(\epsilon_0, 0)^2 \\
&\approx N_c \epsilon_0 \overline{\delta \epsilon_0}.
\end{aligned}
\tag{5.25}
$$

We thus see that the effects of vibrating walls on a neutron distribution are quite different from the effects of the simple one-dimensional walk in total energy space considered by Kosvinstev et al (1979). If we let $\Delta\epsilon_0(\epsilon_0, N_c)$ denote the energy change experienced by a neutron of initial total energy ϵ_0, after a total of N_c wall collisions, the probability distribution of $\Delta\epsilon_0(\epsilon_0, N_c)$ for large N_c is approximately given by

$$
P(\Delta\epsilon_0(\epsilon_0, N_c)) = \frac{1}{\sqrt{2\pi N_c \epsilon_0 \overline{\delta \epsilon_0}}} \exp - \left(\frac{(\Delta\epsilon_0(\epsilon_0, N_c) - N_c \overline{\delta \epsilon_0})^2}{2 N_c \epsilon_0 \overline{\delta \epsilon_0}} \right).
\tag{5.26}
$$

5.5.3   Observations of quasi-elastic heating of UCN

The first search for quasi-elastic heating of UCN (Kosvinstev et al 1979) has been described in the previous section. The experiment set an upper limit for ϵ_0(RMS) of 0.07 neV per bounce. A more sensitive technique has been developed by Richardson (1989). The set-up used by Richardson to look for evidence of this quasi-elastic scattering mechanism is shown in figure 5.21. The procedure relies upon the possibility of using a UCN monochromator (see section 6.3.3) to analyse the UCN total energy as well as to select the UCN energy and it can be described in seven stages

(i) The monochromator vessel is filled with UCN. The absorbing plane is positioned within the monochromator vessel (A), at height Δh_1 above the base plane.

(ii) The UCN are stored for a sufficiently long time period to cleanly define the upper cut-off ($\epsilon_0 = mg\Delta h_1$) on total UCN energy.

(iii) The monochromatic UCN are then allowed access to the entire experimental set-up (A), (B) and guide (C).

(iv) The initial UCN density in the secondary bottle (B) is allowed to reach its maximum value.

(v) The UCN are then stored in the secondary vessel for a time period T. During this storage period all other UCN remaining in (A) and (C) are emptied into the detector.

(vi) At the end of the storage period the neutron door V3 is opened and the neutrons allowed to flow back through (C) and (A) to the detector, recording counts $N(\Delta h_1)$.

(vii) The measurement sequence (1) to (6) is then repeated with one modification; during the UCN storage in the secondary bottle (B), the absorbing

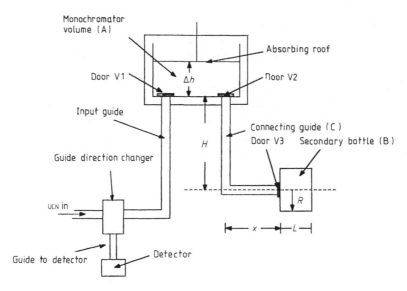

Figure 5.21   Schematic diagram of the apparatus used by Richardson (1989) in searching for evidence of quasi-elastic heating of UCN gas within a liquid walled bottle (Richardson 1989).

plane in the monochromating vessel is set to a new height Δh_2 above the base plane, where $\Delta h_2 > \Delta h_1$. The counts reaching the detector at the end of the second storage period of time T are denoted by $N(\Delta h_2)$.

The probabilities of transmitting UCN through the device depend strongly on whether the UCN have sufficient energy to rise to the height of the absorbing roof Δh_2. Subsequently changes in the neutron energy distribution close to $mg\Delta h_1$, resulting from quasi-elastic scattering, manifest themselves as a difference in the recorded count rates. Richardson (1989) developed a theory to describe the UCN transmission properties of his device, which coupled to the theory previously described and experimentally determined initial UCN spectrum, was used to derive the following limits on $\overline{\delta\epsilon_0}$ (equation (5.27)) for both Fomblin oil and grease,

$$\overline{\delta\epsilon_0}(\text{oil}) = 2(1.5) \times 10^{-14} \text{ eV} \qquad (\delta\epsilon_0(\text{RMS}) = 0.035(0.020) \text{ neV})$$

$$\overline{\delta\epsilon_0}(\text{grease}) < 6 \times 10^{-14} \text{ eV} \qquad (\delta\epsilon_0(\text{RMS}) = 0.065 \text{ neV}).$$

The results correspond to a spectral broadening of approximately 4 neV for 60 to 80 neV UCN after 8000 wall collisions and clearly indicate the high degree of elasticity associated with the UCN reflection process. Although these results should for the present be considered provisional, further more controlled and extensive measurements are under way. It should be noted that Mampe *et al* (1989a see sections 7.2.3 and 5.5.1) required a degree of quasi-elastic scattering consistent with these results to describe his Fomblin oil storage results. Measurements by Richardson *et al* (1991, see section 5.5.1) also provide evidence for the existence of quasi-elastic scattering at this level.

It is fair to conclude this chapter with the statement that UCN storage measurements have come of age. The problems of the past are now well resolved and we are now able to look in detail at the UCN loss processes. It is to be hoped that progress will continue and that UCN storage experiments will be able to yield their full potential in the realms of both fundamental and condensed matter physics.

6

UCN spectrometers, microscopes and monochromators

6.1 INTRODUCTION

UCN are generated by sources with energies spanning the entire UCN energy range $0 < E < 200$ neV. As we have seen in Chapter 4 these UCN can be transported within guides in a gas-like manner to external experiments. For some experiments working with UCN within a wide band of energy is not a problem. However, if we wish to examine processes such as UCN reflection and transmission through thin films, which depend strongly on the UCN energy or, to be more exact, on the normal component of velocity at the surface, we need to be able to define and measure the UCN velocity accurately within the UCN energy range. In this chapter we describe devices constructed to produce and detect UCN with well-defined velocity characteristics and discuss their various applications. The instruments can be roughly divided into two main classes: those that define or measure a single UCN velocity component—one-dimensional UCN spectrometers—and those that determine the total UCN kinetic energy—UCN monochromators.

6.2 UCN SPECTROMETERS DEFINING ONE-COMPONENT OF VELOCITY

6.2.1 Time-of-flight (TOF) spectrometers

The first real possibility of measuring material properties as a function of energy right down to the UCN regime occurred with the construction of a time-of-flight TOF system, coupled to a curved vertical neutron guide at the FRM reactor Munich (Steyerl 1972c).

The device, shown in figure 6.1, consists of a 12.8 m long curved electropolished guide tube fitted with a graphite converter (placed 1 cm from the reactor core). The neutron chopper system, 1.9 m above the converter,

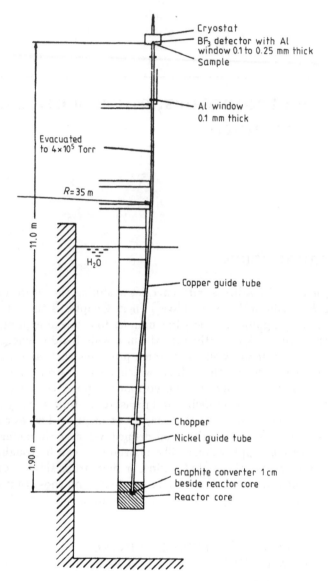

Figure 6.1   Design of the beam tube and TOF spectrometer for UCN, located at Munich (Steyerl 1972c).

is designed to pass VCN. The target and detector system are situated at the top of the guide tube.

The axial (vertical) velocity component of VCN generated in the converter and successfully passing the chopper gradually decreases as they rise against gravity in the guide tube. Their transverse velocity component remains con-

stant providing that the reflection at the guide surfaces between the chopper
and detector can be considered to be specular. This is a fair assumption
for neutrons reaching the detector. Any non-specular reflection within the
guide will almost invariably lead to VCN being scattered into trajectories
with angles of incidence greater than the guide's critical angle for confinement at their next wall collision, resulting in the loss of almost all non-
specularly reflected neutrons before they reach the detector. The degree of
non-specularity of neutron reflection on guide-like surfaces has been considered by Steyerl (1972b) and Ignatovich (1973), and is discussed in sections
2.4.5 and 4.4.5. Although it does not lead to much loss in resolution in the
TOF spectrometer, it does result in neutron loss within the guide. The TOF
from chopper to detector therefore gives a clear measurement of the neutron
axial velocity component at the target position. The spectrometer has a
neutron velocity resolution of 20% in the velocity range 5 to 100 m s^{-1}. A
plot of measured neutron intensity as a function of axial velocity is shown
in figure 6.2.

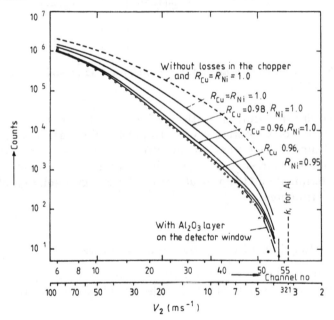

Figure 6.2   Neutron count rate as a function of axial velocity component for
the Munich TOF spectrometer (see figure 6.1). The measurements (points) are
compared with theoretical predictions based on a Maxwellian distribution emerging
from the converter and calculated with various assumptions on the guide tube
reflectivities (Steyerl 1972c).

A TOF spectrometer for UCN has also been constructed at the pulsed UCN
source at Argonne (see figure 6.3). In this instance the UCN are generated

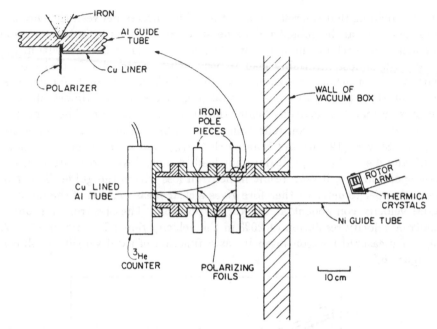

Figure 6.3 Schematic diagram of the Argonne rotating crystal TOF spectrometer (Lynn *et al* 1983).

by Doppler shifting neutrons with a velocity of 400 m s^{-1} into the UCN range by means of a rotating crystal Doppler shifter (see section (3.4.1)). The arrival time of the UCN at the detector after traversal of a short section of highly polished nickel coated guide is monitored and provides for axial velocity determination (Lynn *et al* 1983). An energy resolution of 40 neV at 10 m s^{-1} is quoted.

6.2.2 Applications of UCN TOF spectrometers

The UCN TOF spectrometers have been used in the following experimental investigations; more details on many of these applications are given in Chapter 8.

(i) Neutron transmission as a function of normal velocity component was measured for a variety of thin films (Steyerl 1972c), giving a precision of 1% on the coherent scattering length. Interference effects arising from the partial waves emanating from the two surfaces of the foil were also observed. Similar measurements but with polarized UCN have also been reported on magnetic thin films (Lynn *et al* 1983).

(ii) Total cross section measurements at low neutron energy in a variety of homogeneous and inhomogeneous substances were performed (Steyerl 1969,

Steyerl and Vonach 1972). Amongst the materials investigated were gold, aluminium, copper, glass, mica and air. A value for the Debye temperature of aluminium was derived from the temperature dependence of its inelastic scattering cross section. In addition the fact that the absorption and inelastic scattering cross sections are proportional to $1/v$ where v corresponds to the velocity of the neutron within the foil was verified experimentally by these authors.

(iii) Measurements of total loss cross sections of inhomogeneous samples in the low energy range can be used to yield information relating to microscopic structure, e.g. scattering zone size and density, on a distance scale of 100–10000 Å ((Steyerl 1974) see also section 8.3.2). The accuracy and validity of the technique was tested by performing total scattering cross section measurements on a mixture of D_2O and H_2O containing a known amount of SiO_2 spheres (diameter 130–140 Å) (Lengsfeld and Steyerl 1977). A sphere radius of 72(21) Å was obtained experimentally in excellent agreement with the known diameter. The method was subsequently employed to investigate domain size and wall thickness in polycrystalline iron, nickel and cobalt samples (Lermer and Steyerl 1976).

6.2.3   The UCN diffractometer

The UCN diffractometer (Scheckenhofer and Steyerl 1977, 1981) was constructed to study the neutron optical properties of UCN in detail. Both the reflection and diffraction of low energy neutrons have been investigated with the device. Energy analysis is based upon determining the maximum height $H_{max}(= E/mg)$ to which a UCN can rise in the earth's gravitational field. However it should be noted that the technique only defines, or measures, the vertical component of neutron velocity, i.e. $E = mV_z^2/2$.

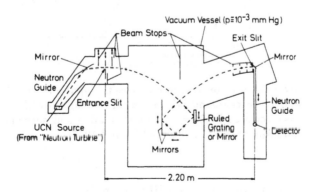

Figure 6.4   Schematic view of the UCN gravitational diffractometer (Schekenhofer and Steyerl 1977).

A schematic diagram of the diffractometer is given in figure 6.4. The combination of mirrors and beam stops at the device input are arranged so that only UCN with a horizontal velocity at the entrance slit are admitted into the vessel. The UCN with an initial horizontal velocity of approximately 3 m s^{-1} hit the first vertical mirror (or grating). The neutrons reflect from the mirror and, after further reflection from two more nickel coated mirrors (all mirrors have adjustable horizontal and vertical positions), are sent on a trajectory towards a further energy analysis unit consisting of beam stops, exit slit and mirror. UCN successfully passing through the analysing stage are reflected from the final mirror and collected at the detector. The whole detector system can be moved up and down to scan the beam profile. The UCN are confined laterally during the course of their motion by vertical plane mirrors fabricated from float glass and separated by 10 cm. UCN typically make 10 to 50 reflections with these vertical mirrors on their trip through the system and so the mirror quality and alignment of these surfaces has to be very good to avoid loss of intensity and to obtain optimum resolution. By positioning the sample appropriately the device permits examination of both the reflectivity/transmission of polished surfaces/films as a function of incident neutron energy and of the reflection/transmission processes involving energy transfer in the vertical velocity component. The resolution curve obtained by measuring the UCN throughput as a function of relative vertical slit separation, from a set-up incorporating all nickel mirrors, and with input and exit slit widths of 2 cm ($\approx$ 2 neV), is shown in figure 6.5.

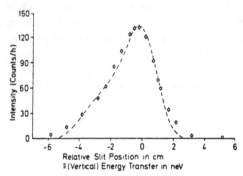

Figure 6.5 UCN diffractometer resolution curve for input and exit slit settings of 2 cm: measured points (background subtracted) in comparison with the theoretical curve (broken), as calculated by Monte Carlo simulation and normalized with the measured peak intensity (Scheckenhofer and Steyerl 1977).

6.2.4 Applications of the UCN diffractometer

The diffractometer has been used in a variety of experimental investigations of neutron reflection and transmission processes.

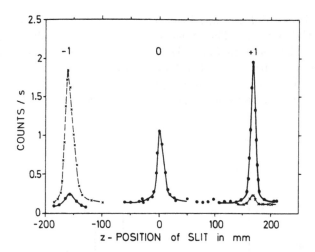

Figure 6.6 UCN diffraction data for a blazed ruled grating. The orders ±1 and 0 were measured with the grating in the normal position (full lines) and in a position reversed by 180° (broken lines) (Steyerl *et al* 1988a).

(i) Momentum changes associated with diffraction. When a neutron is diffracted from a grating the component of the wavevector parallel to the surface changes by $2\pi n/t$ where n is the order of diffraction and t is the line spacing. By placing a diffraction grating at the position of the vertical mirror (see figure 6.4), one can examine the corresponding energy change by scanning the vertical intensity profile resulting from UCN reflection at the grating, with the energy analysing slit. The first experiments were performed at Garching with a slit width of $\Delta Z = 2$ cm (Scheckenhofer and Steyerl 1977); these were repeated with higher resolution $\Delta Z = 0.4$ cm when the device was installed at the more intense ILL source (Steyerl *et al* 1986). Diffraction peaks obtained on the ILL source with a ruled grating of 1200 groove/mm, blazed for first-order UCN reflection, are shown in figure 6.6. Efficiencies of 14% and 8% were observed for the orders 1 and 0 respectively. The FWHM of the peaks was 1(0.1) neV, comparable with the instrument resolution. As pointed out by Steyerl *et al* (1988a), absence of line broadening at this level of resolution would indicate the neutron wave train is at least 2 mm long and would provide a new limit to the speculative nonlinear term in the Schrödinger wave equation of 10^{-17} eV (Bialynicki-Birula and Mycielski 1976). This value is a factor of 100 lower than the present limit (Gahler *et al* 1981). It seems that so far the question of whether the lines actually obtained were broadened has not been satisfactorily resolved (Steyerl 1989a). The less sensitive initial experiments enabled a lower bound on the neutron wave train of 0.1 mm to be set.

(ii) The reflection properties of borosilicate glass were investigated at UCN energies close to and slightly above the Fermi potential cut-off (Scheckenhofer and Steyerl 1977). These experiments yielded information on the surface Fermi potential profiles. The experiments and their implications are discussed in detail in section 5.3. Measurements have also been performed recently on reflectivity from a Fomblin oil surface and have shown that the Fermi potential profile at the surface just above the cut-off is well described by the assumption of an abrupt potential step (Steyerl 1989b, see section 5.5.1).

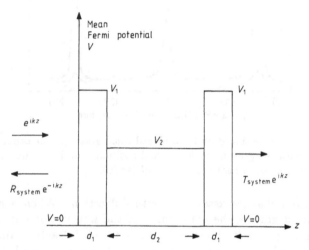

Figure 6.7   Neutron reflection from a one-dimensional potential-well-type barrier as investigated by Steinhauser *et al* (1980).

(iii) Investigations into the transmission and neutron trapping properties of multi-layer films were performed (Steinhauser *et al* 1980). Consider reflection of a plane wave from the Fermi potential distribution composed of three rectangular potentials of height V_1 and V_2 (see figure 6.7). One can follow a procedure similar to that of section (2.4.3), applying continuity of the wavefunction and its derivative at each change in potential. The effects of multiple internal reflections within the sandwiched layer from the potential boundaries also need to be included. The system constitutes the neutron analogue of the optical Fabry–Perot etalon (Born and Wolf 1959). After some effort the following expression for the reflectivity and transmittance of the system are obtained

$$|R_{\text{system}}|^2 = 1 - |T_{\text{system}}|^2 = 1 - \frac{|T|^4}{|T|^4 + 4|R|^2 \sin^2(kd_2 + \phi)} \qquad (6.1)$$

where $|R|^2$ and $|T|^2 = 1-|R|^2$ are the reflection and transmission coefficients for a single barrier of height V_1 and width d_1 sandwiched between vacuum and a semi-infinite potential of V_2 and ϕ is the phase shift on reflection from

the barrier. Expressions for these parameters are easily obtained. Resonance
transmission is seen to occur for

$$k = k_n = \frac{(n\pi - \phi)}{d_2} \qquad \text{with } n = 0, 1, 2, \dots . \qquad (6.2)$$

The resonances correspond to quasi-stationary bound states of the neutron
in the potential well defined by the region of low Fermi potential. The energy
of the state is given by

$$E_n = \frac{\hbar^2 k_n^2}{2m} + V_2 \qquad (6.3)$$

and associated lifetime by

$$\tau = \frac{\hbar}{2\Gamma} \approx \frac{m(d_1 + d_2)|R|}{\hbar k_n |T|^2}. \qquad (6.4)$$

For a suitable choice of system parameters $V_1 > E_n > V_2$ this lifetime can
be made reasonably long. Even more complicated resonance phenomena can
be expected when multiple well potential distributions are considered.

Transmission and reflection from such systems were examined by Steyerl
and collaborators at the FRM reactor at Garching (Steinhauser *et al* 1980,
Steyerl *et al* 1981). Reflectivity measurements from a single potential well
are shown in figure 6.8(*a*). The data are well described by the simple theory outlined here: the lifetime of the quasi-stationary state is estimated to
be 0.1 μs, corresponding to four round trips by the neutron in the cavity.
Transmission through a double well structure is shown in figure 6.8(*b*). In
this instance the individual single well states are coupled by the possibility of
tunnelling through the central barrier. This leads to a splitting of the single
resonator state into two separate states, one symmetric, the other asymmet-
ric. This behaviour is well borne out by the data. Reflection from systems
of multiple coupled resonators have been performed with 4 Å neutrons at
grazing incidence (Steyerl *et al* 1985). These experiments demonstrate most
elegantly the quantum mechanics of scattering from a one-dimensional potential.

(iv) The idea of using the diffractometer as a source of monochromatic
neutrons for interferometric UCN measurements has been discussed by Steyerl *et al* (1979). UCN interferometry allows longer pathlength differences,
larger spatial path separation and increased observation times relative to
conventional crystal and single slit interferometers; these factors can be important in certain experiments such as the search for a possible net charge
on the neutron (Baumann *et al* 1988) and Aharanov-Bohm-type effects with
neutrons (Cimmino *et al* 1989). There is, however, an inherent intensity
problem due to the fact that only low UCN fluxes are available. The diffractometer accepts a wide bandwidth of horizontal UCN velocity components,

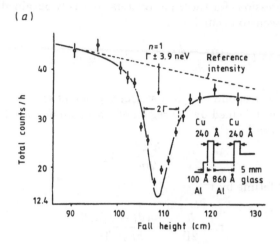

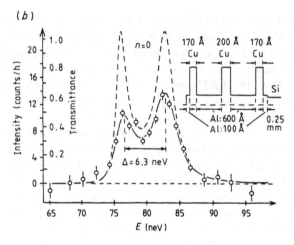

Figure 6.8 (a) Reflection data for a sample with the indicated film sequence which corresponds to a double hump potential. (b) Level splitting observed in UCN transmission through a sample with the indicated coupled resonator structure (Steyerl *et al* 1988a).

whilst accurately defining and analysing the vertical velocity component with which one can perform interferometry. This makes for efficient use of the incident spectrum.

Several possible Michelson-type interferometer configurations, employing semi-transparent mirrors as beam splitters, have been proposed for use within the diffractometer (Steyerl *et al* 1979). The engineering tolerances of the mirrors and beam splitters required to realize such a scheme are quite stringent and so far no experiments have been performed.

6.3   UCN MONOCHROMATORS

6.3.1   Introduction

In the previous sections we have been concerned with UCN velocity selectors which provide UCN with an accurately defined or analysed velocity component. This is ideal for examination of those reflection and transmission properties of UCN, such as mirror reflection, which are solely defined by the neutron's normal component of velocity. In fact, in such instances, we wish to have as wide a bandwidth in the transverse velocity component as possible to make the best possible use of the available flux. However, if we wish to make the following measurements then we need to produce UCN with accurately defined total energies, i.e. monochromatic UCN

(i) Energy-dependence measurements with UCN gas, where the velocity components are readily scrambled on multiple reflection from surfaces. UCN gas measurements offer a whole new range of experiments as they offer the possibility of allowing a single neutron to interact with a target many times (Richardson 1989). For example one can easily examine the average probability of neutron loss on reflection $\mu(E) \approx 10^{-5}$, which is so small as to be impossible to measure directly in a single interaction experiment.

(ii) Measurements on intrinsically rough surfaces, such as a grease, on which one cannot accurately define the surface normal.

(iii) Conventional inelastic scattering measurements with UCN. In such an application the final total energy of the neutron also requires measurement.

In this section we describe devices designed to simply produce monochromatic UCN with which one can address experiments of types (i) and (ii), before closing the chapter with a discussion of UCN total energy spectrometers and microscopes.

6.3.2   UCN gas monochromators

The first attempts to obtain monochromatic UCN were made in the early days of UCN physics and made use of the neutron's gravitational interaction. Neutrons emanating from a UCN source were forced to pass through the crank-style device illustrated in figure 6.9 before reaching a detector (Groshev et al 1971). The height of the horizontal guide section could be adjusted to lie anywhere between $0 < h < 1.5$ m above the input guide. UCN with insufficient energy to reach the height of the horizontal guide were unable to get through the crank to the detector, thus the device passed only UCN within the energy range $mgh < E < 165$ neV where 165 neV is the Fermi potential of the guide material. This form of crank was used to measure the SM-2 source spectrum (see section 3.3.3 and figure 6.10).

A similar device was employed to measure spectra by Kosvintstev et al (1977), however a modification was employed in one instance. A good UCN

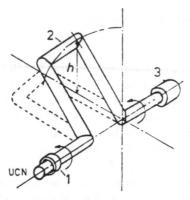

Figure 6.9 The rotating crank monochromator: 1, rotating vacuum joint; 2, elbow; 3, detector (Kosvinstev *et al* 1977).

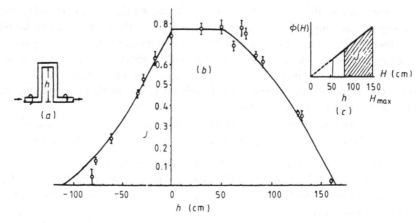

Figure 6.10 UCN spectrum of the SM-2 reactor as measured with a crank monochromator: (*a*) monochromator; (*b*) UCN transmission through the elbow on channel 4; and (*c*) assumed UCN spectrum (Groshev *et al* 1971).

absorber, polythene which has a slightly negative mean Fermi potential and large inelastic scattering cross section, was placed along the upper surface of the horizontal guide, see figure 6.11.

In this instance only UCN in the range $mgh < E < mg(h + \Delta h)$ are passed by the system; UCN with sufficient energy to rise to the height of the polythene are likely to be absorbed in attempting to traverse the horizontal guide. The exact resolution of the device is difficult to assess since it relies on non-specular reflection in the horizontal guide section to effect monochromatization; if all reflections were specular the device would merely velocity-select in the vertical direction. A resolution of 10 neV was quoted. A similar set-up was used to make energy-dependent measurements of the neutron decay time in material traps (Groshev *et al* 1976) and a resolution of

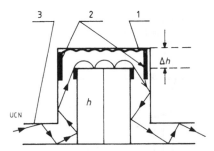

Figure 6.11 Differential crank monochromator designed to pass UCN with an energy in the range $mgh < E < mg(h + \Delta h)$: 1, UCN absorber (polythene); 2, roughened guide inserts; 3, input guide (Kosvinstev *et al* 1977).

25 neV in the range 0 to 180 neV was obtained. Crude energy selection (20–30%) was obtained by Steyerl and Trustedt (1974) by passing UCN through thin-film neutron filters; the UCN were then used to make storage measurements (see section 5.3). Storage measurements using monochromatic UCN in the energy range 0 to 30 neV have also been reported in UCN bottles fitted with neutron absorbing roofs (Kosvinstev *et al* 1982, see section 5.4) but as of yet only one system for producing and supplying monochromatic UCN for external experimentation has been constructed (Richardson 1989). The device and its applications are discussed in detail below.

6.3.3 The UCN gravitational monochromator

As with the previously described crank devices, energy selection within the UCN monochromator shown in figure 6.12, is performed by allowing only those UCN with a maximum possible rise height $H = E/mg$ within a well-defined range to pass through the system. The apparatus consists of a special cylindrical storage vessel equipped with two doors. One door leads to the UCN source and the other to the secondary experiment which is situated at a vertical height H below the monochromator base plane; exactly what is meant by secondary experiment will become clearer in the sections that follow. The monochromator vessel is equipped with a polythene roof which can be lowered or raised relative to the base plane. The height of the roof is defined as Δh. The side and base walls of the vessel are coated in a low UCN loss Fomblin grease coating ($\bar{\mu} \approx 10^{-5}$) (section 5.5.1). In addition to being low-loss the grease is a macroscopically rough surface ensuring that the UCN velocities are effectively randomized at each wall collision.

For the sake of this discussion we shall describe UCN by their total energy ϵ_0 at the base plane of the vessel; we define the potential energy to be zero at this level. When UCN are admitted into the vessel their fate depends almost entirely on whether their total energy is greater than or less than $mg\Delta h$. If

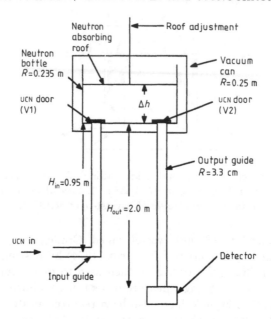

Figure 6.12   The UCN gravitational monochromator configured to measure the device's throughput as a function of resolution setting Δh (Richardson 1989).

we consider the storage of UCN within the vessel, we can use (4.16) to write the following expression to describe the decay of UCN of total energy ϵ_0:

$$\frac{\mathrm{d}}{\mathrm{d}t}N(\epsilon_0, t) = -\frac{N(\epsilon_0, t)}{\tau_{\text{tot}}(\epsilon_0)} \qquad (6.5)$$

where $\tau_{\text{tot}}(\epsilon_0)$ is given by

$$\frac{1}{\tau_{\text{tot}}(\epsilon_0)} = \frac{1}{\tau_{\text{side}}(\epsilon_0)} + \frac{1}{\tau_{\text{base}}(\epsilon_0)} + \frac{1}{\tau_{\text{abs}}(\epsilon_0, \Delta h)} + \frac{1}{\tau_\beta} \qquad (6.6)$$

where we have split the surface integral in (4.16) into discrete components corresponding to the various bottle surfaces. Substituting $A(h) = \pi R^2$, where R is the bottle radius, into expression (4.11) for the effective real space volume, and inserting the appropriate expressions for the individual component surface areas into (4.16), we obtain the following expressions for the individual loss rate components

$$\frac{1}{\tau_{\text{side}}(\epsilon_0)} = \frac{3g}{2Rv(\epsilon_0, 0)} \int_0^{\epsilon_0/mg} (1 - mgh/\epsilon_0)\bar{\mu}(\epsilon_0, h)\,\mathrm{d}h \qquad (6.7)$$

$$\frac{1}{\tau_{\text{base}}(\epsilon_0)} = \frac{3g\bar{\mu}(\epsilon_0, 0)}{4v(\epsilon_0, 0)} \qquad (6.8)$$

$$\frac{1}{\tau_\beta} \approx 885 \text{ s}$$

and

$$\frac{1}{\tau_{abs}(\epsilon_0)} = \begin{cases} 0 & \text{if } \epsilon_0 < mg\Delta h \\ \dfrac{(\epsilon_0 - mgh)v(\epsilon_0)\pi R^2 \bar{\mu}_{abs}(E(\Delta h))}{4\epsilon_0 \gamma(\epsilon_0, \Delta h)} & \text{if } \epsilon_0 > mg\Delta h. \end{cases}$$

(6.9)

$v(\epsilon_0, 0)$ is, as in Chapter 4, the velocity of a neutron of total energy ϵ_0 at the level of the base plane, $\bar{\mu}(E(h))$ is the averaged loss probability per bounce for a neutron of kinetic energy $E(h) = \epsilon_0 - mgh$ (equation (2.70)), $\gamma(\epsilon_0, \Delta h)$ is the previously defined effective real space volume (4.11) with the upper limit Δh on the integral, i.e.

$$\gamma(\epsilon_0, \Delta h) = \pi R^2 \int_0^{\Delta h} \sqrt{\frac{(\epsilon_0 - mgh)}{\epsilon_0}} \, dh.$$

(6.10)

$\bar{\mu}_{abs}(E(\Delta h)) \approx 1$ is the averaged loss probability per bounce on the neutron absorbing roof for a neutron of incident kinetic energy $E(\Delta h) = \epsilon_0 - mg\Delta h$. $\bar{\mu}_{abs}(E(\Delta h))$ can be evaluated for a given absorber from (2.67) (2.69) and (2.71). For materials such as polythene with zero or slightly negative real components of Fermi potential, the loss function is close to unity. Rather suprisingly (2.71) shows that materials such as cadmium and gadallinium, with enormous loss cross sections such that $W \approx V$, can make reasonable reflectors of UCN with energy $E \ll W - V$. Reflection in the non-existent regime $W \gg V$ has been considered by Gurevich and Nemirovskii (1962). In this regime the large complex component of the mean Fermi potential gives rise to strong reflection of low energy neutrons. This effect has become known as 'metallic reflection of UCN' after the well known optical analogue when strong absorbers of light start to reflect once the attenuation length of the light in the medium approaches the optical wavelength regime (Landau and Lifschitz 1960). Reflection of UCN from strong absorbers has been investigated experimentally by Morozov et al (1987) and Bates (1978). For a reasonable bottle geometry and choice of wall coating, $1/\tau_{abs}$ can be made several orders of magnitude larger than any of the other loss-rate contributions. A graph of the total bottle lifetime as a function of energy for a Fomblin-grease-coated monochromator bottle, with an assumed 0.4 cm^2 area of gaps, through which UCN are able to leak is shown in figure 6.13.

Storage of UCN in such a bottle thus offers great potential for monochromatization; neutrons with energy $\epsilon_0 > mg\Delta h$ are lost rapidly from the system whilst those with lower energy are oblivious to the existence of the roof and suffer few losses. The monochromator thus produces UCN with energies in the range $0 < \epsilon_0 < mg\Delta h$. By emptying the bottle and allowing these neutrons to fall a further distance H to the secondary experiment we can supply UCN in the range $mgH < E(H) < mg(H + \Delta h)$ in the experiment. If we assume the bottle to have been filled to its equilibrium density originally and neglect

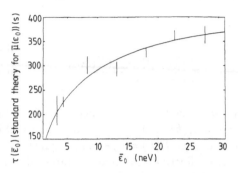

Figure 6.13   Experimentally determined energy dependence of the UCN loss rate within the UCN monochromator, monochromatization vessel (see figure 6.12), fitted with the theory of section 6.3.3. The predominant loss during these measurements was due to a small gap around the neutron valves. In later experiments the gap was reduced and lifetimes of the order 500–600 s were obtained (Richardson 1989).

losses occurring during the monochromatizing storage period, we can easily derive an expression for the expected energy spectrum of UCN in the range $0 < \epsilon_0 < mg\Delta h$. We use (4.11) and (4.10) and setting $a = 0, b* = \epsilon_0/mg$ and $A(h) = \pi R^2$ we obtain

$$N(\epsilon_0, t)\, d\epsilon_0 = \int_0^{\epsilon_0/mg} n(\epsilon_0, t, h)\pi R^2\, dh\, d\epsilon_0$$

$$= \pi R^2 n(\epsilon_0, t, 0)\int_0^{\epsilon_0/mg} \sqrt{\frac{(\epsilon_0 - mgh)}{\epsilon_0}}\, dh\, d\epsilon_0$$

$$= \gamma(\epsilon_0)n(\epsilon_0, t, 0)\, d\epsilon_0 \qquad (6.11)$$

where $\gamma(\epsilon_0)$, the effective real space volume for this geometry, is

$$\gamma(\epsilon_0) = \frac{2\epsilon_0\pi R^2}{3mg}. \qquad (6.12)$$

The effective real space volume for a given UCN is seen to be linearly dependent on ϵ_0 and equal to $\frac{2}{3}$ of the real space volume actually accessible to that neutron. If we assume a Maxwellian source density i.e. $n(\epsilon_0, 0, 0)\, d\epsilon_0 = \rho_0\sqrt{\epsilon_0}\, d\epsilon_0$ we obtain the following expression for the total number of UCN trapped in the bottle

$$N(\epsilon_0 < mg\Delta h, 0) = \int_0^{mg\Delta h} \rho_0 \epsilon_0^{1/2}\gamma(\epsilon_0)\, d\epsilon_0$$

$$= \frac{2\pi R^2 \rho_0}{3mg}\int_0^{mg\Delta h} \epsilon_0^{3/2} d\epsilon_0$$

$$= \frac{4\pi R^2 \rho_0 (mg\Delta h)^{5/2}}{15mg}. \qquad (6.13)$$

A plot of the UCN yield as a function of Δh obtained using the monochromator on the VCN beam on the ILL UCN source (section 3.3.2) is shown in figure 6.14; the data are well described by this theory.

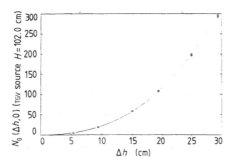

Figure 6.14   Monochromator throughput as a function of resolution setting as obtained on the ILL TGV UCN source (direct VCN beam). The data are well fitted (full line) by the assumption that the source produces a Maxwellian UCN spectrum (see equation (6.13)) (Richardson 1989).

In this storage mode the monochromator can be used to produce UCN within narrow bands of total energy of full width in the range $1 < \Delta\epsilon_0 = mg\Delta h < 30$ neV. These can be accelerated to yield UCN with kinetic energies within the same bandwidth centred anywhere within the range $1 < E(H) < 350$ neV. The device can also be operated in a continuous mode when ultrahigh resolution is not required and it behaves in a manner similar to the earlier crank devices. The energy definition is far better because of the rough nature of the grease and the long time (≈ 30 s) a UCN takes to pass through the system. The resolution can be further controlled by varying the diameter of the input and output apertures. These effectively control the mean time a given UCN remains in the bottle before 'finding' the exit aperture ($\propto$ aperture area) and subsequently the probability it does so before being absorbed or lost from the system. This probability is highly dependent on whether the neutron has sufficient energy to reach the absorbing roof and can be used to optimize throughput($\propto$ aperture area) for a required resolution (Richardson 1989).

A plot of transmission probability against neutron energy as a function of port radius for the device operating in continuous mode is shown in figure 6.15; for small aperture settings the device has an extremely well-defined passband. The transmission for $\epsilon_0 < mg\Delta h$ is seen to be 0.5; this is because neutrons entering the vessel within this energy range have such a long lifetime that they are almost certain to leave the bottle through the input or output port, if the aperture areas are the same both scenarios are equally probable.

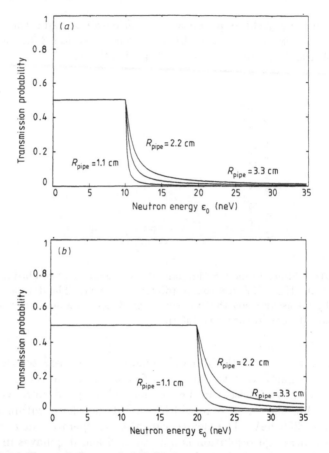

Figure 6.15 Transmission probabilities for UCN entering the UCN monochromator energy selection vessel as a function of energy ϵ_0 and port radius R_{pipe} for a resolution setting of (a) $\Delta h = 10$ cm and (b) $\Delta h = 20$ cm based on the assumption of a Fomblin-grease-coated monochromator vessel fitted with a polyethylene roof $V_{\text{F}}^{(\text{abs})}$ (absorber's Fermi potential) $= -9.0$–0.5i neV: R (bottle radius) $= 25$ cm; V_{F} (wall's Fermi potential) $= 106$ neV; $2f = 2 \times 10^{-5}$; $A_{\text{gap}} = 0.4$ cm^2; $\tau_\beta = 885$ s (Richardson 1989).

6.3.4   Applications of the UCN gravitational monochromator

The UCN monochromator, although still a relatively new device, has been employed in a variety of experiments.

(i) Measurement of the energy dependence of UCN loss rates within material bottles and the determination of $\bar{\mu}(E)$. The UCN monochromator has been employed in a storage mode to make measurements of the decay rate of UCN in material traps as a function of energy. The device is used to select UCN within narrow bands of total energy $0 < \epsilon_0 < mg\Delta h \approx 8$ neV and the output used to fill a secondary bottle situated at a height H below the monochromator. H can be varied during the course of the experiments. Storage measurements are then performed in the secondary vessel which enable the decay rate and subsequently the loss function $\bar{\mu}(E(H))$ to be determined. Although the energy of the UCN changes across the vertical extent of the bottle the effect this has on the determination of the loss function can be minimized by selecting a bottle with small vertical dimensions and a symmetry axis perpendicular to the gravitational field as discussed in section 4.3.1. Loss functions have been determined for both Fomblin oil and grease surfaces. Data from the experiments and their interpretation and implications are discussed in section (5.5.1). The technique is likely to be extended to a variety of materials in the future (Richardson 1989).

(ii) The monochromator has been used in a selector/analyser configuration to search for spectral broadening during the course of neutron storage experiments, resulting from quasi-elastic heating of UCN as they reflect from bottle walls (Richardson 1989). Heating can be caused either by mechanical vibration or, in the case of liquids, by Rayleigh-type surface waves; their effect on UCN spectra is discussed in section (5.5.2). The experimental configuration, measuring procedure and initial experimental results are described in Chapter 5. Preliminary data indicate possible quasi-elastic heating effects at the level of $\delta\epsilon_0(\text{RMS}) = 3.5(2.0) \times 10^{-11}$ eV/collision for UCN on a Fomblin oil surface, although further measurements are required to substantiate this result. By placing a suitable target within the secondary vessel it may eventually prove possible to study low energy exchange, inelastic scattering, from materials by analysing the spread in spectral shape of the confined UCN due to their repeated interaction with the sample. If as an example one considers the interaction of UCN with a target constituting a two-level system—a collision in which a UCN can either gain or lose a unit δE of energy, with equal probability p—the effect of N collisions will be a random walk by that neutron in total energy space. If one starts with a monochromatic spectrum of neutrons then after N collisions one will have a Gaussian distribution centred about the same mean energy but of width $\Delta E = \sqrt{Np\delta E}$. The possibility of allowing a single neutron to interact with a target, thousands of times, coupled to the inherent sensitivity of the device to total neutron energy ($\Delta E = 1$ neV) could allow one to reach sensitivities of 10^{-10} to 10^{-11} eV/interaction (Richardson 1989).

(iii) Fermi potential determination of thin films.   The monochromator produces a UCN spectrum with a very sharply defined upper cut-off ($\approx \pm 0.1$ neV). This feature can be used to determine the Fermi potential of thin films. The technique requires a thin sample of the test material 1–100 μm, but does not require smooth surfaces, as one does with conventional reflectivity-type Fermi potential determination techniques (interferometry, reflectometry). The set-up for mean Fermi potential measurements is shown in figure 6.16. The sample is placed perpendicular to the gravitational field,

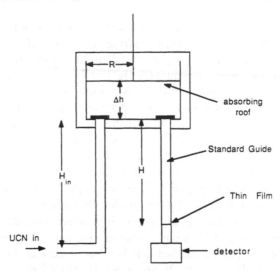

Figure 6.16 A schematic diagram of the apparatus used to determine the mean Fermi potential of thin films (Richardson 1989).

across a section of guide at a height H below the monochromator, the detector is positioned directly beneath the foil. The procedure for Fermi potential determination is to use the monochromator in its storage mode to accurately define the maximum kinetic energy of the neutron energy group at the level of the foil, i.e. $E_{\max}(H) = mg(H + \Delta h)$. Once the spectrum cut-off has been well defined the output valve is opened and the foil becomes, in effect, an integral part of the bottle. If any UCN has an energy at the foil greater than its mean Fermi potential i.e. $E(H) > V_{\text{foil}}$ there is a finite probability that the UCN will strike the foil at such an angle to the surface normal to be transmitted, pass through the foil and be detected at the detector. An individual UCN may make many bounces on the foil before being either transmitted or lost from the system. By gradually increasing $E_{\max}$ until UCN are first detected at the detector one can establish the mean Fermi potential. Figure 6.17 shows data obtained on the UCN source at ILL for a 100 μm Al foil. The value obtained $V(\text{Al})=54.2(1)$ neV is in good agreement with the accurately known value of 54.1 neV, so illustrating the validity of the technique.

The ultimate sensitivity of the method depends on a wide range of parameters, source intensity, foil thickness, loss cross section etc, but 1 neV is about the limit due to the fact that the loss cross section ($\propto 1/v'$ where v' is the neutron velocity in the foil) becomes extremely large for UCN with energies slightly above V_{f}. The real advantage of the monochromator over other conventional techniques is the possibility of analysing intrinsically rough surfaces—the Fermi potentials of Fomblin grease (107.5(+1,-2) neV (Richardson 1989)) and deuterated polystyrene (170.3(+1,-2) neV (Lamoreaux 1988)) have been determined in this manner. The transmission

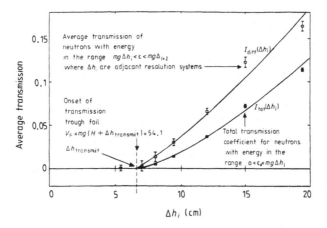

Figure 6.17 Determination of the Fermi potential of a thin aluminium film using the apparatus illustrated in figure 6.16. The full squares are the values of the average total transmission through the foil for UCN with kinetic energies at the foil in the range $mgH < E < mg(H + \Delta h_i)$. The points with error bars correspond to average transmission values of UCN with kinetic energies at the foil in the range $mg(H + \Delta h_i) < E < mg(H + \Delta h_{i+1})$, where Δh_i and Δh_{i+1} are adjacent resolution settings. UCN transmission through the film is seen to begin at a monochromator resolution setting of $\Delta h_{\mathrm{transmit}} \approx 6.5$ cm, giving a value for the mean Fermi potential of alminium of 54.0($\pm$1) neV (Richardson 1989).

properties of polarizing foils have also been investigated with monochromatic polarized UCN (Crampin 1989). The Fermi potential for the two spin states $V_{\pm} = V_f \pm \mu B$ were determined for Fe and Fe/Co polarizing foils as a function of applied magnetic field. A polarizing foil and adiabatic spin flipper was placed in the system between the first foil and the monochromator. A small area collimator made of neutron-absorbing material could also be inserted into the system when required, to perform transmission measurements, under conditions of a single neutron pass. This is required when determining the upper cut-off $V = V_f + \mu B$ since there is a finite probability ($\approx 5\%$) of spin-flip as a neutron reflects from a polarizing foil (Miranda 1987, Crampin 1989). From the determined nuclear potential $V_f = (V_+ + V_-)/2$ the composition of the Fe/Co polarizers could be established and compared with the Fe/Co mixture from which they were evaporated. The results showed that the foils were more Fe-rich than had been previously supposed. Further calculation incorporating data concerning the relative evaporation rates of Co and Fe agreed well with the measured value. This should in future lead to improved polarizer design for which a value $V_- = 0$ neV is desired.

(iv) Richardson and Golub (1989) have proposed using the monochromator as a UCN spectrum analyser in the study of the UCN spectrum produced by the helium superthermal source at ILL (see section 3.5.2). By sampling the spectrum as a function of vertical distance H one is able to locate the upper and lower cut-offs on the spectrum. Blocking or absorption of the generated UCN by contaminants condensing on the systems cold windows (e.g. N_2) is a possible cause of the low UCN production rates observed in the first reported operation of the source (Golub *et al* 1983). The postulate of source emptying problems has been confirmed by recent up-scattering experiments which show that UCN are being produced within the interior of the device at the predicted rate (Kilvington *et al* 1987). In addition the energy dependence of the UCN decay during storage within the ^{4}He filled generation chamber can be determined by passing the UCN remaining in the chamber, after a given storage time, through the monochromator (as a function of height above source) on emptying to a detector.

6.4   UCN TOTAL ENERGY SPECTROMETERS

6.4.1   Introduction

The principles of UCN scattering are covered in detail in Chapters 2 and 8. At the present only one UCN spectrometer, NESSIE, is in existence. The letters stand for 'neutron gravitational inelastic spectrometry' in German. This gravitational spectrometer designed and built by Steyerl (1978) is based upon the idea of energy selection and determination by measurement of the maximum horizontal reach of UCN trajectories in the earth's gravitational field. The device permits measurement of inelastic processes 10 neV$< \Delta E <$ 100 neV at a momentum transfer of $q \approx 0.03$ Å^{-1}, and quasi-elastic processes down to a resolution of 0.1 neV. The scattering function $S(Q, \omega)$ is measured directly with the technique.

6.4.2   The principle of reach analysis

Consider the situation depicted in figure 6.18 where a neutron of energy E is launched from a given point X at angle α to the horizontal. We are interested in the resulting trajectory of the neutron and in particular with finding its maximum horizontal displacement before collision with the base line plane, tilted at angle β to the horizontal. For convenience we consider motion confined to the ZX plane.

The height of the neutron at time t after launch is given simply by

$$Z(t) = V_z(0)t - \frac{gt^2}{2} \tag{6.14}$$

and the corresponding horizontal position by

$$X(t) = V_x(0)t \tag{6.15}$$

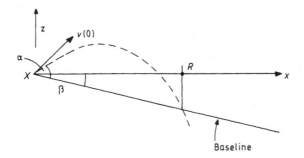

Figure 6.18   Diagram illustrating the principle of reach analysis.

where $V_x(0) = (\sqrt{2E/m})\cos\alpha$ and $V_z(0) = (\sqrt{2E/m})\sin\alpha$ are the initial horizontal and vertical velocity components respectively. The time and point of impact with the baseline can be obtained directly from equations (6.14) and (6.15) yielding the following expression for R:

$$R = \frac{4E}{mg}(\tan\alpha - \tan\beta)\cos^2\alpha. \tag{6.16}$$

For a given value of β we can maximize R and obtain the value of $\alpha = \alpha_{\max}$ yielding the maximum reach. Then

$$\tan\alpha_{\max} = \frac{(1 + \sin\beta)}{\cos\beta} \tag{6.17}$$

which, using (6.16), gives the following expression for $R_{\max}$:

$$R_{\max} = \frac{2}{mg}E\frac{\cos\beta}{(1 + \sin\beta)}. \tag{6.18}$$

Since $dR/d\alpha = 0$ at $\alpha_{\max}$ a small change in α only leads to a second-order change in R and, one is subsequently able to focus a fairly divergent beam of mono-energetic neutrons at their point of maximum reach.

For small values of β

$$\frac{dR}{dE} = \frac{2}{mg} = 1.96 \text{ cm neV}^{-1}. \tag{6.19}$$

Note also that the reach technique applies to total UCN energy and not a single velocity component. This high resolution can only be attained in practice if the second-order beam divergence effect is kept under control, the criterion for this being that

$$\Delta\alpha \leq \sqrt{(\Delta R/2R_{\max})}. \tag{6.20}$$

For a resolution of 1%, at $\beta = 0$ a total beam divergence $2\Delta\alpha < 0.14$ rad is permissible.

The principle of reach analysis is not viable for neutrons outside the UCN range because of the necessarily long trajectory lengths involved. A further advantage of UCN lies in the fact that they are specularly reflected from neutron mirrors at large angles of incidence enabling compactification of the trajectories in real space. This feature can significantly reduce the size of a practical device and has been used to great effect in NESSIE.

6.4.3   NESSIE: A practical application of reach analysis

A simplified scheme illustrating the essential features and operating principles of NESSIE is shown in figure 6.19. For convenience the source sample and detector are chosen to lie in the same plane. Monochromatization of the incident neutrons is effected by reflection on a movable vertical plane mirror. Movement of the mirror causes a change in position of the virtual UCN source. The characteristic energy of the incident UCN is determined by R_m and the degree of monochromatization by the effective area of source sample and monochromating mirror. It should also be added that, in any practical device, the monochromated input beam would also require lateral channelling by vertical mirrors along its flight path, since the acceptance angle of the monochromator with respect to this lateral velocity component could, if made sufficiently large, significantly reduce the monochromator's resolution.

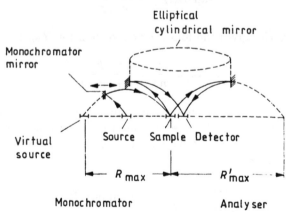

Figure 6.19   A simplified scheme for a total energy UCN spectrometer (Steyerl 1978).

The analyser system invokes the same principle of reach analysis; however, in this instance the scattered UCN trajectories are compactified by reflection on a cylindrical mirror of elliptical cross section. The sample and detector are placed at the focal points of the elliptical mirror and the analyser energy setting corresponds to the horizontal distance from sample to detector R'_{max}. The resolution of the analyser is defined by the effective area of the elliptical mirror, sample and detector. The range of momentum transfer accepted by the device is determined by the position and area of the elliptical mirror relative to the sample.

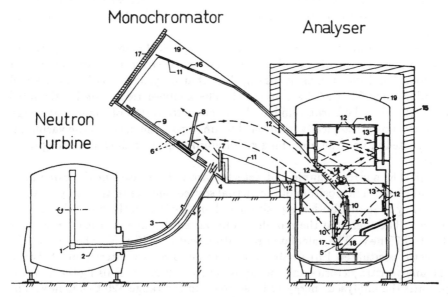

Figure 6.20   Design of the gravity spectrometer NESSIE. The UCN are provided by a neutron turbine. Monochromatization and energy analysis of scattered neutrons are performed by analysis of the maximum reach of the neutrons' flight parabolae. The spatial focusing accomplished by this technique is augmented on the monochromator side, by the focusing properties of two curved mirrors (10). 1, blades of neutron turbine; 2 and 3, neutron guides; 4, source area defined by the guide exit cross section; 5, sample, surrounded by an absorber (polythene); 6, virtual source; 7, first deflecting mirror; 8, monochromator mirror; 9, carriage and spindle for monochromator mirror translation; 10, focusing mirrors; 11, glass plates for lateral beam confinement; 12, beam stops; 13, cylindrical mirrors; 14, two chamber detector with shielding and collimation elements; 15, water shielding; 16, boron containing plastic for shielding; 17, polythene shielding; 18, nitrogen cryostat; 19, vacuum vessel ($p \approx 10^{-5}$ mm Hg) (Steyerl 1978).

Although such a device is simple in principle, in reality high quality mirrors, precision engineering and much careful positioning of beam stops and blockers is required to obtain anything like reasonable resolution. In spite of the enormous technical difficulties, the gravitational spectrometer NESSIE operating on the maximum reach principle was designed and constructed at Munich. A schematic diagram of the device is given in figure 6.20. The principal components and features of the simplified spectrometer of figure 6.19 are the same although there are additional features.

(i) Two additional mirrors are used to focus down the incident monochromatic beam (40 cm by 20 cm) onto the target (15 cm long by 5 cm wide). The focusing in the lateral dimension is effected by tapering the lateral guiding mirrors from 40 to 26 cm above the sample. The divergence and intensity of the beam at the target is thus considerably increased relative to the incident beam.

(ii) Two elliptical mirrors are used to compress the analysed trajectories. The mirror system divides the total analysed trajectories into five shorter sections. The lower mirror is arranged so that only neutrons in the take-off angular range $52.5(\pm 4)°$ are accepted. The analyser resolution is calculated to be 10 neV. The background at the detector from UCN of the wrong energy is reduced by strategically positioned beamstops. The inside of the vacuum is also coated in neutron absorbing gadolinium paint to reduce the background further.

(iii) The source, sample and detector are not situated in the same horizontal plane. The position of the virtual source relative to the sample can be varied by changing both the position and angle of tilt of the monochromating mirror. The energy at the sample can be varied throughout the energy range 390 to 575 neV. The calculated resolution width is 15 neV. The base of the detector is situated 1.67 m above the sample to reduce interference between the analyser trajectories and the input neutron channel. Raising the detector permits the analysed energy to be scanned in the range 420 to 440 neV. The NESSIE detector is actually a conical ^{3}He two-chamber proportional counter (Gmal 1981). The conical shape ensures that the impinging UCN enter the chamber at an angle close to the surface normal thus improving the transmission efficiency of the windows. The two chambers of the detector enable some information on the dependence of the scattered intensity on azimuthal angle to be obtained. As the incoming beam is inclined at a mean angle of $10°$ to the vertical, the count rate in the two chambers gives some indication of the Q dependence of the scattering process. Although this is only true if multiple scattering within the sample can be neglected and that the focusing effects of the elliptical mirrors with regard to launch azimuthal angle can be assessed. The relative Q value resolution so obtained is claimed to be 10% for $Q \approx 0.03$ Å^{-1} .

NESSIE was initially sited at the FRM reactor in Munich which has a thermal flux of only 10^{13} n cm^{-2} s^{-1}. An energy resolution of 17.0(0.6) neV was obtained in spite of the low count rates (≈ 30 n hr^{-1}) (Steyerl et al 1983). The device has recently been installed at the TGV source at ILL where the $\times 10^4$ increase in UCN flux should lead to improved measurements. The overall energy resolution of NESSIE as measured with a beryllium oxide sample at the ILL reactor is 15.8(0.6) neV FWHM (Ebeling 1990). The energy resolution curve is shown in figure 6.21. The background from stray neutrons at the detector is determined by making measurements with a good neutron absorber (Gd_2O_3) positioned at the target position; figure (6.21) has been corrected for background in this manner. The experimentally determined resolution is in good agreement with the expected value.

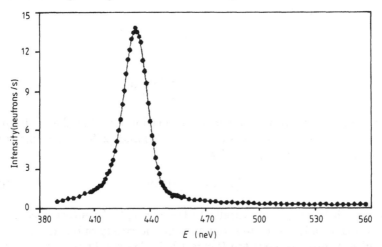

Figure 6.21 NESSIE resolution curve measured with a BeO sample on the TGV source at ILL (Ebeling 1990).

The interest in NESSIE for neutron scattering studies of condensed matter, in terms of sensitivity and the ranges of Q, ω in which it can operate, relative to other neutron scattering devices, is discussed in Chapter 8. The experiments performed so far on diffusion processes in biological systems and the dynamics of polymers are also to be found described in Chapter 8.

Another gravitational spectrometer has also been proposed by Utsuro (1983) who proposes a system for focusing UCN scattered by a target into 4π onto a vertical detector array. An egg-shaped three-dimensional arrangement of mirrors is used to perform the focusing. The energy transfer is determined by TOF and the corresponding vertical height of detection. Such a system would seem difficult to implement in practice.

6.5 UCN OPTICS

6.5.1 Introduction

In Chapter 2 we considered the interaction of neutrons with matter and derived expressions for the reflection and absorption coefficients for neutrons incident on material surfaces. The results were obtained by assuming matter to be represented by a complex potential step relative to vacuum. This approach is most convenient in the case of UCN where the total neutron energy is of the same magnitude as the mean Fermi potential. An alternative approach to the problem of low-energy neutron scattering is to account for the effects of matter by associating a neutron refractive index with the medium, as is conventional in ordinary optics. The expression for the appropriate refractive index, relative to vacuum, in terms of a material's mean Fermi potential $V(r)$ is given by

$$n^2(r) = k_i(r)^2/k_0^2 \tag{6.21}$$
$$= 1 - V(r)/E \tag{6.22}$$
$$= 1 - \lambda^2 \frac{2m}{h^2} V(r) \tag{6.23}$$

where $k_i(r)$ is the neutron wavevector at position r within the medium and k_0 is the vacuum wavevector as shown in (2.43) and (2.44). The Schrödinger equation can thus be re-expressed as

$$\nabla^2 \Psi = -n^2(r) k_0^2 \Psi. \tag{6.24}$$

This approach leads quite naturally to the concept of neutron optics. Neutron optics is usually associated with VCN and a variety of books and reviews on the subject have been written (Sears 1989). We shall not digress here to a long discussion of the subject but will concentrate on the particular features that relate to UCN focusing and microscopy.

The idea of constructing a UCN microscope was first proposed by Frank (1972). Neutrons interact with matter in a different way than light and electrons do. Neutrons are sensitive to both nuclear composition and magnetic field distribution and offer a unique tool for examining matter. Neutron transmission contrast offers the possibility of directly imaging magnetic structures or regions of highly neutron absorbing materials (e.g. H atoms) within samples. Isotope and chemical contrast (e.g. substitution of hydrogen by deuterium) offers additional possibilities and applications. The wavelengths of UCN are in the optical wavelength range and so resolutions equivalent to those of optical microscopes (1 μm) might ultimately be attained. However, unfortunately, focusing UCN in three dimensions is not as simple as the focusing of light because of gravity's strong effect on UCN trajectories. The field of neutron microscopy has been reviewed recently by Frank (1987).

6.5.2 Focusing elements for UCN

We first consider various focusing elements for UCN and show how the gravitationally induced chromatism affects their focusing abilities. Lenses, the most commonly used focusing for conventional optics, are not suitable for focusing UCN beams because of the large absorption cross section for UCN in matter. However, UCN are reflected specularly from mirror surfaces with a negligible probability of loss and so mirrors would seem to be the natural focusing element for UCN.

If a UCN strikes a concave mirror of radius R at a small angle θ to the optical axis it is easily shown by classical mechanics and the assumption of specular reflection that its trajectory will intersect the z-axis (the mirror's optical axis) at a height $R(1 - gR/v^2 \cos^2 \theta)$ above the mirror surface. The paraxial focus is thus at a distance F from the centre of the mirror given by

$$F = \frac{R}{2}(1 - gR/2v^2). \qquad (6.25)$$

The corresponding expression in the case of light is $F = R/2$, the displacement in focal point is caused purely by the effect of gravity on the UCN trajectories. Gravitational chromatism is also apparent in the value of the magnification from a concave mirror, the expression for which (Frank 1987) is given by

$$M = \frac{1}{\{1 - (2Z/R)[1 + (gZ/2v^2)]\}} \qquad (6.26)$$

where Z is the position of the source above the mirror. The corresponding equation for light is given by $M = 1/(1 - (2Z/R))$. In principle, if we had an intense source of highly monochromatic UCN, chromatism would not be a problem. However in reality we will always need to accept a considerable spread in neutron energy and incident angle when working with UCN in order to obtain sufficient count rates. Chromatism is thus a severe problem. Sharp focusing of UCN spanning a range in energy, by a single mirror, is thus not possible, although it should be noted that there is one configuration in which the chromatic effect of gravity on image position disappears completely. This occurs when a point source of UCN is placed at the optical focus of a parabolic mirror. In this instance all UCN irrespective of their initial velocity are focused back onto the source point (see figure 6.22). The system is, however, highly chromatic with respect to magnification as can be seen by setting $Z = R/2$ in (6.26). An additional unique feature of this geometry is that the flight time and phase between source and image for a given classical neutron trajectory is independent of the original flight direction†. Such a system has been called the 'neutron fountain' by Steyerl *et al* (1988a).

---

† In conventional optics Fermat's principle $\delta \int_a^b k \, dl = \int_a^b c \, dt = 0$ would require that this statement be true. However, as pointed out by Steyerl (1984) and Frank (1987), in the case of the optics of a massive non-relativistic particle $k \, dl = mv^2 \, dt/\hbar$. Fermat's principle thus requires $\delta \int_a^b v^2 \, dt = \delta \int_a^b n^2 \, dt = 0$ in this instance. Therefore, although the neutron wavefront is, as in optics, the surface of equal phase with the phase determined by the integral along the classical ray or 'trajectory', the time that a classical particle would take to reach the surface by different trajectories can and generally will differ. This feature of neutron optics has been analysed with respect to the effect of gravity on interference pattern stability in neutron interferometers (Frank 1983).

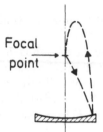

Parabolic mirror

Figure 6.22   Special system for UCN imaging with a parabolic mirror ('the neutron fountain') (Steyerl *et al* 1988a).

Methods of calculating the UCN focusing properties of systems of vertically arranged mirror systems have been developed (Steyerl *et al* 1988a, Frank 1987). In the latter scheme, the neutron's gravitational interaction is included in the expression (6.21) for the refractive index. This leads to the following expression for the refractive index of vacuum for a neutron with vertical velocity component v at $z = 0$

$$n^2(z) = (1 - 2gz/v^2). \tag{6.27}$$

UCN optics calculations can thus be carried out by applying standard geometrical optics matrix theory, generalized to cover the case of layerwise inhomogeneous media (Frank 1979a). It is found that the formulae for geometric neutron optics contain quantities proportional to the propagation time between two points $T = \int_1^2 dl/n(l)$ instead of linear dimensions $L = |r_1 - r_2|$ common to conventional optics (Frank 1987). Several multiple mirror systems have been designed which can simultaneously eliminate magnification and position chromatism to various orders. Such mirror systems have been used in all of the UCN microscopes so far designed. Specific UCN focusing mirror arrangements are described in later sections.

The first direct experimental results of UCN imaging were obtained by focusing UCN with a carefully designed zone plate (Schutz *et al* 1980). As previously shown the focal length of a concave mirror arranged with its optical axis parallel to gravity is strongly achromatic. A zone plate, however, by virtue of its design (see figure 6.23) has a focal length which varies as

$$f = \frac{r_m^2}{m\lambda} \tag{6.28}$$

where r_m is the radius of the largest zone and m is the total number of zones. Since the focal length is proportional to velocity in this instance, the chromatism can be used to compensate for the gravitational chromatism of the concave mirror by arranging for the optical axis to be vertical (Steyerl

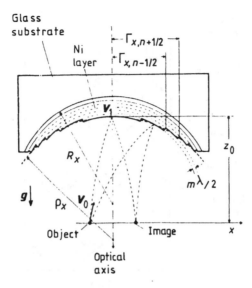

Figure 6.23 Diagram of the zone mirror for UCN. Neutron waves from a point object are bought to a sharp focus by efficient diffraction at the circular zone rings with appropriate profiles. In the case of equal object and image, which is shown, two possible neutron trajectories are indicated. They are parabolic because of gravity which acts along the optical axis. The zone mirror is prepared by electro-deposition, on an aspherical concave glass substrate, of about 50 circular nickel sheets with successively diminishing radii (Schutz *et al* 1980).

and Schutz 1978). Images of a 2 mm slit situated at a small distance from the optical axis of an elliptical zone plate were formed using such a device. UCN with wavelengths in the range 600–800 Å were used and images were obtained for a variety of magnifications in the range 1 to 6. A schematic diagram of the apparatus and typical results obtained so far are given in figure 6.24.

6.6 CURRENT UCN MICROSCOPE AND IMAGING SYSTEMS

6.6.1 Two-mirror microscope system

The first experimental results reporting the successful operation of a UCN microscope were reported in 1985 (Hermann *et al* 1985). A two-mirror system consisting of one convex ($R = 1209.8$ mm and diameter $= 20$ cm) and one concave ($R = 585.2$ mm, diameter $= 4.5$ cm) mirror arranged with a common vertical axis were used to obtain an image of a 1.06 mm object slit, positioned slightly above the light optical focus of the first mirror, at a magnification of $M = 50$. The device is illustrated in figure 6.25.

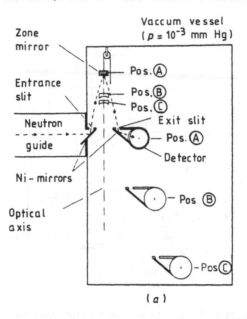

(a)

Figure 6.24   ((a) is shown here and(b) is shown on the next page.)Neutron imaging using a zone plate. (a) Zone plate scheme for UCN imaging. (i) Magnification ×1 (zone mirror and detector position (A); (ii) ×3 (position (B)) and (iii) ×6 (position (C)). A possible UCN trajectory from the object (the entrance slit) to the zone mirror and the exit slit is indicated. Inclined nickel mirrors are used for the desired beam deflections of about 90°. The zone mirror position can be varied in a vertical range of 65 mm, which allows precise adjustment for the desired magnification. The exit slit and detector assembly may be moved within a plane of 40 cm × 20 cm for scanning the image produced by the zone mirror.

The particular mirror configuration employed cancelled out the first-order chromatic aberration of the image and magnifications for input UCN with speeds in the range $5.5 < v < 6.7$ m s^{-1} were obtained. The correct velocity range for optimum resolution and suppression of background was obtained by careful arrangement of beam stops. The device was designed for object scanning, the imaging mirror was a 17 mm by 17 mm spherically concave mirror fixed at an angle of 15° to the vertical at the focus of the system. The count rate obtained as a function of slit position, on the ILL PN5 UCN source, is shown in figure 6.26(b); the authors deduced a resolution of ≈ 0.1 mm from the data. Also shown in figure 6.26(a) is the slit image obtained from the single concave mirror. An image mirror, of dimensions 1.5 cm by 1.5 cm, was placed 8 cm above the image slit and mounted at an angle of 45° to the horizontal in order to deflect the focused UCN into the detector. The image is considerably blurred because of the chromatism of the magnification and wide spectrum of the UCN used. The microscope was moved to the turbine UCN source at ILL (section 3.3.2) in 1986 where it profited from a ×125

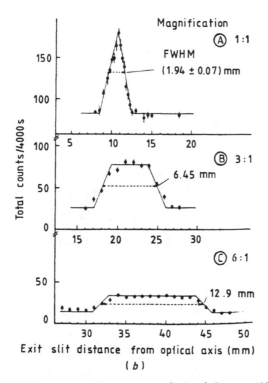

Figure 6.24   (b) The measured count rates (points) for magnifications (A) of 1 (entrance slit width, 2 mm); (B) of 3; and (C) of 6 (entrance slit width, 2.15 mm), plotted against the distance of the exit slit from the optical axis, are compared with the theoretical line shapes. The intensities of measurements (B) and (C) are normalized with respect to the initial intensity incident on the zone mirror in the case of magnification. The half-widths of the measured curves agree with the expected image sizes (Schutz *et al* 1980).

increase in flux. By slightly optimizing the mirror parameters and, more importantly, by accepting a narrower band of input UCN energy, an image of a 0.4 mm slit at a magnification of ×243 was obtained. An impressive aberration limit of $\approx 10 \ \mu$m was deduced from the data (Steyerl *et al* 1988b).

6.6.2   Four-mirror microscopes/imaging systems

A four-mirror 'microscope' was constructed at the Kurchatov Institute (Arzumanov *et al* 1984). A schematic diagram of the arrangement is shown in figure 6.27. The system can be considered to be composed of two sets of plane/convex mirror units placed back to back; the inherent chromatism of the single system is then compensated by the chromatism of the other. In reality the system was more an imaging system than a microscope as

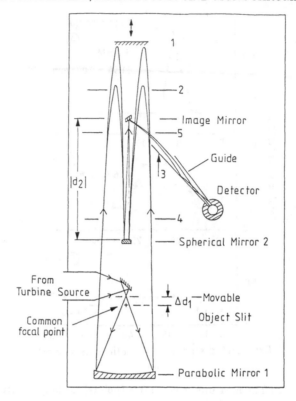

Figure 6.25   The two-mirror UCN microscope (Hermann *et al* 1985).

only very low magnifications 0.7 to 1.4 were obtainable. In the first experiments the authors formed an image of a object slit by scanning an image slit across a simple detector system; however, in a second set of experiments the first two-dimensional direct optical representation of a neutron image was obtained (Arzumanov *et al* 1986).

The optics were slightly different for the second experiment as is seen from figure 6.28. The system is found to be achromatic to first order in terms of both magnification and position when

$$R_1 + R_2 = 6d \tag{6.29}$$

where R_1 and R_2 are the radii of curvature of the two mirrors and d is their separation. In this experiment the image slit was replaced by a nickel figure (thickness 200 nm) evaporated onto a 0.35 mm thick silicon wafer, placed at the focus of the top mirror. Total reflection by the nickel provides a strong UCN transmission contrast at the object for the range of UCN speeds used (3.2 to 5.7 m s^{-1}). The conventional UCN detector was replaced with a position-sensitive UCN scintillation counter consisting of a thin layer of ^{6}Li/^{7}Li on

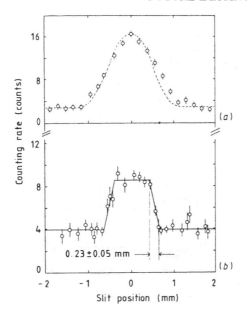

Figure 6.26   Data for scans of a 1.06 mm wide object slit for (*a*) a one-mirror system where (ideally) object and image coincide at the light optical focus, for any arbitrary neutron velocity; and (*b*) the two-mirror microscope set-up of figure 6.25. In (*a*) the magnification is strongly chromatic, and this leads to the observed edge blurring for the wide UCN spectrum used. The trapezoidal image in (*b*) implies a resolution of $\approx$ 0.1 mm. It was limited by coarse-grained image detection rather than aberrations. The measured data are consistent with calculations (broken and full curves). Measurement (*a*) was performed at a reduced primary intensity (factor 6) (Hermann *et al* 1985).

top of a thin sheet of a ZnS scintillator. Photons emitted from the ZnS via α-particles created in the reaction $^6\mathrm{Li}(n,\alpha)$ are collected with a coherent optical fibre bundle; the resulting image is amplified with a micro-channel plate and the image recorded on film. The optical image obtained and the object specimen used in the course of these experiments are shown in figure 6.29. The resolution of the instrument was estimated to be 200 μm.

6.7   FUTURE IMAGING SYSTEMS

It has also been proposed by Skachkova and Frank (1981) to remove gravitational magnification chromatism from a mirror system by effectively removing gravity in the region of the focusing element. Gravity is cancelled by direct application of a magnetic field gradient, i.e.

$$\nabla(\mu_0 B) + mgk = 0. \tag{6.30}$$

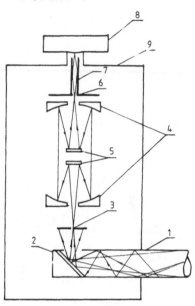

Figure 6.27   Multi-mirror optical instrument for UCN: 1, neutron guide; 2, auxiliary mirror; 3, slit object; 4, concave mirrors; 5, plane mirrors; 6, analysing slit; 7, neutron guide to detector; 8, detector; 9, vacuum chamber (Arzumanov *et al* 1984).

This requires the use of polarized neutrons and necessitates a uniform field gradient of 1.7×10^{-2} T cm^{-1} over the focusing element (see figure 6.30). To eliminate gravity over the entire system would clearly be preferable but the large field gradients required render this impossible. Although image distance chromatism does not disappear in this instance it can be made tolerable (see Frank 1987). By keeping the dimensions of the system low, resolutions of a few microns are thought to be achievable.

Steyerl *et al* (1988b) have proposed a four-mirror, aplanatic, achromatic microscope. The device is a combination of the Frank optics shown in figure 6.27 and the double-mirror system of the two-mirror microscope. By choosing suitable mirror shapes the system can be designed to eliminate longitudinal aberration to third order, and transverse aberrations to second order; the system is aplanatic. The aberration limit of the device is expected to be at the level of a few microns. However, note that the highest UCN current densities presently available (at the ILL UCN source section 3.3.2) give 10^4 cm^{-2} s^{-1} in a velocity band $\Delta v = 1$ m s^{-1} at 6 m s^{-1}. This current corresponds to 10^{-4} UCN s^{-1} μm^{-2}, so a resolution of 1 μm implies a count rate of 1 UCN/3 h/resolution element, a very low count rate indeed.

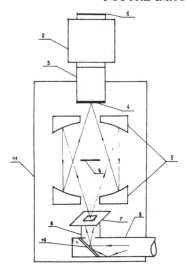

Figure 6.28 Two-mirror optical instrument: 1, photographic film; 2, image converter; 3, optical fibre; 4, UCN sensitive scintillator; 5, concave mirrors; 6, shielding from direct rays; 7, specimen; 8 and 9, neutron guides (Arzumanov *et al* 1986).

Figure 6.29 A nuclear contrast image obtained using the apparatus illustrated in figure 6.28. The dark regions correspond to the areas on the specimen coated in nickel (Arzumanov *et al* 1986).

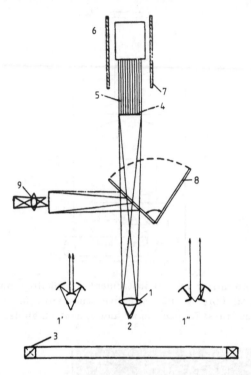

Figure 6.30 Possible design for a neutron microscope with magnetic compensation for the gravitational field: 1, objective (1' and 1" are possible variants of mirror objectives); 2, object; 3, coil of superconductive magnet; 4, image plane, in which the sensitive layer of the detector is placed; 5, fibre light guide; 6, light amplifier; 7, magnetic shielding; 8, hinged mirror; 9, ocular (Frank 1987).

7

Applications of UCN to fundamental physical tests and measurements

7.1 INTRODUCTION

Although the generation, transportation and storage of UCN are in themselves profound tests of fundamental physical principles, the development of more intense UCN sources, improved guides, etc, is currently motivated by the usefulness of UCN as a tool to determine the properties of the neutron and to study fundamental physical interactions. UCN are useful because they can be readily stored in material bottles for times approaching the β-decay lifetime and can be spin-polarized. In a suitable magnetic environment, the spin-polarization lifetime can easily exceed the storage lifetime. In addition, the low velocity (long wavelength) of UCN and the nearly ideal potentials presented by magnetic fields, gravitational fields and material surfaces make a number of interesting experiments possible.

In this chapter we will describe a number of experiments using UCN, some of which are still in the speculative stage. Experiments to determine the properties of the neutron will be described first followed by a discussion of experiments to determine fundamental interactions and to test quantum mechanics. As will become evident, the main advantage of using UCN in most experiments is that such long interaction times (approaching the β-decay lifetime) are possible.

7.2 NEUTRON β DECAY LIFETIME MEASUREMENTS WITH BOTTLED UCN

7.2.1 Neutron β-decay

Neutron β-decay, as described in section 2.1, is the prototype semi-leptonic (involving both leptons and hadrons) weak decay. The study of neutron

β-decay has been an important source of information concerning weak interactions. An accurate value for the neutron lifetime is important for the theory of semi-leptonic weak interactions.

Neutron β-decay is described by the interaction of two left-handed lepton currents with two hadron currents; through the $V-A$ current–current weak Hamiltonian density (γ_μ, or vector current coupling to $\gamma^\mu\gamma_5$, or axial vector current, where the γ represent Dirac matrices) (Marshak *et al* 1969)

$$H = \frac{g_F}{\sqrt{2}}[\bar{\psi}_e\gamma_\mu(1+\gamma_5)\psi_{\bar{\nu}}][\bar{\psi}_p\gamma^\mu(1+\lambda\gamma_5)\psi_n] \tag{7.1}$$

where g_F is the weak interaction vector (Fermi) coupling constant and $\lambda = g_A/g_F \approx -1.25$; both of these are experimentally determined quantities. λ can be determined directly through angular correlation in the decay of polarized neutrons (and other hadrons). Although one could determine both g_F and λ separately using UCN, bottled UCN experiments have so far been restricted to the measurement of the total decay rate which is a combination of g_F and λ.

g_F can be derived from the measured β-decay $f\tau$ (f is the phase space factor, τ the lifetime) values for allowed $0^+ \rightarrow 0^+$ 'Fermi' transitions (for example, $^{14}\text{O} \rightarrow {}^{14}\text{N}$), where the $\gamma^\mu\gamma_5\lambda$ term (the 'Gamow–Teller' transition) is suppressed (Wu and Moskowski 1966). After taking Coulomb and radiative corrections into account, one obtains the 'bare' β-decay weak coupling constant averaged over various nuclear $0^+ \rightarrow 0^+$ transitions (Wilkinson 1982). Comparing g_F for various semi-leptonic processes (neutron decay, pion decay, hyperon decay, etc.) is an important test of the universality of the current–current interaction, the conserved vector current (CVC) hypothesis and the Cabibbo model (Siebert 1989).

Weak-decay data give a value for the bare coupling constant of

$$g_F = 1.026 \times 10^{-5}m_p^{-2} = 89.6 \text{ eV fm}^3. \tag{7.2}$$

This is true only for $q^2 \approx 0$, that is for low energy scattering (Commins and Bucksbaum 1980). What sets the energy scale is the mass of the charged intermediate vector boson m_W as described by the Weinberg–Salam–Glashow (WSG) model of the weak interaction. The Feynman graph for neutron decay, shown in figure 7.1, leads to a structure for the second-order exchange

$$g\frac{1}{m_W^2 - q^2}g \xrightarrow{q^2\to 0} \frac{g^2}{m_W^2} = \frac{g_F}{\sqrt{2}} \tag{7.3}$$

and justifies writing the interaction Hamiltonian in the current–current point contact form. The idea of the WSG model is that electromagnetic and weak interactions are manifestations of the same force with $g \approx e$. The weak force appears weak because it involves the exchange of a massive boson (W)

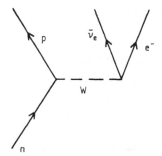

Figure 7.1   Feynmann graph describing the weak decay of the neutron.

while the electromagnetic force—involving exchange of a zero-mass photon—is much stronger and has an 'infinite range' ($1/r^2$ dependence). The WSG model sets $m_W^2 \approx e^2/g_F \approx 4\pi\alpha/10^{-5}m_p^{-2} \approx 100$ GeV.

Using the Hamiltonian equation (7.1), it is straightforward to show that the total unpolarized neutron decay lifetime, in terms of $(f\tau)_{0+\rightarrow0+}$, is (Wu et al 1957)

$$\frac{(f\tau)_n}{(f\tau)_{0+\rightarrow0+}} = \frac{2}{1+3\lambda^2}. \tag{7.4}$$

The neutron decay rate provides no information on the sign of λ.

The values of λ from various neutron decay experiments (including angular correlation determinations of λ) can be compared to determine the validity of the $V - A$ interaction. Specifically, one can test whether there are currents proportional to other combinations of the Dirac matrices (scalar or tensor forces parameterized by $s = g_s/g_F$ and $t = g_t/g_F$). These experiments complement similar limits from searches for atomic electric dipole moments which set limits on s and t which are orders of magnitude smaller (from atomic electron–nucleon interaction) (Lamoreaux et al 1987, Lamoreaux 1989). In addition, it is possible to determine if there is an admixture of right-handed $V + A$ currents in the Hamiltonian, which would be evidence of a right-handed heavy W boson. So far, experiments constrain the mass of the right-handed boson, $M_{\text{right}}^2 \geq 8M_{\text{left}}$ and the mixing parameter $-0.13 \leq \xi \leq 0.03$ (Erozolimskii 1989).

The value of the neutron lifetime is also important to cosmology where it affects the predictions of the ratio of He to H abundances in the early universe. By comparing the predicted ratio with the primordial ratio inferred from experimental observation, quite interesting and remarkable limits may be set on the number of weakly interacting particle families; that is the number of low-mass ($m_\nu \leq 10$ MeV) neutrino families (Schramm 1989, Steigman 1989). Such limits are complementary to those determined from high energy accelerator experiments which are also sensitive to very massive particles ($\leq m_W$) provided they have nearly full strength weak interaction. Both high energy experiments (width of the Z_0 resonance) and cosmological limits indicate that there are three lepton families (Denegri *et al* 1990).

7.2.2   Principles of τ_β measurement and limitations to accuracy

From the earliest days of UCN physics it was hoped that UCN storage techniques would provide experimenters with a new and more accurate means
of determining the neutron β-decay time constant, τ_β. Despite the problem
of the anomalous surface loss rates apparent from the very first UCN experiments it is now fair to say that neutron bottle β-decay measurements have
come of age and are competing favourably with neutron beam experiments.
At present three different UCN storage experiments are being performed, one
of which has reported the lowest ever error for the lifetime $\tau_\beta = 887.6 \pm 3$ s
(Mampe *et al* 1989). In this section we will outline the general principle of
such measurements and describe the current experimental techniques.

Neutron beam measurements of τ_β generally involve determining the neutron decay rate in a well defined volume traversed by a cold or thermal neutron beam. The decay rate within the volume is determined by detection
of one of the (charged) products of the decay. If N_0 is the mean number of
neutrons within the detection volume at a given instant and N_d is the mean
number of decays detected per second then $\tau_\beta = N_0/N_d$. To determine the
decay time constant one thus needs an accurate determination of the absolute neutron flux, the trap volume and the detector efficiency. In addition,
one usually has a large background in the charged particle detector from
the presence of the intense neutron beam. These problems make neutron
beam decay experiments notoriously difficult to perform; Byrne *et al* (1990)
report results from a beam experiment indicating accuracy at the 0.6% level.
Achieving an accuracy of about 0.1% seems feasible, so it is an important
complement to bottled UCN experiments.

Neutron storage experiments bypass many of the problems inherent in
beam measurements. In storage experiments, a bottle is filled with N_0 neutrons and the number of neutrons which survive after waiting a given time
is determined. The number of neutrons which survive as a function of time
is approximately given by

$$N(t) = N_0 e^{-t/\tau} \tag{7.5}$$

where τ^{-1} is the loss rate (see Chapter 4 for a more complete discussion of
the problem of UCN storage in material bottles). In general, there will be
several contributions to τ:

$$\tau^{-1} = \tau_\beta^{-1} + \sum \tau_i^{-1} = \tau_\beta^{-1} + \tau_{\text{walls}}^{-1} + \tau_{\text{gas}}^{-1} + \tau_{\text{holes}}^{-1} \tag{7.6}$$

τ_{walls}^{-1} is the wall loss rate, τ_{gas}^{-1} is the loss rate due to neutron absorption and
scattering by the residual gas, and τ_{holes}^{-1} describes leakage of UCN through

holes in the bottle. In most experiments, τ_{gas} and τ_{holes} are negligible. Determination of the various loss rates and isolation of the one of interest is
the main experimental problem. It is evident that, in order to maximize
the accuracy and minimize systematic effects, $\tau_i \gg \tau_\beta$. The principle of
neutron storage β-decay measurements is to measure the total decay rate of
UCN within a closed storage volume and to separate the wall loss contribution from the total decay rate (equation (7.6)) either by making these losses
negligible relative to τ_β or by careful variation of the bottle geometry and/or
temperature so that the contribution can be accurately determined.

As the determination of the decay constant only requires a relative determination of the count rates, the problems associated with absolute flux
determination and detector efficiency are avoided. Also, since the UCN beams
are low flux and can be turned off during the neutron counting procedure,
detector background ceases to be a problem. Neutron storage measurements
thus avoid all of the problems inherent in conventional beam techniques.

Unfortunately, UCN storage experiments have a few of their own particular
problems, all of which are associated with eliminating the contribution of
wall loss to the decay time. If one fills a bottle of given geometry with
monochromatic UCN (those with a single velocity) and makes a measurement
of the number of neutrons remaining in the bottle after periods of storage
$N(t_i)$, then the counts remaining after each time period are given quite
accurately by equation (7.5). If a wide range of neutron energies are used in
the experiment then it will be difficult to extract an accurate value for τ_β
because the wall loss rate is, in general, a function of UCN energy (velocity)
with average rate given by

$$\langle \tau_{\mathrm{wall}}^{-1} \rangle = \langle \mu(v)v \rangle / \lambda \qquad (7.7)$$

where $\mu(v)$ is the probability of loss per bounce, v is the velocity, $\langle\rangle$ represents
the average over the velocity spectrum, and λ is the mean free path, as in
kinetic theory

$$\lambda = \frac{4V}{S} \qquad (7.8)$$

where S is the surface area of the storage vessel and V is the volume. The
loss probability is given by equation (2.70) (Antonov et al 1969a, Golub and
Pendlebury 1979)

$$\mu(v) = 2f \arcsin(y)/y^2 - \sqrt{1 - y^2}/y \qquad (7.9)$$

where $y = v/v_{\mathrm{crit}}$ (v_{crit} is the maximum velocity of the storable UCN as
determined by the Fermi potential) and f is the ratio of the real to the
imaginary parts of the wall potential.

Thus, one finds that the decay time constant is a function of time, becoming longer at longer times (the wall loss rate increases with energy) as the

neutron velocity spectrum 'softens' or shifts to lower average velocity. The net result is that the decay can no longer be described by a simple exponential. Such a process is extremely difficult to model accurately. However, there are several ways to get around the problem.

(i) The use of an ultra low loss surface, low energy UCN and a geometry give a long mean free path between wall collisions which makes the wall loss rate negligible.

(ii) By using approximately monochromatic UCN and making measurements of total bottle decay constants as a function of UCN energy, the decay times can be determined as a function of energy using equation (7.5) and an appropriate expression for $\langle \mu(v) \rangle$ to extract a value for τ_β. This clearly requires the use of assumptions regarding loss mechanisms within the bottle.

(iii) One can make measurements of UCN decay curves in varying geometries but with identical surface coatings and make use of the mean free path scaling technique of Pendlebury (1988b).

We will now determine the fundamental limit (shot noise) to the accuracy of a general bottled neutron lifetime experiment. The optimum storage time T for determining τ can be readily determined when the only noise source is counting statistics, that is when the fluctuations in $N(T)$ between measurements is given by

$$\delta N(T) = \sqrt{N(T)}. \qquad (7.10)$$

This gives a fluctuation in τ between measurements

$$\delta \tau = \frac{\partial \tau}{\partial N(T)} \delta N(T) = N_0^{-1/2} \frac{\tau^2}{T} e^{T/2\tau} \qquad (7.11)$$

where we have used equation (7.5) in the form

$$\tau = -T \left(\log \frac{N(t)}{N_0} \right)^{-1}.$$

Note that this is the fluctuation in τ per measurement; if we do many (n) measurements over an extended time period $t \gg T$, then

$$n = t/T \qquad (7.12)$$

and the final uncertainty in τ is

$$\sigma(\tau) = \frac{\delta \tau}{\sqrt{n}} = N_0^{-1/2} \tau^2 e^{T/2\tau} \frac{1}{\sqrt{tT}}. \qquad (7.13)$$

This has a minimum, for fixed t, when $T = \tau$.

It is evident that, for maximum sensitivity, the storage time should be equal to the total loss time, all spurious losses should be minimized, and N_0 should be large. The arguments presented here are highly idealized and most experiments will have to incorporate many systematic checks.

7.2.3   Neutron lifetime from fluid-walled bottle experiment

The fluid-walled bottle was invented and developed by Bates (1983) at the Risley reactor. In such bottles, the UCN reflecting surfaces are thin liquid oil films on glass. The oil is a hydrogen-free perfluorinated polyether $(F_3CCF_2OCF_2CF_5)_n$ with average molecular weight 2650. Such oil is used in diffusion pumps (trade name Fomblin Y Vac18/8) and has an extremely low vapour pressure. Fomblin is quite viscous at room temperature and easily wets glass surfaces; the oil remains as a thin stable film on the glass surface for quite long times.

Fomblin has a very low UCN reflection loss probability at room temperature; of the order 10^{-5} per bounce, depending on UCN energy. The mean Fermi potential on the surface is 106.5 neV, corresponding to a critical velocity of 4.55 m s^{-1}. Another advantage of the use of the Fomblin-on-glass combination is that a microscopically flat and reproducible surface can be created. In addition, the fluid readily seals the holes and gaps around the neutron input and exit valves used to fill and empty the bottle, provided that the mechanical parts fit rather closely.

A fluid-walled bottle experiment was recently completed at the ILL turbine UCN source (Steyerl et al 1986 and Chapter 3) with the result $\tau_\beta = 887.6 \pm 3$ s, which is currently the most precise value (Mampe et al 1989, Ageron et al 1986). A schematic diagram of their apparatus is shown in figure 7.2.

The UCN are stored in a rectangular vessel of height 0.3 m, width 0.4 m, and length adjustable between 0.5 and 0.01 m. The fixed walls and the movable piston-like rear wall are made of glass. The bottom is an aluminium plate which is covered with a puddle of Fomblin about 1 mm deep. A spray head allows the walls to be coated while the system is under vacuum; this is usually done after changing the bottle length to make sure the surface is pristine.

Two holes 65 mm in diameter can each be sealed with a sliding Fomblin-coated glass plate. Stainless steel guides connect to theooo valves; one guide to the neutron turbine, the other to a ^{3}He proportional-counter UCN detector.

The length of the box can be controlled to better than 0.05 mm. To make sure that the UCN velocity distribution rapidly becomes isotropic (the UCN from the guide are more or less moving in a forward direction) the rear movable wall is corrugated glass with an approximately sinusoidal surface. These corrugations are about 2 mm deep and 4–5 mm wide. This surface

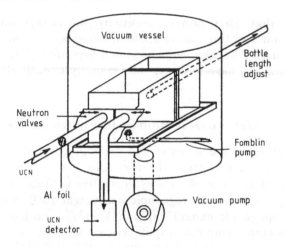

Figure 7.2   Fluid-walled bottle neutron lifetime apparatus (Mampe *et al* 1989).

'roughness' spoils the specularity of the Fomblin film and randomizes the velocity distribution; the angular velocity distribution must be isotropic for statistical mean-free-path arguments to be valid.

The method of operation is: a particular bottle length is chosen, the bottle is filled (for a time long enough to ensure saturation of the UCN density) and then the neutrons are stored for periods of time ranging from 100 to 3600 s, after which they are emptied into the UCN detector. The numbers N_1 and N_2 counted after storage periods t_1 and t_2 give an effective lifetime (see equation (7.5))

$$\tau = (t_2 - t_1)/\log(N_1/N_2) \tag{7.14}$$

which is associated with these survival times and bottle length, the bottle length determining the mean free path λ through equation (7.8).

To extract τ_β from the wall losses, a scaling technique due to Pendlebury (1988b) is used. One chooses combinations of storage times and mean free path lengths so that the total number of collisions with the surface is the same at particular times for the different mean path lengths; one can then determine the total loss rate for measurements with the same number of wall collisions and, in principle, determine the neutron decay rate by extrapolating to zero wall collisions.

Since the mean number of collisions in a time t is tv/λ, if i and j represent sets of measurements at λ_i and λ_j, we choose, for example

$$\frac{t_2(i)}{t_2(j)} = \frac{t_1(i)}{t_1(j)} = \frac{\lambda_i}{\lambda_j} = \frac{t_2(i) - t_1(i)}{t_2(j) - t_1(j)} \tag{7.15}$$

where typical time settings for a given mean free path are chosen so that $t_n(i) = 2^n t_1(i)$ where i represents a given mean free path and $t_1(i) = 112.5$ s.

This means that the velocity spectrum will always be the same at corresponding points in the different measurements i, j if the initial spectra are identical. This latter assumption is only approximately true as different volumes will load the source differently resulting in a small bottle size dependence of the initial spectra. In addition the effects of gravity on the trajectories mean that the time between wall collisions does not scale exactly with the geometrical mean free path. The total UCN lifetime within the bottle, determined between adjacent time settings as previously described, as a function of inverse mean free path, is shown in figure 7.3. The scaling principle permits the extrapolation of the curve to infinite mean free path as a method of obtaining τ_β, that is

$$\tau_i^{-1} = \tau_\beta^{-1} + f(t_1, t_2)/\lambda_i. \tag{7.16}$$

Such groups of measurements give τ_β with a reproducibility to within 1 in 900.

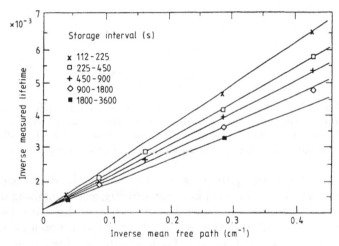

Figure 7.3   Results for the fluid-walled bottle experiment. The measured inverse lifetimes are plotted as a function of the mean free path for different storage intervals. The data are from about one week of running; the error bars are smaller than the data points (Mampe *et al* 1989a,b).

A number of systematic checks were performed. It was demonstrated that the background gas was not contributing to the neutron decay even for the longest storage time (the vessel interior cannot be pumped while neutrons are being stored). In addition, the loss rate is highly temperature-dependent; over a set of measurements, the temperature was held to 0.1 K. Varying the temperature allowed the wall loss to be changed for a set of measurements; no change in the derived value of τ_β was observed.

Unfortunately, there was some discrepancy in the values of τ_β, particularly for short storage times and in completely smooth-walled vessels, as shown in figure 7.4. The problem is that the scaling technique does not work exactly in the presence of gravity. For neutrons without enough energy to reach the top, the relation between mean free path and bottle length is different. Also, there were some differences in the initial neutron spectra due to source loading by the bottle volume as previously mentioned. In all, corrections of the order of 1% were required. The uncertainty in the total correction is the major source of the quoted error—887.6 ± 3 s.

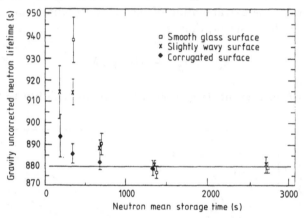

Figure 7.4   Gravitational effects on the neutron lifetime for different fluid-walled bottle surface structures: open square, smooth glass surface; cross, slightly wavy surface; full square, corrugated surface. The smooth-wall short-time effects disappear after including corrections for gravitational effects (Mampe *et al* 1989a,b).

The corrections are quite sensitive to the energy dependence of the UCN/Fomblin loss process. $\mu(v)$ has been directly measured using monochromatic UCN produced by the gravitational UCN monochromator (see section 6.3.4) and agrees well with that expected from a step potential (see figure 5.16). In addition, a small amount of quasi-elastic heating (see section 5.5.2) was required to describe the individual decay curves adequately (for fixed mean free path) and was used in the derivation of the correction. Evidence of such an effect of the order $\delta E(\text{RMS}) = 6 \times 10^{-11}$ eV/bounce has been seen in experiments with monochromatic UCN (Richardson 1989). Such heating could be due to diffraction of the neutrons from surface capillary waves. Measurements by scattering laser light from the Fomblin surface, viscosity measurements, surface tension and density measurements are necessary to model the surface and make theoretical determinations of the microscopic process of heating; comparison of such predictions with the experimental data will be most interesting. These measurements and calculations should help in the understanding of the steep temperature dependence of the loss function.

7.2.4 The 80 K aluminium/heavy water ice bottle experiment

This experiment presently being performed in Dmitrovgrad at the SM-2 reactor (section 3.3.3) is similar to the Fomblin oil measurement of τ_β and relies upon making measurements of total bottle decay rates as a function of mean free path (Kosvintsev *et al* 1986, 1987, Morozov 1989). A diagram of the experimental arrangement is given in figure 7.5. To vary the mean free path, four packets of aluminium plates, the plates separated by 1 cm, are suspended by aluminium filaments. These packets can, by means of solenoids, be raised high enough within the chamber so that UCN cannot reach them, or be lowered into the vessel, to change the net wall collision frequency. The storage vessel itself is hermetically sealed within a larger vacuum vessel, the storage vessel having a pressure of 10^{-6} Torr and the outer vessel, 10^{-5} Torr.

The lifetime of a UCN of speed v_0 at $z = 0$ due to interaction with the bottle walls is given by the following expression.

$$
\begin{aligned}
\tau^{-1} &= \eta\gamma(v_0) \\
&= \frac{\eta \int_{(S)} v_c^2 \left(\arcsin \sqrt{(v_0^2 - 2gz)/v_c^2} - \sqrt{(v_0^2 - 2gz)(v_c^2 - v_0^2 + 2gz)} \right) \, \mathrm{d}S}{2 \int_{(\Omega)} \sqrt{v_0^2 - 2gz} \, \mathrm{d}\Omega}
\end{aligned}
$$

$$(7.17)$$

where η is the ratio of the imaginary-to-real parts of the neutron–wall interaction, S is the full surface area of the vessel and plates, z is the height above the bottom, $\gamma(v)$ is a geometrical factor, defined by (7.17), v_c is the critical velocity of the wall material, and Ω is the volume of the vessel. This expression, which is equation (7.9) averaged over the velocity, is easily derived from the theory of Chapter 4 where the loss function for a step change in potential is assumed.

In the presence of a broad velocity profile,

$$
\tau^{-1} = \eta\gamma(t) = \frac{\int_0^{v_{\max}} f(v_0)\gamma(v_0) \exp[-\eta\gamma(v_0)t] \, \mathrm{d}v_0}{\int_0^{v_{\max}} f(v_0) \exp[-\eta\gamma(v_0)t] \, \mathrm{d}v_0}
\tag{7.18}
$$

where $f(v_0)$ is the initial UCN velocity spectrum and $v_{\max}$ is the upper limit of the spectrum.

The time evolution of the total UCN population is closely described by

$$
\frac{1}{\tau} = \frac{1}{\tau_\beta} + \eta\langle\gamma\rangle = \lambda
\tag{7.19}
$$

where $\langle\gamma\rangle$ is the time average of $\gamma(t)$ over the storage period. $\langle\gamma\rangle$ is controlled experimentally by insertion of a set of plates, with a well defined area and

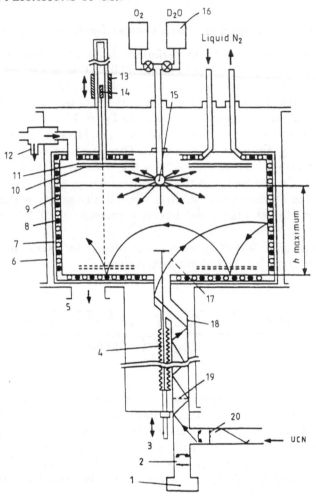

Figure 7.5   80 K aluminum–heavy water ice neutron lifetime apparatus: 1, UCN detector; 2, detector gate; 3, shaft; 4, bellows system; 5, heater; 6, outer vacuum jacket; 7, heat shield; 8, inner vacuum vessel; 9, liquid nitrogen cooling coil; 10, lower plate; 11, upper plate; 12, hermetically sealed valve for evacuating the vessel; 13, solenoid; 14, armature; 15, O_2 and D_2O vapour inlet; 16, O_2 and D_2O reservoirs; 17; disc-shaped gate; 18, vertical guide; 19, aluminum membrane; 20, UCN inlet gate (Morozov 1989).

geometrical arrangement, into the vessel, as already described. The volume of the bottle does not change during the course of the measurements. By extrapolating the plot of $\langle \tau^{-1} \rangle$ against $\langle \gamma \rangle$ to $\gamma = 0$ a value for τ_β can be deduced. The major methodological errors arise from representing the bottle losses by an averaged decay constant (0.2%), from possible deviation in surface characteristics over the bottle wall (0.1%), and from experimental error in the determination of the initial spectrum (0.2%). Eliminating all other bottle losses due to gaps and residual gas scattering and capture can also be a problem at the 0.4% level.

UCN in the energy range $0 < E < 49$ neV are used in the experiment. The shape of the initial UCN spectrum at the start of the storage cycle, required to evaluate $\langle \gamma \rangle$, is determined experimentally by inserting an absorbing plane into the vessel at a series of different heights. The principle and theory behind this kind of measurement were discussed in section 6.3.3. A typical spectrum derived this way is shown in figure 7.6. Experiments were initially performed using a carefully prepared aluminium bottle. The vessel could be heated (to 750 K) and cooled (to 80 K) as required *in situ* and under vacuum. Measurements were made at several temperatures. A plot of λ as a function of $\langle \gamma \rangle$ is given in figure 7.7. As can be seen, loss rate due to wall interaction is of the same order as that due to β-decay. The last published result of these measurements is $\tau_\beta = 903(\pm 13)$ s (Kosvintsev *et al* 1986).

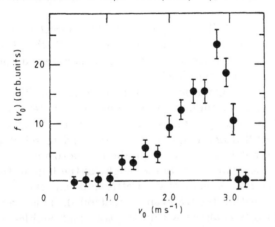

Figure 7.6   Spectrum of the UCN stored in the vessel of figure 7.5 (Morozov 1989).

The device has recently been altered slightly to permit the freezing of D_2O ice onto the bottle walls. The wall loss rate now only contributes 3–4% of the total bottle loss rate (Kosvintsev *et al* 1987). The latest results from these measurements give $\tau_\beta = 893(\pm 20)$ s.

A combined best value for sets of data with both wall materials is $\tau_\beta = 900(11)$ s. This error does not yet represent the ultimate limitation to the method.

7.2.5   Gravitational UCN trap

The idea behind this experiment is to reduce the UCN bottle wall losses to a level well below that of free β-decay and thus to observe the β-decay directly

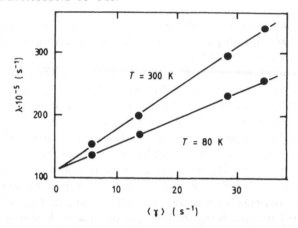

Figure 7.7   The UCN loss probability as a function of the geometrical factor γ for different temperatures in the bare aluminum bottle (Morozov 1989).

(Kharitonov *et al* 1989). The design also allows a direct measurement of the energy dependence of the bottle lifetimes as a check that loss mechanisms other than those due to β-decay have been fully suppressed and, if they have not, allows them to be determined.

The experimental apparatus is shown in figure 7.8. The bottle itself is a hollow sphere of aluminium 75 cm in internal diameter, internally coated with 3000–5000 Å of beryllium. There is a hole in the side of the sphere to allow it to be filled and emptied with UCN, and to select the maximum UCN energy. The bottle can be rotated about a horizontal axis through the centre of the bottle. This permits the height of the hole in the vessel to be set relative to the lowest point of the bottle and thus the maximum energy of UCN confinable within the trap can be controlled. This manoeuvre can be carried out with UCN confined within the trap and enables one to tip UCN within certain energy bands out of the bottle at will.

All possible steps have been taken to minimize UCN wall losses; the bottle walls can be coated in frozen oxygen which has an extremely low UCN surface loss, only low energy UCN are used, and the combination of a spherical vessel with the effective bottle roof defined by gravity ensures a long mean free path.

The lifetime of a UCN in the trap has the following form:

$$\tau_{\text{st}}^{-1} = \tau_{\beta}^{-1} + \frac{1}{4}\frac{3\cdot5\cdot7}{2\cdot4}\frac{\eta g}{v_{\text{c}}}\frac{\left[(4x-3)\arcsin x^{1/2} - (2x-3)y^{1/2}\sqrt{1-x}\right]}{x^{5/2}(3.5 - 2x/x_{\max})} \quad (7.20)$$

where $x = h/H_{\text{c}}$, $x_{\max} = H/H_{\text{c}}$, h is the distance from the lower part of the sphere to the lower edge of the filling hole (this is also the maximum energy of UCN stored in height units), H is the sphere diameter, H_{c} is the critical energy of the wall material in height units, v_{c} is the critical velocity

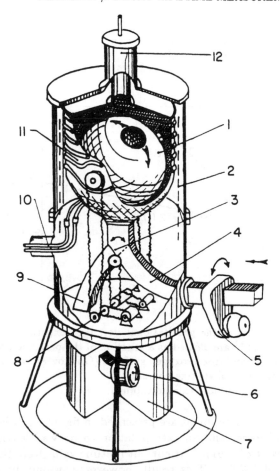

Figure 7.8 The gravitational UCN trap neutron lifetime measurement apparatus: 1, storage volume; 2, liquid nitrogen heat shield; 3, distribution valve; 4 and 9, neutron guides; 5, injection valve; 6, UCN detector; 7, detector shielding; 8, mechanism for rotating storage volume; 10, cryogenic pipes; 11, cryostat; 12, gas inlet system (Kharitonov *et al* 1989).

of the wall material, g is the gravitational acceleration, and η is the ratio of imaginary-to-real parts of the wall scattering amplitude. In the limit of small energy, $\tau_{\text{wall}} \to \eta g / v_{\text{c}}$.

The experimental method is to fill the bottle directly from the source and quickly rotate the trap until the hole is at the top of the bottle. The UCN are then stored for a certain time period before the bottle is rotated to a new height setting. UCN with sufficient energy to rise to the height of the hole now leak out of the bottle and are counted at the detector. The procedure is then repeated until the trap hole is at its lowest position and contains no

UCN. The measurement is performed again with a new time period before commencement of the rotation phase. In such a way the energy dependence of the total lifetime is determined.

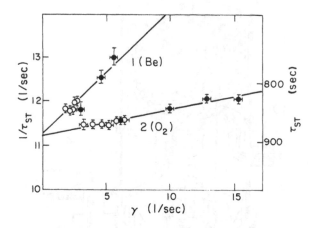

Figure 7.9   Results of the gravitational UCN trap experiment. The inverse storage lifetimes $(\tau_s)^{-1}$ are plotted as a function of the parameter γ showing the results for beryllium and frozen oxygen surfaces (Gudkov *et al* 1991).
o — results from spherical storage volume;
● — results from a cylindrical storage volume.

Results from the experiment are presented in figure 7.9, where the effects of the UCN wall losses and an extrapolation to no wall losses ($\gamma = 0$) are shown (Gudkov *et al* 1991). The current lifetime value from this experiment is quoted as 884 ± 2.9 s. This result, however, can only be considered as a preliminary value. Additional work is required to determine the energy dependence to higher precision and to look for effects arising from inhomogeneous wall coating and bottle vibrations. The method looks very promising.

7.2.6   Measurement of τ_β by magnetic storage

Confining neutrons in a magnetic hexapole torus field (Kügler *et al* 1978, 1985) has been used to measure the neutron lifetime $\tau_\beta = 876.7 \pm 10$ s (Anton *et al* 1989).

Hexapole focusing magnets have long been used in atomic beam experiments where they both select the spin state and focus the neutral atomic beam (Friedburg and Paul 1951, Ramsey 1956). A magnetic hexapole is formed by a combination of poles as shown in figure 7.10(*a*); when used with atomic beams, one spin state has a force directed toward the centre, while the other is directed away. A hexapole can also be generated by a set of wires carrying a current as shown in figure 7.10(b). If these wires are then formed into loops, as in figure 7.10(c), a hexapole torus trap results; the appropriate spin state has a restoring force directed toward the centre of the torus. A neutral particle with the appropriate spin state will orbit where the centripetal force is balanced by the magnetic force.

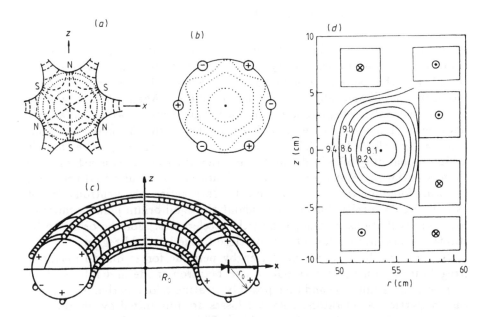

Figure 7.10   (a) Cross section of an ideal linear hexapole. The broken lines represent the magnetic field while the dotted lines represent magnetic equipotentials. (b) Linear hexapole generated from six current lines. (c) Currents lines bent into loops, generating a hexapole torus trap. (d) The current configuration used in the experiment with equipotential lines (units of 10^{-7} eV) for neutrons of velocity 15 m s^{-1}; the potential includes both magnetic and centrifugal components (Kügler *et al* 1978, 1985).

It can be shown that the magnetic field at the centre of a hexapole as shown in figure 7.10(a) can be written (Kügler 1985):

$$B_x = -2B_0(xz)/r_0^2 \qquad \text{and} \qquad B_z = B_0(x^2 - z^2)/r_0^2 \qquad (B_y = 0)$$

which gives $B = B_0 r^2/r_0^2$ where r is the displacement from the centre. In the case of the torus, the radius r_0 is much less than the large radius, R_0.

The force on a magnetic moment in the radial direction can be found from the gradient of the potential energy

$$F_r = \frac{\partial}{\partial r}\boldsymbol{\mu}\cdot\boldsymbol{B} = \pm\mu(\partial B/\partial r) \qquad (7.21)$$

where μ is the neutron magnetic moment, $\pm$ refers to the spin state, and B is the magnetic field. The equilibrium point for the orbit is found by balancing the magnetic and centripetal forces: $F_c(R) = mv_\theta^2/R$. For the hexapole torus, stable orbits can be formed since $F_R = -2B_0\mu(R - R_0)/r_0^2$, for the appropriate spin state, which gives the condition

$$R = (R_0 + \sqrt{R_0^2 + 2mv_\theta^2 r_0^2/B_0\mu})/2$$

where R is the distance of the neutron from the origin in figure 7.10(c). Only neutrons within a limited velocity range defined by the trap size r_0 may be stored, and the average velocity must be low since the neutron magnetic moment is small and laboratory fields are limited. Again, it is interesting to note that only one spin state is trapped. One problem with such a trap, however, is that the spin has to be fixed parallel to the magnetic field, that is $\omega_L = 2\mu B/\hbar \gg |dB/dt|/B$, as seen by the neutron, otherwise there is a high probablility of spin flip and losing the neutron. This is achieved by slightly deviating from the hexapole configuration. Note also that there is a restoring force for vertical displacements. In practice, the neutrons do not exhibit circular orbits, but oscillate about the equilibrium point, similar to betatron oscillations in charged particle accelerators. For the orbits to be stable, the oscillations cannot get too close to the low-field region.

Betatron oscillations can be driven by periodic forces due to magnetic irregularities. Energy comes out of one betatron mode and into another increasing its amplitude and thus possibly creating a loss mechanism. In the case of particle accelerators, betatron losses are minimized by limiting the range of momentum stored, $\delta p/p \approx 10^{-3}$. This is impractical with UCN for they are not monochromatic and defining the momentum to within such a small range would severely limit the number stored. For the neutron storage ring, this problem is dealt with by having the ratio of the radial to vertical betatron frequencies of the order of two, by limiting the initial phase space by using 'beam scrapers' (boron-carbide-coated knife edges which define the outer edges of the storage cross section) when filling and hence limiting the initial betatron amplitudes, by limiting the momentum range $\Delta p/p \approx 2$, and by generating a slight decapole term which makes the betatron oscillations nonlinear and their frequencies amplitude-dependent.

The arrangement of coils used in the experiment is shown in figure 7.10(d). The conductors are made from superconducting Nb–Ti, 1 mm in diameter. The current is of the order of 200 A, generating a usable field of 3.5 T and a gradient of 1.2 T cm^{-1}. Neutron velocities between 7 and 20 m s^{-1} can be trapped; these are actually VCN. A plot of the equipotential lines for VCN of 15 m s^{-1} is given in figure 7.10(d), indicating the depth of the potential well.

The apparatus was initially used with the PN5 source at ILL where the VCN density was such that 10 neutrons were left after waiting 500 s. The apparatus was installed in 1987 on the TGV source at a special VCN guide where the density was such that about 400 neutrons were left after 500 s, a factor of 40 improvement.

To fill the trap with neutrons, a double-bend guide brings neutrons to the trap volume along the inside radius. This guide then feeds two curved guides which can be inserted into the storage volume and extracted in a time much faster than the circular transit time of a neutron. Each of these then directs the neutron velocities tangentially around the storage ring in opposite directions, so neutrons circulate both clockwise and counterclockwise around the ring. After filling the ring, a set of mechanically controlled beam scrapers are inserted to set the limits of the outer edges of the neutron beam, after which they are withdrawn. After waiting a given time, a ^{3}He proportional counter is moved into the beam and the remaining neutrons are counted. This sequence is repeated for different storage times, ranging from about 20 to 3500 s. All of the timing in the experiment was carefully controlled, and effects due to rates of movement of the mechanical parts were carefully checked. Fluctuations in the reactor power are removed by normalizing the counts to a monitor counter. The number of neutrons remaining as a function of time gives the lifetime, apart from systematics. Figure 7.11 is a plot of their results, normalized to 1000 counts at $t = 0$.

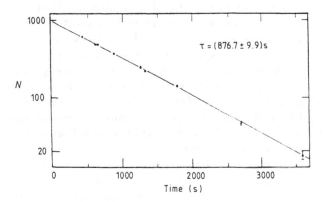

Figure 7.11   Neutron lifetime in the hexapole torus trap; all data are included, normalized to 1000 at $t = 0$ (Anton *et al* 1989).

A number of systematic tests were performed. Varying the scraper positions, thus changing the storage volume and position in the torus, and varying the magnetic field did not significantly alter the result, particularly if only storage times greater than 450 s are considered. The authors could not estimate the spurious loss rate; if one compares their result of 876.7±10 s (where the error is purely statistical) with that due to Mampe *et al* (1989b) of 887.6±3 s, this implies a 'bottle loss' time t for the torus of $t > 4 \times 10^4$ s, a long time indeed.

7.3   MEASUREMENT OF THE NEUTRON ELECTRIC DIPOLE MOMENT USING UCN

7.3.1   Introduction and history

Purcell and Ramsey (1950) suggested that nuclear forces need not conserve parity; one consequence of this is that there could be an electric dipole moment (EDM) for the neutron or other elementary particle, in addition to the usual magnetic dipole moment. The first experiment to search for a neutron EDM was performed on a neutron beam at Oak Ridge and produced the result (in 1951) $d_n = -(0.1 \pm 2.4) \times 10^{-20}$e cm which was eventually reported in Smith *et al* (1957).

Lee and Yang (1956) suggested that parity non-conservation could explain the τ–θ puzzle; the most sensitive experimental evidence supporting parity conservation was the neutron EDM. They suggested experiments to test for parity violation, most notably the measurement of the angular distribution of β-rays from a polarized nucleus. If parity were not conserved, there could be electron emission of the form $A = I \cdot p_e$, where I is the nuclear spin and p_e is the electron momentum. Landau (1957) suggested that it would be more satisfactory if the operations of charge conjugation C together with parity P left the theory invariant, that is CP is conserved, otherwise empty space would contain information. He also pointed out that one would not see an EDM unless time reversal T is violated in addition to P. Wu *et al* (1957) demonstrated parity violation in the β-decay of ^{60}Co; subsequent experiments showed that the symmetry conserved was indeed CP.

It had already been recognized that, if a local Lagrangian theory is invariant under the proper Lorentz transformation, then CPT (and its permutations) must be a symmetry of the theory (see, for example, Pauli 1955). Ramsey (1958) argued that time-reversal symmetry was an open question which could only be answered experimentally, that is that it was still important to search for the neutron EDM.

In 1964 Christensen *et al* (1964) reported the observation that CP was violated in the decay of the K^0 meson, which implies a T violation as well. Arguments have been given (Casella 1969, Schubert *et al* 1970) that the K^0 experiments show T violation directly. Observation of CP violation is the main motivation behind the on-going search for the neutron EDM, and searches for EDMs in other systems, such as atoms (Lamoreaux *et al* 1987, Lamoreaux and Golub 1989).

The origin of the CP violation, discovered in the K^0 system now 26 years ago, remains an enigma (He *et al* 1990, Shabilin 1982, 1983). Historically, experimental limits on the neutron EDM (and more recently atomic EDMs) have placed strong constraints on theoretical models that are used to explain CP violation (see He *et al* 1990, Ellis 1990 for reviews of the theory). The most recent experimental values for the neutron EDM are those of Altarev *et al* (1986b), $d_n = (-14 \pm 6) \times 10^{-26}$ e cm, which they interpret as an upper limit of $|d_n| < 26 \times 10^{-26}$ e cm and Smith *et al* (1990), $d_n = -(3 \pm 5) \times 10^{-26}$ e cm, which implies $|d_n| < 12 \times 10^{-26}$ e cm. Both of these experiments made use of bottled UCN.

It is fair to say that the neutron EDM has ruled out more theories (put forward to explain K_0 decay) than any experiment in the history of physics. For example, in Ramsey (1982), roughly 24 out of 35 theoretical estimates limits are incompatible with the experimental results, while about 16 out of the 17 listed in Golub and Pendlebury (1972) are incompatible. At present, there are 5 or 6 tenable theories (Ellis 1989, He *et al* 1990); continued improvement in the neutron EDM limit will be crucial in determining which of these are correct.

The neutron ground state, having spin $I = \frac{1}{2}$, is completely specified by the spin projection quantum number $m_I = \pm\frac{1}{2}$. In external electric and magnetic fields E and B, the field-dependent part of the Hamiltonian is

$$H = -(d_n I \cdot E + \mu_n I \cdot B)/I \qquad (7.22)$$

where d_n and μ_n are the electric and magnetic dipole moments of the neutron (Golub and Pendlebury 1972). The EDM must lie along I otherwise additional quantum numbers would be necessary to describe the neutron ground state; in addition, any component perpendicular to I would be unobservable. This Hamiltonian manifests P and T violation; under P, $E \to -E$ and B is unchanged; under T, $B \to -B$ and E is unchanged; however, CPT is conserved.

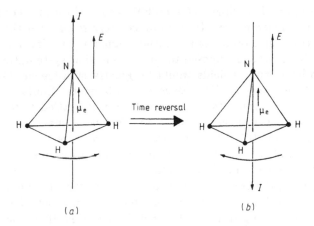

Figure 7.12   Behaviour of an ammonia molecule under time reversal.

Under time reversal, equation (7.22) is altered; the time-reversed state is different from the original state which implies time reversal is not a symmetry of the Hamiltonian. Why is it then that molecules which have electric dipole moments are not evidence of time-reversal violation? Figures 7.12(a)

and (b) show the normal and time-reversed states of an ammonia molecule. These two states are degenerate and it is this degeneracy which allows the ammonia molecule to have an EDM without violating time-reversal symmetry. However, we assumed that the neutron ground state was completely described by the spin projection quantum number m_I and the spin states are non-degenerate. That the ground state is non-degenerate is also supported by the observed fact that neutrons obey the Pauli principle. One can also understand the time-reversal violating character of equation (7.22) through Kramer's theorem (Messiah 1966) which implies that, for a Hamiltonian that depends on an external field which is time-reversal invariant (an electric field, for example) times the angular momentum, the eigenvalues of the system are at least twofold degenerate and the degeneracy is of even order. Since there are only two eigenstates of the neutron, under application of an electric field, these states remain degenerate. The degeneracy is removed by application of a magnetic field.

The neutron EDM is measured by comparing the Larmor frequency of the neutron (measured, for example, by use of Ramsey's method of separated oscillatory fields (Ramsey 1956)) for parallel and for antiparallel magnetic and electric fields. It follows from equation (7.22) that the shift in Larmor frequency between the two field configurations is $\delta\omega_0 = -4d_n E/\hbar$; the minus sign is necessary because $\mu_n < 0$.

7.3.2   Neutron EDM measurements with bottled neutrons

Shapiro (1968) and Lushchikov et al (1969) suggested that UCN could be used to search for a neutron EDM. It was immediately recognized that a storage experiment could give orders of magnitude higher sensitivity due to a longer interaction time, 10^2 s as opposed to 10^{-2} s, and that systematic effects due to non-parallel E and B fields would be greatly suppressed (Okun 1969). Golub and Pendlebury (1972) discussed fundamental limits to sensitivity for a variety of neutron EDM experiments; use of bottled UCN is the clear winner. They further suggest that 'if no effort were spared a limit for the dipole length of 5×10^{-27} e cm might ultimately be reached'. The present experimental limits are not too far from this and this level is expected to be reached in the next round of measurements.

An important advantage to the use of bottled UCN over beam experiments is the elimination of the systematic effect due to the magnetic field generated by the neutron moving through the field $E \approx v \times E$; if E and B are not exactly parallel (Sandars and Lipworth 1964)

$$\delta B = \frac{v}{c} E \sin \theta_{EB} + \sqrt{B^2 + \left(\frac{v}{c}E\right)^2} \qquad (7.23)$$

where it is assumed that the velocity, v, is approximately perpendicular to E. E and B are roughly parallel and $\theta_{EB} \approx 0$ is the angle between the magnetic and electric fields. In the case $\theta_{EB} \neq 0$, there is a change in magnetic field associated with the application of the electric field which generates a shift in Larmor frequency indistinguishable from an EDM shift. In addition, even if the fields are parallel, there is a shift in magnitude proportional to E^2; thus it is necessary for the magnitude of the electric field to be reversed exactly in any case, something which can be difficult in the presence of dielectrics. Since UCN stored in a bottle have an average velocity approximately zero, $v \times E$ effects are substantially reduced.

We will now derive the fundamental limits to a general class of bottled UCN experiments, which include the neutron EDM search, where a shift in energy between the two spin states (a shift in Larmor precession frequency ν) due to the application of an external field is measured using magnetic resonance techniques. The magnetic resonance linewidth, which determines the accuracy with which a shift can be measured, is given by

$$\Delta \nu \propto 1/T \qquad (7.24)$$

where T is the time that the neutron spin was in a coherent superposition (precessing about the field). This is a restatement of the uncertainty principle, that $\delta E \delta t \geq 1$ (energy measured in hertz). Again, there is a finite neutron survival time τ with total number at time T given by equation (7.5). The fluctuations in ν between measurements due to counting statistics can be readily determined (assuming that the total storage time is equal to t, the coherence time):

$$\delta \nu \propto \frac{1}{\alpha T} N_0^{-1/2} e^{T/2\tau} \qquad (7.25)$$

where α represents the polarization efficiency. Assuming again that we do many (n) experiments over a time $t \gg T$, we find, using equation (7.12), that

$$\sigma(\nu) \propto N_0^{-1/2} \alpha^{-1} e^{T/2\tau} / \sqrt{tT}. \qquad (7.26)$$

This has a minimum when $t = \tau$. Thus, it is evident that we want a long survival time, α nearly one, and N_0 to be large.

As has already been mentioned, the use of Ramsey's method of separated oscillatory fields is a convenient way to measure the magnetic resonance frequency (Ramsey 1956). In this method, we start with the spin along a static magnetic field. An oscillating magnetic field nearly at the Larmor frequency and perpendicular to the static field is turned on for a time τ' such that the spin (as viewed in the rotating frame) precesses through $90°$, that is a $\pi/2$ pulse; the magnitude of the field $2b$ and τ' satisfy the relation $\gamma b \tau' = \pi/2$, where γ is the magnetic moment. After the oscillating (or RF) field is turned off, the spin precesses about the static field for a time $T \gg \tau'$, at which time a second pulse is applied; note that the oscillator has been running in the background and has complete phase coherence with the first pulse. However, if the Larmor frequency and RF frequency are not exactly equal, a phase difference builds up, $\phi \approx (\omega - \omega_0)T$, where ω is the RF frequency and $\omega_0 = \gamma B$ is the Larmor frequency. Thus, in the rotating frame, the spin is not at right angles to the RF field and its spin, after a second $\pi/2$ pulse, will end up at an angle ϕ to the static field. The final polarization is given by $-P_0 \cos \phi$. As the RF is tuned away from resonance, the initial spin flip probability is reduced, thus the oscillations die away as shown in figure 7.13.

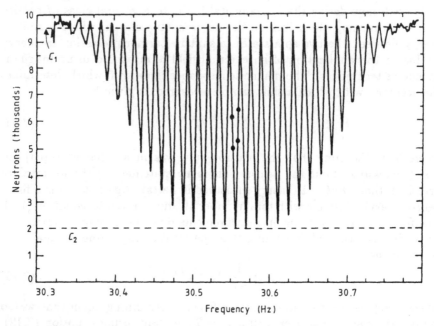

Figure 7.13 A neutron magnetic resonance curve generated through Ramsey's method of separated oscillatory fields. The polarized neutrons were stored for a total of 80 s; the time between the 4 s $\pi/2$ pulses was 68 s, giving a linewidth of 7 mHz for the central fringe. Taking data sequentially at the four points shown allows the determination of the resonance frequency (Smith *et al* 1990).

An important advantage of bottled UCN over beam experiments is that the timing conditions are the same for all neutrons. In a beam, faster neutrons spend less time, slower ones more time in the fields, hence the Ramsey fringes get washed out. However, in the neutron bottle, if the RF field is sufficiently homogeneous and, if τ' is longer than the mean collision time so that the RF field is sufficiently averaged, all the neutrons see the same pulse length, and the same time between pulses, thus the beautiful pattern as in figure 7.13. Since the time between pulses, T, is the same for all neutrons, all the neutrons have $\cos \phi = 0$ at the same frequency and all the maxima in figure 7.13 are the same height (transition probability = 0).

When separated oscillatory fields are used, the determination of the effective linewidth is not so obvious. If we use equation (V.37) given in Ramsey (1956) in the limit $b \gg \Delta\omega$ where $\Delta\omega = \omega_0 - \omega_{\mathrm{RF}}$ and b is the RF field strength, and if $\pi/2$ pulses are used (that is the pulse length τ' and the RF field strength satisfy the condition $\tau'\gamma b/2 = \pi/4$), we find that the probability to flip the neutron spin is

$$P \approx -\cos \Delta\omega T \qquad (7.27)$$

where $T \gg \tau'$ is the time between pulses, which is the same result as our qualitative argument previously given. If we use the same polarizer for both initial polarization and for analysis after the two pulses, and take into account the polarizer inefficiency, we find that the number of neutrons which get through the polarizer is

$$N(\Delta\omega) = N_0[1 - \alpha \cos \Delta\omega T]/2 \qquad (7.28)$$

where the visibility $\alpha = (C_1 - C_2)/(C_1 + C_2)$, $C_1 = N_0$, as in figure 7.13.

Since we will be looking for a small change in frequency with application of the electric field, we want to sit on the side of the resonance where the sensitivity is highest, that is where the slope of the plot of the number of counts against $\Delta\omega$ is highest, and look for a change in counts with application of the electric field:

$$\frac{\partial N(\Delta\omega)}{\partial \Delta\omega} = N_0 \frac{\alpha}{2} T \sin \Delta\omega T. \qquad (7.29)$$

We need to minimize

$$\sigma(\Delta\omega) = \frac{\partial \Delta\omega}{\partial N(\Delta\omega)} \sqrt{N_0} \qquad (7.30)$$

since we count N_0 neutrons total for both spin states. This has a minimum at $\Delta\omega T = \pi/2$ where we find the frequency noise due to counting statistics

$$\sigma(\Delta\omega) = \frac{2}{\alpha T} \sqrt{N_0} \qquad (7.31)$$

N_0 is the number of counts for both spin states. Using our Hamiltonian, equation (7.22), leads to

$$\sigma(d_n) = \frac{\hbar}{2\alpha E T \sqrt{N_0}} \qquad (7.32)$$

since $\delta\omega_0 = -4d_n E/\hbar$ on reversal of the electric field.

7.3.3 Bottled UCN EDM experiment at the Institut Laue–Langevin

Figure 7.14 is a schematic diagram of the experimental apparatus which is described more fully in Miranda (1987) and Pendlebury *et al* (1984). The apparatus was initially used on the old ILL PN5 UCN source and produced the result $0.3 \pm 4.8 \times 10^{-25}$ e cm (Pendlebury *et al* 1984). In 1986, the experiment was moved to the ILL neutron turbine (Steyerl *et al* 1986, and described in Chapter 3) where the UCN flux is two orders of magnitude higher; the UCN density at the turbine output is 90 cm^{-3}. The new result of this experiment is $-(3 \pm 5) \times 10^{-26}$ e cm (Smith *et al* 1990). A description of the experiment follows.

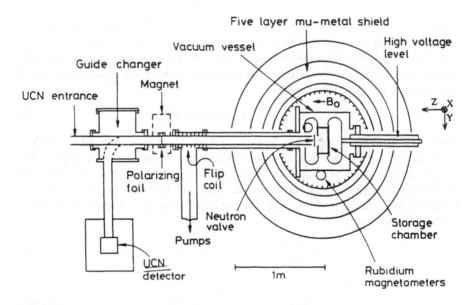

Figure 7.14   The neutron EDM apparatus used at the Institut Laue–Langevin. The UCN are stored for a total of 80 s; the static magnetic field is about 10 mG and the electric field is about 10 kV cm^{-1} (Smith *et al* 1990).

The UCN are transported from the turbine to the experiment through a stainless steel guide. In the apparatus, the UCN are polarized by transmission through a magnetically saturated 1 μm thick iron-cobalt foil. The neutron bottle, consisting of two beryllium electrodes 25 cm in diameter separated by a 0.1 m long cylindrical beryllium oxide tube (1 cm thick wall) which serves as an insulator, has a net critical velocity 6.9 m s^{-1}. The BeO tube rests in grooves, about twice the tube thickness deep, in the Be plates. This arrangement gives better high voltage stability. The ideas and technology behind the BeO/Be bottle are described by Golub (1986). Neutrons enter the bottle through a hole, which can be sealed with a Be door, in the grounded electrode.

The bottle is inside a five-layer Permalloy shield (shielding factor of 10^5) (Sumner *et al* 1987) with the bottle axis perpendicular to the cylinder axis of the shield, the orientation such that the magnetic shielding is maximum.

A 10 mG magnetic field B, parallel to the axis of the bottle, is produced by a cylindrical coil with a constant number of turns per unit distance perpendicular to the axis of the shield (cosine distribution), to produce a uniform field inside the magnetic shield.

The magnetic field between the polarizer and the storage bottle was carefully tailored so that the adiabatic condition, $\omega_L = \gamma B \gg |\mathrm{d}B/\mathrm{d}t|/B$, where γ is the gyromagnetic ratio, and so that $B \neq 0$ anywhere; thus there is a gradual change from the approximately 1 kG polarizer field to shield, and through the shield to the bottle field of 10 mG; thus no loss of polarization occurs.

The experiment is operated as follows. The 5 litre storage volume is filled for 10 s (three filling time constants), after which the door is closed. Immediately after filling, the density of the polarized UCN is about 10 cm^{-3}. After waiting 6 s to allow the neutron velocities to become isotropic within the bottle, the first pulse of a Ramsey separated (in time) oscillatory field magnetic resonance sequence (Ramsey 1956) is applied for 4 s. This turns the neutron spins perpendicular to the magnetic field. The neutrons are allowed to precess for 70 s (the neutron storage lifetime) after which a second 4 s Ramsey pulse is applied. The neutron valve is opened and those neutrons in the appropriate spin state pass through the polarizing foil (which now serves as an analyser), are diverted to a detector, and are counted for 10 s. The spin-flip coil, which consists of about 5 turn/cm wound on a NiCo-coated glass guide (so that the 3 kHz RF field can penetrate) 20 cm long, is then turned on and adiabatically reverses the spin, thereby permitting the remaining neutrons to pass through the polarizing foil and be counted, also for 10 s. The two counting periods give approximately 12 000 and 8000 neutrons. Including filling and emptying, each measurement cycle takes 124 s.

An electric field, E, with magnitude up to 1.6 MV m^{-1}, follows a 32 measurement cycle (about 1 hour) sequence: 8 cycles applied parallel to B, 4 cycles off, 8 cycles antiparallel, and 4 cycles off, with 2 cycles taken to change each state (the magnitude of E is constant over each approximately 3 day set of measurements). The direction of E is reversed by simply changing the polarity of the voltage applied to the upper plate. The neutron bottle is maintained under vacuum (10^{-6} Torr) or with 10^{-4} Torr of nitrogen or helium to help quench sparking; it was found that helium works much better. Leakage currents across the bottle are monitored and are kept below 30 nA by operating at an appropriate voltage. For most of the data presented here, the leakage current was less than 5 nA.

Over the course of a reactor cycle (about 6 weeks), the neutron storage and high voltage properties of the bottle gradually deteriorate. This is probably due, in part, to hydrogen contaminating the surface (Lanford and Golub 1977). The background system vacuum is rather poor, of order 10^{-6} Torr. Running a discharge of argon and deuterium *in situ* (about 150 V, 50 Hz for 10 min at 1-5 Torr) restores the storage time (Mampe *et al* 1981) and low leakage currents. Such discharge cleaning is repeated approximately once every reactor cycle. There is some evidence that the presence of the high voltage accelerates the degradation.

The magnetic field within the shield is monitored using three optically pumped rubidium magnetometers placed around the neutron bottle (Pendlebury *et al* 1984) at a maximum distance of 40 cm from the bottle. The field at each magnetometer is averaged over the neutron storage time. In addition, there is a flux gate magnetometer placed between the outer two layers of the shield to check for possible externally generated systematic signals.

Between measurements, the frequency of the RF field for Ramsey pulses is changed so that there can be a continuous calibration of the apparatus; the frequency is varied sequentially through four points around the central fringe of the resonance pattern (see figure 7.13). The two points on each side of the resonance centre are separated by one-tenth of a linewidth. The four points are used to determine both the neutron spin polarization and the resonance frequency. To determine the neutron resonance frequency, ν_n, a first pass is made through the data to extract the visibility of the resonance curve, $\alpha = (C_1 - C_2)/(C_1 + C_2)$, which is typically 0.64 (see figure 7.13). A second pass uses α and a combination of the counts for the two neutron spin states to yield a single resonance frequency ν_n for each measurement cycle, as implied by equation (7.28). An important advantage of this technique is that it suppresses non-statistical fluctuations in the neutron flux (due to reactor power fluctuations, for example), since one works with the ratio of spin up to spin down counts. Over the course of the measurements, the frequency is changed so that the resonance is tracked as the magnetic field drifts.

A run lasts about three days and has about 1000 measurements of the resonance frequency. The electric field is reversed about every 10 measurements as described previously.

Figure 7.15 shows ν_n as a function of time for a typical 1-day run, and the same data after the average of the magnetometer readings (normalized to the neutron frequency) is subtracted from ν_n. Jumps in the magnetic field, as shown in figure 7.15, can be due, for example, to movements of the reactor crane, relaxation movement of the magnetic shields or the switching of magnets on neighbouring experiments.

An EDM would appear simply as a shift in resonance frequency with application of an electric field. The question is how to extract the EDM frequency shift from other systematic effects due to sparks, leakage currents or externally generated fields associated with switching the high voltage.

A number of techniques have been used to extract the neutron frequency shifts which change sign with the applied electric field direction, $\Delta\nu_n = [\nu_n(E) - \nu_n(-E)]/2$. Generally, a drifting background is removed by fitting terms linear and quadratic in time for the individual measurement sets. Cycles around clear discrete jumps in the magnetic field are

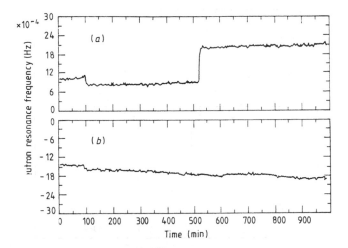

Figure 7.15   Data from a typical 1-day run showing (*a*) the measured neutron resonance frequencies and (*b*) the same data but corrected by the average of the magnetometer readings. It should be noted that about 90% of the large jump was corrected for, but only about 50% of the smaller jump was taken out. This demonstrates the usefulness of this technique for suppressing disturbances generated external to the shields (Smith *et al* 1990).

discarded. Similar analyses are performed for the three magnetometers yielding $\Delta\nu_{m1}$, $\Delta\nu_{m2}$, and $\Delta\nu_{m3}$. Frequency shifts quadratic in E (voltage on/off effects) and shifts between the zero electric field groups (due for example to the shields being magnetized by leakage currents or sparks) are also extracted. To help identify systematic effects, the direction of the fixed magnetic field is reversed every few weeks; a true EDM frequency would have its sign reversed by this. In addition, the high voltage was varied between measurements sets; a true EDM would scale with the electric field strength.

The individual measurement set results can be combined into 15 groups, one for each 6-week reactor cycle of measurement time. Figure 7.16 shows the apparent (uncorrected) neutron EDM for each reactor cycle. The weighted average of these data yields an uncorrected neutron EDM of $\tilde{d}_n = (-1.9 \pm 2.2) \times 10^{-26}$ e cm with a normalized $\chi^2 = 3.1$. As is apparent in figure 7.16, there is evidence for non-statistical (systematic) variations in the uncorrected EDMs.

The magnetometer readings have been used to correct for systematic errors in two independent ways. The first makes use of the observation that for perturbations external to the magnetic shields, the average value of the three magnetometers usually tracks the neutron resonance frequency to within 20% (as illustrated in figure 7.15). For each measurement set, a corrected $\Delta\nu_n$ is obtained by subtracting the average of $\Delta\nu_{m1}$, $\Delta\nu_{m2}$ and $\Delta\nu_{m3}$. The

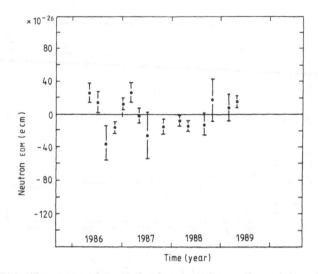

Figure 7.16 The measured neutron EDM per reactor cycle, without any corrections. The weighted average is $-(1.9 \pm 2.2) \times 10^{-26}$ e cm with $\chi^2 = 3.1$ (Smith *et al* 1990).

data in figure 7.15, after correcting by the averaged magnetometer results, yield $d_{n1} = (-3.4 \pm 2.6) \times 10^{-26}$ e cm with normalized $\chi^2 = 2.2$. The reduction in χ^2 for the corrected data comes primarily from the additional noise contributed by the magnetometers. Although this technique does not eliminate the non-statistical variations in the data, it reduces contributions to d_n from external magnetic field fluctuations.

The second magnetometer correction method allows for the possibility that systematic frequency shifts may be different for the neutrons and magnetometers (such as those due to magnetic fields from leakage currents across the storage bottle). As described previously, the frequency shift with the electric field $\Delta\nu_{ki}$ over each run (set of 1000 measurement points) where k represents the neutrons or any of the magnetometers and i represents the run number, are determined. A true neutron EDM is a frequency shift $\delta\nu_n$ which is correlated with the electric field and direction of the magnetic field, that is EB where E is the electric field magnitude and B is the sign of the magnetic field relative to the electric field, and *that is not correlated with a shift on any of the magnetometers* . Any part of $\Delta\nu_n$ correlated with the magnetometers is a systematic effect.

There are two cases to consider.

(i) Systematic effects which do not follow EB, e.g. they could come from ground loops or pressure sensitive sparks, etc. In this case $\Delta\nu_n$ would be correlated with the magnetometers but not with EB.

(ii) Systematic effects which are proportional to EB, e.g. helical leakage currents. In this case, it is not possible to distinguish between an EDM and systematic effect. Instead, a correlation between the magnetometers and EB indicates that a systematic shift is present. If such a correlation exists, the error bar on $\Delta\nu_n$ must be increased.

For consecutive data sets between major alterations to the experimental apparatus (several reactor cycles), linear correlation coefficients between $\Delta\nu_n$ and $\Delta\nu_{m1}$, $\Delta\nu_{m2}$ and $\Delta\nu_{m3}$ were evaluated. A multiple linear regression routine (Bevington 1969) was then used to fit $\Delta\nu_n$ with a term linear in the magnitude of E and linear terms in any $\Delta\nu_{mi}$ if the mi correlation with $\Delta\nu_n$ was statistically significant (greater than 80% probable). There was essentially no systematic-free data as indicated by the cross-correlations between the magnetometer signals and the electric field.

Before the discovery of the correlation between the magnetometers and electric field, selection of seemingly 'clean' (no magnetometer EDM signals) data gave a 2σ result in agreement with that of Altarev et al (1986a) (Pendlebury 1988a). The correlation technique subsequently indicated that all of the data were contaminated with systematic effects. The point is that, even though the average of the magnetometers gave a zero result, there were fluctuations about zero highly correlated with the neutrons. This indicated that the systematic error has contributions from both inside and outside the magnetic shields, with different relative strengths for the neutrons and each magnetometer. It is important to note that, in order to see the correlation in the presence of the usual noise, it was necessary to look at groups of about 20 runs, that is data taken over entire reactor cycles.

Because of the presence of correlations in the independent variables (magnetometers and electric field), one cannot say which is responsible for the observed response (shift in the neutron's Larmor frequency). Subsequently, the final error bar had to be increased by a factor of 2.5 in amplitude to account for the correlations between the magnetometer signal (the 'background' signal by which the neutron EDM was corrected) and the electric field.

The weighted average of the four sets of data by this analysis yields a second corrected value for the neutron EDM, $d_{n2} = (-3.3 \pm 3.5) \times 10^{-26}$ e cm with a normalized $\chi^2 = 1.5$. The larger error bar for d_{n2} compared with d_{n1} comes from the uncertainties generated by cross-correlations in the linear regression and from the use of a smaller data set (tighter cuts were used to determine which runs would be included in the correlation analysis). Most importantly, the regression made the data sets between major alterations of the apparatus consistent.

A similar analysis was done for 'quadratic' frequency shifts (voltage on/off effects) in the applied field, E: $\Delta\nu_Q = [\nu(|E|) - \nu(0)]/E$ gives $|d_Q| < 2 \times 10^{-26}$ e cm. Again, there were highly significant correlations between the various channels which were compensated for by the regression fit. This was also true for the zero-field shifts, that is at $E = 0$: $\Delta\nu_0 = [\nu(E =$

$0\ after\ +\boldsymbol{E})-\nu(\boldsymbol{E}=0\ after\ -\boldsymbol{E})]/E$ gives $|d_0| < 3 \times 10^{-26}$ e cm.

The data analysis rejected measurement cycles on either side of a detected spark. When the cycles with sparks were included in the analysis, no significant changes in the results are found. Effects from small, undetected sparks are therefore expected to be negligible. It is also interesting to estimate the size of a systematic error generated by the leakage current if that current were to flow in a loop around the BeO cylinder circumference. For a current of 10 nA and radius of 10 cm, the magnetic field is 50 pG, or a shift in frequency of 0.15 μHz. This corresponds to an EDM of roughly 3×10^{-26} e cm. Although this is an upper limit, one should be aware that the shift at the magnetometers will be different in magnitude and of opposite sign for they are outside of the current loop. So this could be a systematic effect.

Components of the neutron velocity, $\boldsymbol{v}_n$, that circulate around the bottle circumference can generate a frequency shift proportional to $\langle \boldsymbol{v}_n \times \boldsymbol{E} \rangle$. Such a circulation may persist for times on the order of the bottle-emptying time constant (3 s). No evidence for such systematic effects was observed when the time interval (usually 6 s) between closing the bottle door and applying the first Ramsey pulse was varied.

In summary, the regression technique both identified and reduced the systematic errors. The authors are confident that there is very little systematic error in the final number quoted because the results of the two analysis methods give corrected EDM values d_{n1} and d_{n2} which are in good agreement and differ little from the uncorrected value, d_n. The conclusion is that the best result for the neutron EDM comes from the average of d_{n1} and d_{n2} with the larger of the two error bar (times $\sqrt{\chi^2}$):

$$d_n = (-3.3 \pm 4.3) \times 10^{-26} \text{ e cm}$$

$$|d_n| < 12 \times 10^{-26} \text{ e cm} \quad 90\% \text{ CL.}$$

The contribution to this error from neutron counting statistics, using equation (7.32), is 1.9×10^{-26} e cm.

The present experiment is limited by how well the magnetic field within the neutron bottle is monitored by the spatially separated rubidium magnetometers. In section 7.3.5 we will describe a new experiment which is being built from the components of this experiment and incorporating a comagnetometer (a polarized atomic species within the neutron storage volume); this should give a net factor-of-ten improvement.

7.3.4   Bottled UCN EDM experiment at the VVR-M Reactor, Leningrad

Altarev et al (1986b) have recently reported the result $d_n = -(1.4 \pm 0.6) \times 10^{-25}$ e cm which they interpret as an upper limit of $d_n < 2.6 \times 10^{-25}$ e cm.

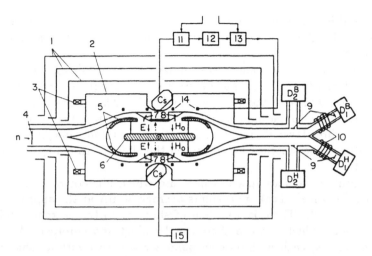

Figure 7.17 The neutron EDM apparatus used at the Leningrad reactor: 1, magnetic shields; 2, vacuum chamber; 3, Helmholtz coils for producing the static magnetic field; 4, UCN polarizer; 5, ground electrodes; 6, high voltage electrode; 7, entrance shutters; 8, exit shutters; 9, analysers; 10, spin flippers; 11, caesium magnetometers; 12, frequency divider; 13, control for oscillating field pulses; 14, coils for producing oscillating field; 15, caesium magnetometer; D labels the UCN detectors with B and H specifying upper and lower chambers, 1 and 2 specifies the two polarizations (Altarev *et al* 1986b).

This should be compared with their earlier result (Altarev *et al* 1980) $d_n <$ 1.6×10^{-24} e cm. The major improvements for this experiment were in the UCN source (see Chapter 3) and an increased neutron storage time.

A schematic diagram of the experimental apparatus is shown in figure 7.17. In many ways the apparatus is similar to that of the ILL experiment; however, there are important differences. As can be seen in figure 7.17, there are two neutron storage chambers with oppositely directed electric fields (relative to the magnetic field) in each chamber. The high voltage is applied to the plate separating the two chambers while the outer plates are held at ground potential. They typically run at 15 kV cm^{-1}, somewhat higher than the average of the ILL experiment.

Using two bottles with oppositely directed electric fields essentially doubles the sensitivity to a neutron EDM while reducing the background magnetic field noise; an EDM-generated shift will be of opposite sign for the two chambers. In addition, since the two chambers are located quite close spatially, one would expect high discrimination from background magnetic field changes since the EDM shift is given by the difference in the resonance frequency between the chambers as a function of electric field direction; this difference is sensitive only to changes in spatial gradients of the magnetic field. Such gradients could be due to locally generated fields such as leakage currents within the bottle. However, even for local fields, the close place-ment of the chambers should give a fairly good cancellation of net magnetic field change.

The position of the resonance is stabilized by synthesizing the RF pulses for the neutrons from the output of two caesium magnetic Zeeman atomic oscillators (located near the storage chambers). (The ratio of neutron res-onance frequency to caesium is about 120.) This was necessary in part because the three-layer Permalloy shield did not provide adequate stability and shielding. The stabilization improved the effective shielding by a factor of about 15.

The chamber walls are quartz rings onto which films of BeO and Be_3N_2 have been deposited. Leakage currents average 30 nA at the approximately 15 kV cm^{-1} used. There is a system of shutters to the entrance and exit guides through which the chambers can be filled and emptied. A system of four detectors, each with its own polarizer and two with adiabatic spin flippers allow the spin-up and spin-down neutrons to be counted for each chamber simultaneously, as opposed to the ILL experiment where the spin-up and spin-down are counted in sequence.

The experiment had previously been operated in a 'flow-through' mode; that is the neutron shutters were continually opened and neutrons were continuously counted; a shift between the spin-up and spin-down counting rate as a function of electric field direction would be an indication of an EDM. More recently, they began using a pulsed technique much like the ILL experiment, making full use of the chamber lifetime of 50 s, as implied by equations (7.26) and (7.32). The increase in sensitivity gained here along with the improved source of UCN resulted in a daily statistical uncertainty of approximately 2.5×10^{-25} e cm, comparable with the ILL experiment.

There is still a question of whether the nearly non-zero effect is due to systematic problems or to a neutron EDM. In the data analysis, they have analysed each bottle separately for common-mode (overall magnetic field change) compared with a differential shift. It seems that the differential change with application of the electric field leads to a non-zero effect of 3.5σ in one chamber while only 1σ in the other. It should be noted that there existed the possibility of a $v \times E$ shift in the flow-through system. There has been additional work on the apparatus and they are continuing to take data.

7.3.5   UCN EDM experiment with a ^{199}Hg comagnetometer

Since the ILL experiment described in section 7.3.3 is no longer limited by counting statistics but only by systematic errors, it was decided to rebuild the apparatus and include a comagnetometer, that is a polarized atomic species within the same storage volume as the neutrons. This provides a nearly exact spatial and temporal average of the magnetic field as seen by the neutrons over the storage period. The use of polarized ^{3}He had already been considered (Ramsey 1984), but the extreme difficulty in detecting the ^{3}He polarization makes its use impractical.

The use of ^{199}Hg had been suggested (Lamoreaux 1986). The advantage here is that ^{199}Hg can be readily directly optically pumped and its polarization optically detected with 254 nm resonance radiation (Lamoreaux 1989). Since ^{199}Hg is a $^1S_0$ atom, its ground-state polarization is specified by the nuclear angular momentum, which is $\frac{1}{2}$ for ^{199}Hg.

An important feature of a spin-$\frac{1}{2}$ system is that its Larmor frequency cannot be affected by electric fields other than through an EDM. This is a statement of Kramer's theorem which says that the energy levels of a T-even Hamiltonian are doubly degenerate (Messiah 1966). Since there are only two levels which describe the ground state, these levels are degenerate and there is no observable effect. This is to be compared with higher-spin species which were considered, such as the alkali atoms caesium and rubidium. In these cases, the total spin is greater than 1 and the atoms have a large electric polarizability; the ground-state Zeeman levels split proportional to E^2. Such shifts would make exact reversal of the electric field imperative, an experimental difficulty in the presence of dielectrics.

Furthermore, it is necessary that the atomic species do not have an EDM of their own which could possibly mask a neutron EDM; in the case of ^{199}Hg, experimental limits have already been set at the level of sensitivity needed (Lamoreaux et al 1987, Lamoreaux and Golub 1989). In these experiments, ground state spin-polarization lifetimes in excess of 100 s, were routinely achieved in cells of about 5 cm^3 volume, even in the presence of electric fields up to 15 kV cm^{-1} However, these cells included 250 Torr of nitrogen to improve the high voltage stability.

An unfortunate disadvantage of ^{199}Hg is that the walls of the container must be specially prepared to have long spin relaxation times. In all previous experiments, hydrocarbon waxes were used; these would be unusable with UCN. In addition, the wall coating has to be stable under the application of high voltage in vacuum since a high-pressure background gas cannot be used with the UCN.

A possible wall coating material, deuterated polystyrene (DPS), has been developed (Lamoreaux 1988). Although the Hg spin-polarization characteristics are not as good as the hydrocarbon waxes (10 s cm^{-1} compared with 100 s cm^{-1} mean free path), it should give a lifetime of about 100 s in the much larger neutron bottle. In addition, thin films of DPS seem to be stable under application of high voltage. This stability is a bit puzzling as every other thin-film (suitable for use with UCN) material tested would break down at relatively low electric fields. These materials included Fomblin and Teflon where electric fields of about 2 kV cm^{-1} would cause breakdown. It has long been known that polystyrene can be formed from an electrical discharge in styrene vapour; perhaps there is some complicated dynamical chemistry which leads to the high voltage stability. In addition, it has been shown that for a material to have high vacuum voltage stability, its vapour pressure must increase slowly with temperature (Trump and van de Graaff 1947). The vapour pressure of Fomblin certainly rises rapidly with temperature, hence its use as a diffusion pump fluid. Teflon breaks down readily upon heating; the breakdown temperature of polystyrene seems to be higher than that of Teflon.

One might expect that the inelastic scattering from deuterium would be significantly worse than that of the fluorine in Fomblin but DPS is found to have excellent UCN storage properties. Preliminary tests show that it is about as good as Fomblin and is about as easy to apply to a surface. The Fermi potential of DPS is about 165 neV, higher than that of Fomblin. However, the Hg polarization storage properties have so far only been demonstrated in relatively small cells, of order 5 cm^3. Whether the proper surface preparation can be achieved in a quite large vessel is still an open question. It should be pointed out here that Teflon has good polarized Hg storage properties and the surface can be adequately prepared for arbitrarily large bottles; its only drawbacks being its instability in high voltage, although this problem might be solved, and its somewhat lower Fermi potential.

A rough schematic diagram of the proposed experimental apparatus is shown in figure 7.18. To increase the sensitivity through storage time and neutron counts, and to account for the loss of neutrons due to the lower Fermi potential of DPS over Be–BeO, a larger (about 10 times) volume storage bottle is being considered. Since there is a considerable shift in the centre of mass between the UCN gas and atomic gas in the gravitational field (due to the difference in temperature) (Ramsey 1984), the experiment is designed so that the shorter axis of the bottle is vertical, thus minimizing the displacement. In addition, there is a safety consideration with the use of Hg, although the miniscule quantities of Hg to be used should present no problem. However, it is necessary to have a gas-tight window which can withstand atmospheric pressure. It has been decided that this will be the polarizer; the design for the polarizer is to evaporate iron onto aluminum. To account for the fairly high Fermi potential of the aluminum, after passing through the foil the UCN will rise about 1 m. Tests are currently in progress to determine the optimum combination of heights to maximize the number of UCN left in a test bottle after a 100 s storage period.

The neutron storage bottle will be an aluminum oxide cylindrical spacer about 60 cm diameter and 20 cm high separating aluminum plates. The entire inner surface will be coated with DPS. The neutron valve will be gas-tight to minimize the loss of the polarized atomic vapour.

Provisions will be included for polarizing the atomic vapour. There is an optical pumping cell connected to an isotopically enriched Hg reservoir. The Hg will be optically pumped to the appropriate spin state, parallel to the

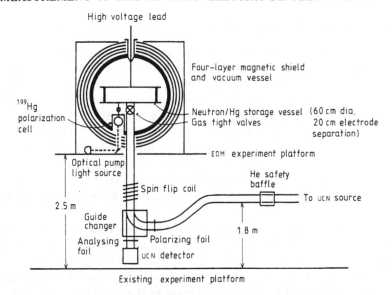

High voltage lead

Four-layer magnetic shield
and vacuum vessel

^{199}Hg
polarization
cell

Neutron/Hg storage vessel (60 cm dia,
Gas tight valves   20 cm electrode
                        separation)

EDM experiment platform

Optical pump
light source

He safety
baffle

Spin flip coil

To UCN source

2.5 m

Guide
changer

1.8 m

Analysing
foil

Polarizing foil

UCN detector

Existing experiment platform

Figure 7.18 Tentative neutron EDM measurement apparatus which includes an
Hg comagnetometer. The heights will be adjusted to maximize the UCN density
after storage for 120 s. The Hg spin will be optically detected by a beam of
circularly polarized resonance radiation propagating perpendicular to the page
and across the neutron storage volume (Lamoreaux 1988).

static field, with circularly polarized light from an Hg discharge lamp. After
the Hg is polarized, and after the neutron bottle is filled with polarized UCN
and the neutron valve closed, the polarized Hg is admitted to the neutron
bottle. $\pi/2$ pulses are applied for both the neutrons and Hg (the ^{199}Hg
magnetic moment is about one-third of the neutron magnetic moment). The
free precession of the Hg spin is observed with a beam of circularly polarized
resonance light which propagates across the bottle diameter, through quartz
windows in the insulating cylinder. The magnetic field over the storage time
can be determined from the free precession signal.

At the end of the storage period, the second neutron pulse is applied,
the bottle door opened, and the neutrons are counted as usual. The Hg
is pumped away. While the storage was in progress, more Hg had been
admitted to the optical pumping cell and polarized; the process is thus ready
to be repeated.

An EDM will be evident from a change in the ratio of the magnetic moments between reversals of the electric field. Although the sensitivity of the
Hg to the magnetic field is only one third that of the neutron, the high
signal-to-noise inherent in the free precession signal is a compensating factor
and in fact preliminary estimates show that determination of the average
field should be a factor of 10 higher in sensitivity that the neutron signal
and hence contribute very little noise.

7.3.6   Superfluid helium neutron EDM with ^{3}He comagnetometer

Golub (1983, 1987) has suggested performing an EDM search directly in the liquid helium of a superthermal source (described in Chapter 3) using a dilute solution of polarized ^{3}He as a polarizer and detector since ^{3}He absorbs neutrons only when the total spin is zero (neutron and ^{3}He spins antiparallel). The polarization and transport of polarized ^{3}He is a well-developed technology (Aminoff *et al* 1989). The reaction between ^{3}He and neutrons should produce ultraviolet scintillation in the liquid helium which should be easily detectable, giving a detection of ^{3}He–n reactions with nearly 100% efficiency. In addition, liquid helium has good dielectric characteristics and it should be possible to establish electric fields at least as high as can be reached in vacuum.

Such an experiment would be sensitive to a neutron EDM by looking at the scintillation rate at the end of a double-pulse sequence as a function of electric field. It has been shown by solving the Schrödinger equation in the presence of a spin-dependent absorption probability that this technique is slightly less sensitive than the conventional bottle technique; however, this loss of sensitivity is more than made up for by elimination of the extraction losses, transport losses, polarization transmission losses, etc. In fact, extremely high UCN densities might ultimately be achieved by locating the helium as close as possible to the reactor core. In addition, the presence of liquid helium suggests the use of superconducting shields which might be better than the usual Permalloy.

Unfortunately, the problem of measuring the magnetic field remains and it has been demonstrated that experiments are presently limited by magnetic systematic effects. It might be possible to use SQUID magnetometers to detect the ^{3}He magnetization, the ^{3}He could then serve as a magnetometer. However, the sensitivity is at best marginal. Lamoreaux and Golub (1989) have suggested using the ^{3}He as a direct comagnetometer by using 'dressed atom' techniques to make the magnetic moments of the neutron and ^{3}He equal. The idea is as follows.

In the presence of a strong oscillating magnetic field, the magnetic moment will be modified, or 'dressed' (Dubbers 1989)

$$\gamma' = \gamma J_0(\gamma B_{RF}/\omega_{RF}) = \gamma J_0(\gamma x) \tag{7.33}$$

where γ is the gyromagnetic ratio, B_{RF} and ω_{RF} are the amplitude and frequency of an applied oscillating magnetic (RF) field, J_0 is the zeroth-order Bessel function, and γ' is the observed gyromagnetic ratio.

In practice, the oscillating field is at right angles to the static field B_0 around which the spins are precessing. It is a happy coincidence of nature that the gyromagnetic ratios of ^{3}He and the neutron are nearly equal; $\gamma_1/2\pi \approx -3$ Hz mG^{-1} and $\gamma_2/2\pi \approx -3.33$ Hz mG^{-1} where subscripts 1 and 2 refer to the neutron and ^{3}He respectively. We will use this notation in the rest of this discussion. In the absence of the oscillating field, one would see scintillation due to reactions occur at a rate

$$\phi(t) \propto (1 - p_1 p_2 s_1 \cdot s_2) = (1 - p_1 p_2 \cos[(\gamma_1 - \gamma_2)B_0 t + \Phi]) \qquad (7.34)$$

where p is the polarization, s ($|s| = 1$) is the spin vector which lies approximately perpendicular to the static field, and Φ is an arbitrary phase.

Thus, one sees oscillations in the scintillation rate occur at the difference in the precession frequencies ($\delta\omega$). If the RF dressing field is now applied, we find

$$\delta\omega = (\gamma_1 J_0(\gamma_1 x) - \gamma_2 J_0(\gamma_2 x))B_0. \qquad (7.35)$$

This has the amazing property that $\delta\omega = 0$ when $\gamma_1 x \approx 1.1$.

If the neutron EDM were non-zero (the ^{3}He EDM is expected to be quite small due to Schiff (1963) shielding), the neutron precession frequency will be shifted by an amount $2d_n E J_0(\gamma_1 x)$ since the dressing dilutes the angular momentum. Thus, the value of $x = x_0$ to give $\delta\omega = 0$ is changed. By measuring the value of x with respect to electric field direction, we will be sensitive to the neutron EDM. The important point is that the effect of magnetic fields is cancelled.

An experimental method might be to keep the neutron and helium spin vectors nearly parallel; as the value of x is varied the scintillation will increase or decrease as x is varied away from the value x_0 such that $\delta\omega = 0$. Over the course of a storage, x could be modulated at a low frequency and the value x_0 inferred from the modulation in the scintillation rate.

As before, there will be a filling and RF pulse sequence, with electric field flips between measurement cycles. It will be necessary to have provisions to remove the ^{3}He at the end of each measurement cycle, possibly through use of the heat flush technique (McClintock 1978). So far an exact experimental procedure has not been determined. There is also the question of what ^{3}He density is optimum; a rough estimate shows that the reaction rate should equal the bottle loss rate, that is the average value of $\phi = 1/\tau$ where τ is the usual bottle lifetime (with β-decay included). There could be some problems associated with the generation of strong homogeneous oscillating fields, particularly near conductors.

Preliminary estimates show that it is possible to achieve a very good sensitivity; perhaps in the not too distant future, this technique could yield a sensitivity 100 times better than the current limits. Such an experiment would be operated with the helium bath as close as possible to a reactor core or target of a spallation source to achieve the maximum UCN density, assuming that the only limitation will be counting statistics, the counting statistics being determined by the total number of scintillation pulses (neutron-^{3}He reactions).

7.4   THE MAGNETIC DIPOLE MOMENT OF THE NEUTRON

Detection of a non-zero neutron EDM when a comagnetometer is used, as described in sections 7.3.5 and 7.3.6, will be through measurement of the ratio of the gyromagnetic ratios of the neutron and the atomic species as a function of electric field. Such measurements will be of the order 5×10^{-8}Hz out of about 30 Hz and thus, in the absence of an EDM (or other systematic), it is a measurement at the one part per billion level, as compared with the current value of the neutron magnetic moment accuracy of 250 parts per billion (Greene *et al* 1979) relative to the nuclear magneton.

For such a measurement to be useful, the neutron magnetic moment must be eventually related to the nuclear magneton or proton magnetic moment. The magnetic moments of the two atomic species being considered for a comagnetometer, ^{3}He and ^{199}Hg, are known only at the part per million level. However, with modest effort, these could be significantly improved upon.

The only major systematic which might be difficult to deal with is the displacement between the centre of mass of the UCN gas and the magnetometer gas due to the Earth's gravitational field giving a different average magnetic field for the two species.

7.5   THE ELECTRIC CHARGE OF THE FREE NEUTRON

The electric neutrality of the neutron is an experimental question with important theoretical implications. For example, a non-zero neutron charge would imply that conservation of baryon number and conservation of electric charge are not independent. A non-zero neutron charge could also eliminate the possibility of neutron–antineutron oscillations.

Most experimental determinations of the neutron charge q_n have been through the deflection of a neutron beam by an applied electric field; among the first use of this method yielded (Shapiro and Estulin 1956) $q_n < 6 \times 10^{-12}q_e$. More recently Baumann *et al* (1988) have reported $q_n < 2 \times 10^{-20}q_e$; a factor of ten improvement seems likely (Baumann *et al* 1989).

There has been a demonstration of the feasibility of using UCN to measure q_n (Borisov *et al* 1988) which in three days' running has produced the result $q_n = -(4.3 \pm 7.1) \times 10^{-20}q_e$. This result compares favourably with the most recent beam determination, particularly after taking into account loss of sensitivity due to the short running time of the UCN experiment. Since the sensitivity of these experiments depends on the square of the time that the neutron spends in the electric field, one gains rapidly with decreasing neutron velocity.

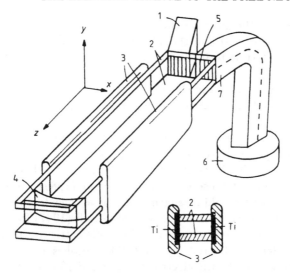

Figure 7.19   The neutron optical camera for measuring the charge of the free neutron: 1, input guide; 2, optical mirrors which also serve as electrode spacers; 3, electrodes with polished titanium surfaces; 4, vertical curved mirror; 5, input and output gratings; the output grating is split into an upper and lower half, the two halves displaced horizontally by a grating spacing; 6, UCN detectors; 7, dual guide for the upper and lower exit gratings (Borisov *et al* 1988).

A schematic diagram of the apparatus, a neutron optical camera, is shown in figure 7.19. The experiments were performed at the Leningrad VVR-M reactor. The measurement is based on the focusing of UCN by a vertical cylindrical mirror.

As shown in figure 7.19, neutrons enter the camera through a grating composed of 40 vertical slits 0.7 mm wide and 50 mm high, separated by 1.5 mm. The camera is made from two plane-parallel mirrors 1065 mm long and 100 mm wide which separate two high-voltage plates. These surfaces are specially prepared optical surfaces. The UCN are reflected from a vertical cylindrical mirror with radius of curvature 1041 mm and produce an image of the entrance grating at the exit grating. The exit grating is split into two equal parts, a top and a bottom, each half feeding its own detector; these are offset by the grating spacing. This allows determination of the displacement of the image through a differential measurement. The exit grating could be translated by a micrometer and was set to the point of maximum change in counting rate with image displacement. The mean neutron transit time (squared) was determined from the UCN velocity spectrum.

An electric field of order 10 kV cm^{-1}, a mean square transit time of 0.3 s^2 and a total counting rate 140 s^{-1} in each detector gives a statistical error of $3.6 \times 10^{-20} q_e$ per day of running; however, the sensitivity of the apparatus is quite sensitive to how level it is. This is because the neutrons used were not monochromatic; the image at the exit grating is shifted according to the neutron velocity (squared) and the tilt angle. The tilt angle of $1'$ reduced the effective transit time to 0.12 s^2 and the sensitivity was reduced to 9×10^{-20}/day.

There were some systematic effects due to electrostatic forces on the optical system; the systematic errors were reduced after everything was glued together.

The sensitivity can be further improved by reducing the tilt of the apparatus, increasing the electric field, and by increasing the precision of the optical elements. It should be noted that the electric fields in this experiment were one-sixth of those of Baumann *et al* (1989). Since the limit set scales as the electric field strength, there is a considerable factor to be gained here. However, as Trump and van de Graaff (1947) found, there is a limit to the maximum potential across a single dielectric in vacuum of about 300 kV, independent of electrode spacing. This implies a maximum field of 30 kV cm^{-1}, as implied by the electrode spacing of 10 cm, which is a factor of three greater than the present field. The authors are confident that an overall factor of ten increase in sensitivity is possible.

7.6 GRAVITATIONAL INTERACTIONS OF THE NEUTRON

The external force which affects us most is the one we know the least about; this unhappy instance is due to the weakness of the gravitational force when considering interactions at the microscopic level. It is observed that the weaker the interaction, the more symmetries it violates with the exception of gravity; however, the symmetry properties of gravity are not so well tested experimentally. General relativity is symmetric under all transformations, but this is a macroscopic theory. Most tests of general relativity have dealt with bulk matter.

Testing for symmetry violation in the neutron gravitational interaction has been suggested where a possible neutron gravitational dipole moment (GDM) is described through the interaction (Okun and Rubbia 1967, Hari Dass 1976, Golub 1978)

$$V(r) = mlG\boldsymbol{\sigma} \cdot \nabla\phi = \frac{1}{(\hbar/mc)} \left(\frac{GM}{rc^2}\right) \left(\frac{\hbar/mc}{r}\right) mc^2 \boldsymbol{\sigma} \cdot \hat{\boldsymbol{r}} \qquad (7.36)$$

where $G\phi$ is the gravitational potential $G\phi = GM/r$ and $G\nabla\phi = g$, at a distance r from a source mass M, and ml is the GDM. This can be rewritten in the form

$$V = \alpha_1 (g\hbar/2c)\boldsymbol{\sigma} \cdot \hat{\boldsymbol{r}} \qquad (7.37)$$

so α_1 is a dimensionless constant which measures the strength of the interaction.

The GDM must lie along the spin as in the case of an EDM. The existence of a non-zero $\alpha_1 \propto ml$ would imply a breakdown of the equivalence principle as l represents a displacement between the inertial and gravitational masses.

Another possible force is the coupling of the spin to its velocity through a gravitational field (Leitner and Okubo 1964),

$$V_2 = \alpha_2(g\hbar/2c)\boldsymbol{\sigma} \cdot \boldsymbol{V}/c \qquad (7.38)$$

where $\boldsymbol{V}$ is the velocity relative to the gravitational source.

Strengths of various sources are as follows:

$$
\begin{aligned}
g\hbar/2c &= 3.7 \times 10^{-29} \text{ (earth's field)} \\
&= 2.6 \times 10^{-26} \text{ eV (sun's field)} \\
&= 2.6 \times 10^{-21} \text{ eV (galactic field).}
\end{aligned}
\qquad (7.39)
$$

$\hbar/mc = 2 \times 10^{-14}$ for neutrons.

Limits on α_1 and α_2 could be set using existing data from the neutron EDM experiment; the Larmor frequency, or energy splitting between the two spin states, has already been measured as a function of the direction of the quantization axis (magnetic field) in space since data were taken over approximately 3-day periods uninterrupted. The effects due to the solar field would occur once per solar day and could be contaminated by diurnal systematics whereas effects due to the galactic field would occur once per sidereal day; over the course of three months one would expect a $\pi/2$ phase shift in diurnal systematic errors. Roughly, the present sensitivity would be at the 1 μHz level (4×10^{-21} eV), implying $\alpha_1 < 10^5$ and $\alpha_2 < 10^9$ in the sun's field. These limits are comparable with the previous best limits set by Lamoreaux et al (1986, 1988) on such couplings to ^{199}Hg and ^{201}Hg.

Although α_1 (T, P violating) and α_2 (P violating) are expected to be of order 1 for maximal breaking of these discrete symmetries, these are nonetheless the best experimental limits.

The ultimate experiment will be to use an EDM apparatus with a comagnetometer as described in sections 7.3.5 and 7.3.6. Such a comagnetometer system will be sensitive to such interactions only insofar as the ratio of the couplings is different from the ratio of the magnetic moments; otherwise the effect will be indistinguishable from magnetic field changes. The main advantage here is that one could change the orientation of the magnetic field relative to the various sources and not have to worry about the magnetic effects. We could expect a factor of 100 improvement in the limits sometime in the foreseeable future.

One could also consider looking at the gravitational acceleration of UCN for spin states oppositely directed to the earth's gravitational field, as equation (7.34) implies that the gravitational acceleration should be slightly different for the two orientations (Puricia 1979), although this probably would not yield such tight limits since there is the background potential of 100 neV m^{-1}. Puricia (1979) also suggests that the β-decay rate could be modified due to the presence of a gravitational field.

UCN could also lend themselves to a test of weak boson coupling to ordinary matter, the so-called 'fifth force'. Puricia (1988) describes a clever technique for cancelling the gravitational acceleration of a test source, with what he calls a 'gravitational lens' which is a combination of source masses configured so that the total potential due to the usual $1/r$ potential from all of the source masses cancels over a region of space. Forces which arise from short-range interactions do not cancel in the configuration. The idea here is to use a spin-echo technique to measure the transit time of UCN with and without the gravitational lens test source in place; a change in velocity would be an indication of forces in addition to gravity which has been cancelled to some degree of precision. Although there is some merit in considering the coupling to bare neutrons, the recent popularity in considering such effects seems to be waning at most institutions.

7.7   NEUTRON–ANTINEUTRON OSCILLATIONS WITH UCN

Mohapatra and Marshak (1980) have suggested that there may exist a baryon number violating interaction ($\Delta B = 2$) capable of transforming neutrons into antineutrons and *vice versa*, with the result that neutrons would change into antineutrons after a finite time, $\tau_{N-\bar{N}}$. Theoretical estimates of this time (Mohapatra 1989) based on superstring and other grand unification theories range from 10^7 to 10^{10} s.

Attempts to detect the oscillation of free neutrons to antineutrons ($n - \bar{n}$) have so far been restricted to beam experiments, Fidecaro *et al* (1985) set a limit at 10^6 s; a new experiment might be sensitive at the 10^8 s level (Puglierlin 1989). The sensitivity of such experiments increase as t^2 where t is the free flight time of the neutrons.

Golub *et al* (1989) have considered the possibility of using UCN stored in a material bottle and looking for the UCN̄ produced; while this is attractive because of the long storage times available in material bottles, collisions with the confining wall lead to phase shifts between the neutrons and antineutrons, and absorption losses of the antineutron. Thus, one would not expect the full T^2 enhancement of the UCN̄ density, where T is the storage time, as in a beam experiment. (Ignatovich and Luschikov (1981) have suggested that UCN–UCN̄ oscillations might account for the anomalously large wall loss rates seen in early bottled UCN experiments.)

However, analysis indicates that this might not be a limitation and can be partially compensated for by a 'bootstrap' technique (Golub *et al* 1989). They show that for lower wall reflectivity and wall phase shifts for UCN̄ relative to UCN, the density of UCN̄ approaches a steady-state value proportional to the UCN density; by applying a magnetic field, the relative phase shifts on wall reflection can be reduced to some degree (the 'bootstrap') and the UCN̄ density enhanced.

The UCN̄ wall interaction is still an open question, but reasonable estimates can be placed on the real and imaginary parts of the potential (see, for example, Kazarnovskii *et al* 1981).

An experiment based on UCN has a number of advantages over a beam-type experiment. The average magnetic field in the space over which the neutron–antineutron oscillation takes place will suppress the transition rate; to see the free-space oscillation time one must have $\gamma BT \ll 1$, where γB is the energy splitting between the antineutron and neutron induced by the magnetic field and T is the observation time. Such a requirement imposes a severe difficulty on beam experiments; for example, a 100 m long magnetic shield has been built for the beam experiment. The magnetic shielding of reasonable size bottles is a well developed technology, as in the EDM searches. In addition, there might be a much lower background rate with UCN since the high flux cold beam is eliminated.

In a UCN-based experiment, one would simply store UCN in a bottle and look for the annihilation of UCN̄ in the wall which is a 2 GeV event and should be readily detectable.

We can now calculate the sensitivity of such an experiment. Golub *et al* (1989) show that in the presence of non-perfectly reflecting walls for antineutrons, the UCN̄ density approaches a constant proportional to the UCN density

$$P_{\bar{N}}(t) \propto P_{N}(t_c/\tau_{N-\bar{N}})^2 \qquad (7.40)$$

where t_c is the time between wall collisions, and P_i represents the probability (density) of neutrons in the appropriate state. There is a finite wall reflectivity $\rho_{\bar{N}} < 1$ for the antineutron; at each wall collision we find

$$P = P_{\bar{N}}(1 - \rho_{\bar{N}}^2) \qquad (7.41)$$

annihilations; these occur at times separated by t_c. We thus find that the annihilation rate is

$$\Phi = P/t_c. \qquad (7.42)$$

We must, however, take into account the loss of UCN; both through the ordinary wall loss and through β-decay. Using equations (7.5), (7.40) and (7.42), and assuming that the equilibrium density of the antineutrons is established in a time much shorter than the bottle lifetime

$$\Phi(t) = \alpha N_0 e^{-t/\tau} t_c/\tau_{N-\bar{N}}^2. \qquad (7.43)$$

The constant of proportionality α is greater than 1 but probably less than 100. The total average counting rate over a storage time T is

$$\langle \Phi \rangle = \frac{1}{T} \int_0^T \Phi(t)\, \mathrm{d}t = \alpha N_0 \frac{\tau t_c}{T \tau_{N-\bar{N}}^2} (1 - e^{-T/\tau}) \qquad (7.44)$$

which has a maximum when $T = \tau$. In a running time T_r, we would expect to see

$$\langle \Phi \rangle T_r = \alpha N_0 \frac{t_c T_r}{\tau_{N-\bar{N}}^2} \qquad (7.45)$$

annihilations, setting an upper limit if none are seen of

$$\tau_{N-\bar{N}} \propto \sqrt{N_0 t_c T_r} \qquad (7.46)$$

which yields $\tau_{N-\bar{N}} \gtrsim 10^7$ s for the modest $N_0 = 600\,000$ and one year of running, with $t_c = \frac{1}{2}$ s and $\alpha = 10$. This might be improved by proper choice of wall material and full use of the 'bootstrap' and could thus be an important complement to the beam experiments.

7.8   DEMONSTRATING BERRY'S PHASE USING BOTTLED UCN

A manifestation of Berry's phase has been demonstrated using bottled spin-polarized UCN subjected to a temporally varying magnetic field, (Richardson and Lamoreaux 1987, 1989). Although Berry's phase for spin-$\frac{1}{2}$ is somewhat trivial, this demonstration is nonetheless interesting as it indicates a use of Berry's phase in determining the sign of magnetic moments.

Berry (1984) discovered a rather remarkable property of the Schrödinger equation; that is if the Hamiltonian H of a system depends on a multi-dimensional parameter (for example, a magnetic field, which has three independent components), if that parameter is temporally varied (adiabatically) around a closed curve C in parameter space, the system will return to its orginal state with the exception of a phase shift $e^{i\gamma(C)}$ in addition to the usual dynamical phase $\exp(-i \int H(t)dt)$. This additional phase factor is called Berry's phase and has a simple form when the system is nearly degenerate: $\gamma(C) \propto \Omega(C)$ where $\Omega(C)$ is the solid angle of C as viewed from the origin of parameter space (degeneracy point).

For spins in a magnetic field B which is varied such that the tip of B traces out a closed curve during a time interval T, $\gamma_m(C) = -m\Omega(C)$ where m is the spin component along B (see section 4 of Berry (1984)). This phase appears in addition to the usual dynamical phase $\phi_m = \kappa m \int_0^T B(t)dt$ where κ is the magnetic moment. For spin-$\frac{1}{2}$ particles, this reduces to $\phi_\pm = \pm(\kappa/2) \int_0^T B(t)dt$ and $\gamma_\pm(C) = \mp\Omega(C)/2$, where the $\pm$ refers to the two spin states and $\Omega(C)$ is the solid angle swept out by the magnetic field.

In the test of Berry's theory, spin-polarized stored UCN were subjected to a slowly varying magnetic field. This experiment complements a recent test using a cold neutron beam (Bitter and Dubbers 1987). In the UCN experiment, in addition to verifying the solid angle dependency, it was possible to demonstrate that $\gamma(C)$ is additive for multiple excursions along the same path, that 'ellipticity' (to be defined) does not affect the outcome, and that the neutron has a negative magnetic moment.

The magnetic field in the neutron storage volume was varied as follows: for $0 \leq t \leq NT$,

$$
\begin{aligned}
B_x &= \alpha B_0 && \text{(for } \epsilon = 0) \\
&= 0 && \text{(for } \epsilon \neq 0) \\
B_y &= \pm(1 + \epsilon)B_0 \cos 2\pi t/T \\
B_z &= B_0 \sin 2\pi t/T
\end{aligned}
\tag{7.47}
$$

and $B_x = B_y = 0$, $B_z = B_z^0$ for all other times. The neutron polarization initially pointed along z. N is the number of rotations of the magnetic field vector, T is the time for one rotation, B_0 is the length of the magnetic field vector, ϵ is the ellipticity parameter which allowed the magnitude of B to be varied over the path (the projection of the tip of B traces out an ellipse in the y–z plane; this parameter is used only in the case of $\Omega(C) = 2\pi$), and α is used to vary the solid angle: $\Omega(C) = 2\pi[1 \pm \cos(\theta)]$ where θ is the apex angle of the cone described by B and $\cos\theta = \alpha/(1 + \alpha^2)^{1/2}$, and $\pm$ depends on the rotation direction and sign of the magnetic moment. All of the parameters were each independently adjustable.

Schrödinger's equation can be solved exactly for a system of non-interacting spin-$\frac{1}{2}$ particles subjected to the magnetic field given in equation (7.47); the solution is the same as for a classical magnetic dipole (Cohen-Tannoudji *et al* 1977). The solution is simplest if one works in the rotating frame (Richardson and Lamoreaux 1987) and for the case $\epsilon = 0$. If one transforms into a frame rotating about $\hat{x}$ at $t = 0$, that is where B ($B_{x'} = \alpha B_0$, $B_{y'} = B_0$, and $B_{z'} = 0$) is constant, an additional field is generated along $\hat{x}$, giving $B_x = \alpha B_0 \mp 2\pi/(\kappa T)$ where $\mp$ refers to the rotation direction. The neutrons are initially polarized along $\hat{z}$ and the spins simply precess about the quadrature sum of the fields and $P_{z'} = \cos[\kappa(B_{x'}^2 + B_{y'}^2)^{1/2}t]$. At $t = T$, the rotating and laboratory frames coincide: $P_{z'}(T) = P_z(T)$. Thus, the final neutron polarization is given by

$$
P_z = \cos\{[(\kappa B_0 T)^2 + (\alpha\kappa B_0 T \mp 2\pi)^2]^{1/2} - 2\pi\}
\tag{7.48}
$$

where the -2π is necessary so that the total phase $\Phi = 0$ when $B_0 = 0$. The total field in the laboratory frame is $B = B_0(1 + \alpha^2)^{1/2}$. In the adiabatic limit ($\kappa BT \gg 1$), the total phase is

$$
\Phi = \kappa BT - 2\pi[1 \pm \frac{\kappa}{|\kappa|} \frac{\alpha}{(1 + \alpha^2)^{1/2}}] = \phi - \Omega(C) = \phi + \gamma(C) = 2(\phi_+ + \gamma_+(C))
\tag{7.49}
$$

which is exactly as predicted by Berry's analysis. For multiple rotations, the total time is NT and the phases in equation (7.49) are simply multiplied by

N. Thus, for multiple rotations, Berry's phase is additive as expected. Note that since the neutron wavefunctions are not interferred with, there is no sensitivity to spinor rotation effects (Bitter and Dubbers 1987, Bonse and Rauch 1979).

The ILL neutron EDM measurement apparatus which has been described in section 7.3.3 was used for the measurements, the only change being the addition of another coil so that a magnetic field could be applied along any axis within the magnetic shield. (The high voltage components of the EDM apparatus and the rubidium magnetometers are unused in this experiment and can be ignored.) As before (see figure 7.14), neutrons were transported from the neutron turbine to the apparatus through a stainless steel guide and polarized by a magnetically saturated Fe–Co foil, and transported to the bottle. The neutron polarization was initially along a 5 mG field in the z-direction. After filling the bottle for about 10 s the neutron density approached its maximum value of about 10 n cm^{-3} and the neutron valve was closed. After waiting 2 s to allow the system to equilibrate, the polarized neutrons were subjected to the temporally varying magnetic field over a time $T = 7.387 \pm 0.005$ s followed by another 2 s wait. The neutron valve was then opened and the two neutron spin states were counted as already described in section 7.3.3.

The residual field averaged about 10 μG in the storage vessel with a gradient across the vessel of the same magnitude. At low values of B_0, the neutron spins were depolarized by the residual fields and thus set a lower limit on B_0 and an upper limit on T. This depolarization is evident in figure 7.20(b) for $B_0 < 100$. In addition, there was a lower limit on T because the storage vessel was enclosed in an aluminium can. Eddy-current-induced field gradients rapidly depolarized the neutrons if the magnetic field was varied too rapidly. It was found that $T \approx 7$ s did not give any significant depolarization at high B_0 and measurements with low B_0 ($\kappa B_0 T/2\pi \approx 1$) were also possible. The plot of maximum rotation rate against the minimum magnetic field to define a quantization axis defined a limited range and it was fortunate that the experiment was possible at all.

For a given set of the parameters N, α and ϵ, the spin-up and spin-down neutron counts were measured as a function of B_0, each datum requiring a fill/store/count cycle. Figure 7.20 shows typical results ($\gamma(C) = 2\pi$). Note the aperiodicity in the neutron count oscillations near $B_0 = 0$. It is this aperiodicity which leads to the Berry's phase shift in the adiabatic limit.

The data were analysed as follows. The spin-up and spin-down neutron counts vary with the phase Φ as

$$n_{\text{up}} = n_{\text{up}}^0 (1 + p_{\text{up}} \cos \Phi)$$
$$n_{\text{down}} = n_{\text{down}}^0 (1 - p_{\text{down}} \cos \Phi) \tag{7.50}$$
$$\Phi = 2\pi (a_1 x^2 + (a_2 x + a_3)^2)^{1/2}$$

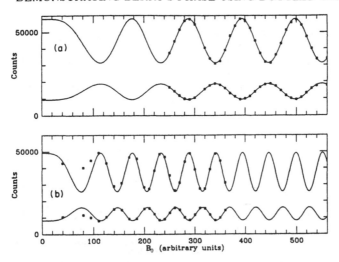

Figure 7.20 Spin-up (upper curves in each plot) and spin-down neutron counts as a function of B_0 in the case $\alpha = 0$, $\epsilon = 0$, and the rotation sense is positive; (a) data for one rotation and (b) data for two rotations. Note the low-field depolarization in (b) (Richardson and Lamoreaux 1989).

where x is proportional to B_0, $p_{up} \approx p_{down} \approx A^2$ where A is the polarizing efficiency of the polarizer which was also used as the analyser, n_{up}^0 and n_{down}^0 are the average detected number of up or down spin-state neutrons. Φ in equation (7.50) is the most general form of the phase as equation (7.48) indicates. Note that an accurate calibration of B_0 is not needed to determine Berry's phase for the data set; one can readily determine the phase in the adiabatic limit by taking the appropriate combination of the fit coefficients (see equation (7.48)). Although the data do not extend to $B_0 = 0$, it was possible to determine the phase unambiguously. The data extend low enough so that at most one oscillation can occur between the lowest B_0 and $B_0 = 0$. Note also that the oscillation period increases as B_0 approaches zero in all cases. Thus, at most the first maximum is missed and its location can be approximately determined.

It was found that typically $n_{up}^0 \approx 34\,000$, $n_{down}^0 \approx 13\,000$, $p_{up} \approx p_{down} \approx 0.5$ and p is lower than is observed in normal operation of the EDM experiment where $p \approx 0.6$. Some loss of polarization probably occurred because the initial field along $\hat{z}$ which guides the neutron polarization into the bottle was 5 mG as opposed to 10 mG as used in normal operation.

The results demonstrate that Berry's phase is indeed additive and that ellipticity does not affect the outcome to high accuracy, and that the sign of the neutron magnetic moment $\kappa/|\kappa| = -1$. This is particularly evident in the data in figures 7.21(a) and (b) since the oscillations are drawn toward the origin for the $-$ rotation. Figure 7.22 is a plot of the solid angle $\Omega(C)$ inferred from the experimental results compared with the expected functional form;

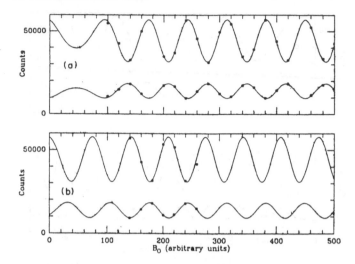

Figure 7.21   Neutron counts as a function of B_0 in the case $\alpha = 1$ ($\theta = 45°$), $\epsilon = 0$: (a) is for the positive rotation sense and (b) for negative rotation sense. The sign of the neutron magnetic moment is evident from this data as the phase oscillations are drawn toward zero for the negative rotation sense (Richardson and Lamoreaux 1989).

the solid angle dependency of Berry's phase is thus experimentally verified with reasonably good precision.

Although this additional phase is accounted for by transforming to the rotating frame (for spin-$\frac{1}{2}$ systems), Berry's analysis suggests a new way to think about magnetic moments precessing in a varying magnetic field. Most interestingly, it allows one to determine the sign of the magnetic moment. While the concept of Berry's phase has profound implications for many fields of physics (Shapere and Wilczek 1989), it appears almost trivial in the spin-$\frac{1}{2}$ system treated here. Hopefully, this example will lead to an improved understanding of its other manifestations.

7.9   THE UCN–MATTER WEAK INTERACTION

7.9.1   Introduction

It is well known that the neutron interacts weakly with both nucleons and electrons. There have so far been no UCN experiments to investigate the weak interaction directly. In these sections we will theoretically investigate the interaction of UCN with an electric current, and the interaction with the storage bottle walls in the presence of a gravitational field.

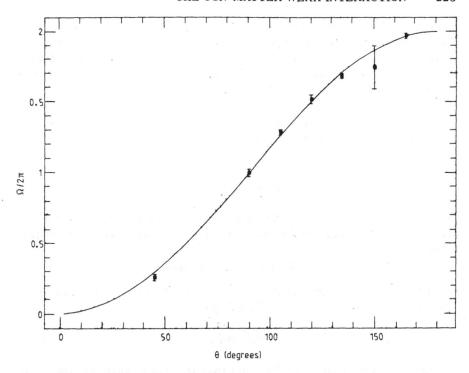

Figure 7.22   The solid angle inferred from the experimental results compared with the geometrically determined solid angle (Richardson and Lamoreaux 1989).

7.9.2   The UCN–electric current weak interaction

We first consider polarized UCN (random or zero velocity) interacting with an electric current (well-defined electron velocity or momentum, p_e). (This interaction was suggested by Khriplovich and Pospelov (1989) between atoms or molecules with electrons in a conducting solution.) In this case, the electron–neutron interaction can be written in the non-relativistic limit (Commins and Bucksbaum 1980, Fortson and Lewis 1984)

$$H_{\text{PNC}} = \frac{G_{\text{F}}}{m_e c\sqrt{2}}\frac{1}{2}(1 - 4\sin^2\theta_{\text{W}})g_{\text{A}}\boldsymbol{\sigma}\cdot\boldsymbol{p}_e \int |\psi_{\text{N}}|^2|\psi_e|^2\,\mathrm{d}V \qquad (7.51)$$

where $\boldsymbol{\sigma}$ represents the neutron spin, ψ_i represents the electron or neutron wavefunction, and the integral is over all space.

Consider a storage bottle similar to that used in the neutron EDM (figure 7.14) of radius R, only with a conducting layer on the inner surface so that when a voltage is applied between the electrodes a current density

$$\boldsymbol{j} = \sigma\boldsymbol{E} \qquad (7.52)$$

is established where σ is the electrical conductivity and $\boldsymbol{E} = (\Delta V/h)\,\hat{z}$, where ΔV is the voltage applied across the bottle and h is the plate separation.

We consider the case where the conducting layer has good neutron storage properties (e.g. pyrolytic graphite) and has thickness $t \approx 100$ Å so that the UCN wavefunction has a high overlap with the electron current.

The current density $\boldsymbol{j}$ is approximately constant over the conducting layer. The non-zero average part of the electron motion which establishes the current is v_{drift}, the drift velocity of the electrons in the applied electric field. So at any instant there are

$$N = \rho_{\text{e}}(2\pi Rth) \tag{7.53}$$

electrons flowing where ρ_{e} is the conduction electron density and $2\pi Rth$ is the volume of the thin conducting layer, with the total current I flowing, satisfying

$$I/(2\pi Rt) = |\boldsymbol{j}| = e\rho_{\text{e}}v_{\text{drift}}. \tag{7.54}$$

The electron wavefunction is assumed constant in the conducting layer; since we must have $N = \int |\psi_{\text{e}}|^2 dV$, we find

$$|\psi_{\text{e}}|^2 = \rho_{\text{e}}. \tag{7.55}$$

Similarly, for the neutron, we find that over the entire bottle we must have $\int |\psi_{\text{N}}|^2 dV = 1$ so

$$|\psi_{\text{N}}|^2 \approx 1/(\pi R^2 h) \tag{7.56}$$

or $1/$(total bottle volume). Inside the conducting layer, the neutron density must decrease as $e^{-(r-R)/\lambda}$ where λ is the usual UCN penetration length. We thus find

$$\int |\psi_{\text{N}}|^2 |\psi_{\text{e}}|^2 \, dV \approx \tfrac{1}{3}\rho_{\text{e}}(2\pi Rth)/(\pi R^2 h) = \tfrac{1}{3}\frac{V_{\text{wall}}}{V_{\text{bottle}}} \tag{7.57}$$

for a conducting layer approximately λ thick.

The electron momentum is given simply by

$$\boldsymbol{p}_{\text{e}} = m_{\text{e}}v_{\text{drift}} = m_{\text{e}}\frac{\sigma \boldsymbol{E}}{e\rho_{\text{e}}}. \tag{7.58}$$

Substituting equations (7.58) and (7.57) into (7.51), we find

$$H_{\text{PNC}} = \frac{G_{\text{F}}}{c\sqrt{2}}\frac{1}{2}(1 - 4\sin^2\theta_{\text{w}}W)g_{\text{A}}\frac{1}{3}\frac{\sigma}{e}\frac{t}{R}\boldsymbol{\sigma}_{\text{N}} \cdot \boldsymbol{E}. \tag{7.59}$$

Note that this is exactly the same form as the time-reversal violating interaction for a neutron EDM, as in equation (7.22). The time reversal in this case is due to the macroscopic process, $\boldsymbol{j} = \sigma \boldsymbol{E}$, which is due to electron–phonon scattering, for example, which is an irreversible process. The point is that under time reversal, $\boldsymbol{j} \rightarrow -\boldsymbol{j}$. This interaction, however, could lead to a systematic effect in the neutron EDM experiments.

We can rewrite equation (7.59) for the shift in energy between the two neutron spin states relative to the current flow I

$$\Delta E = \frac{G_{\mathrm{F}}}{c\sqrt{2}}(1 - 4\sin^2 \theta_{\mathrm{W}})g_{\mathrm{A}}\frac{1}{3}\frac{1}{2\pi R^2}\frac{I}{e} \tag{7.60}$$

where we have used Ohm's law, $V = IR$. The interesting result is that the macroscopic properties of the wall material cancel. Using $\sin^2 \theta_{\mathrm{W}} \approx 0.23$ and equation (7.2), we find a shift in the Larmor frequency

$$\delta\omega_0 \approx 1.7 \times 10^{-21} R^{-2} I \ \mathrm{m}^2 \ \mathrm{Hz} \ \mathrm{A}^{-1}. \tag{7.61}$$

Thus, for a bottle of radius 10 cm with 100 A flowing in the wall, one would find $\delta\omega_0 = 1.7 \times 10^{-17}$ Hz, or, comparable with the signal expected from an EDM (at 10 kV cm^{-1}) of 4×10^{-36} e cm, a very small shift indeed. One could consider making R effectively smaller by trapping the neutrons between two cylinders; however, the shift is small for any imagined geometry. To observe such shifts, one would have to monitor the magnetic field with a comagnetometer; even this might not be sufficient as the neutron penetrates into the wall where the current is flowing, while the atoms will not. If there were perfect cylindrical symmetry, this would give a shift proportional to I^2 since the applied magnetic field is along the cylinder axis; however, maintaining such perfect symmetry in an actual system would be impossible.

7.9.3   Spontaneous polarization of bottled UCN in a gravitational field

As demonstrated by the experiments described by Heckel (1989), terms in the coherent forward neutron scattering amplitude (from bulk matter) of the form

$$f(0) = C\boldsymbol{\sigma}_{\mathrm{N}} \cdot \boldsymbol{p}_{\mathrm{N}} \tag{7.62}$$

lead to a helicity dependence of the neutron index of refraction. Although such effects can be due to electron–neutron interactions as described by equation (7.51), neutron–nucleus interactions are about 1000 times larger.

We consider the case of low energy UCN so that they reflect only from the bottom of a storage bottle (a possible geometry would be a cylindrical bottle of large radius with the cylinder axis parallel to the earth's gravitational field). Since the spin $\boldsymbol{\sigma}$ of each neutron is constant (in the absence of transverse magnetic fields), the helicity ($\boldsymbol{\sigma} \cdot \boldsymbol{p} = \sigma_z p_z$, where $\boldsymbol{p}$ is the momentum of a particular neutron and we have chosen the z-axis as the spin quantization direction) is constant for each neutron at every collision with the bottle surface and is unchanged from collision to collision. If there is a helicity dependence to the wall interaction, one helicity state will be lost more rapidly than the other; i.e. for one orientation of σ and p, the loss rate will be higher than the other. The net result is that the UCN become slightly polarized along the gravitational field. We can estimate the size of this effect.

The wall loss rate is proportional to the imaginary part of the *forward* scattering amplitude (f in equation (7.9)) which is the imaginary part of C in equation (7.62). Thus, from the transmission asymmetry A_z for the two neutron helicity states we can find the difference in the loss rates:

$$A_z = (\sigma_+ - \sigma_-)/(\sigma_+ + \sigma_-) = 4\pi \Im(C)/\sigma$$
$$= \frac{f_+ - f_-}{f} \qquad (7.63)$$
$$= \tau/\tau_+ - \tau/\tau_-$$

where $\sigma_\pm$ refers to the transmission of the two helicity states and σ is the total absorption cross section. As Stodolsky (1982) has shown, $A_z \to 0$ as $p \to 0$; using $A_z \propto p$ and scaling the 7 Å values of A_z to UCN energies, we find

$$A_z \approx 10^{-7} \qquad (7.64)$$

for those elements where A_z is large.

Thus, the fractional difference in the loss rates for the two helicity states is $A_z \approx 10^{-7}$. If one stores UCN for a time t the ratio of spin-up ($-$) to spin-down ($+$) is

$$P = \frac{e^{-t/\tau_+} - e^{-t/\tau_-}}{e^{-t/\tau_+} + e^{-t/\tau_+}} \approx A_z t/2\tau \qquad (7.65)$$

where we assumed that $\tau_+ \approx \tau_-$, and τ^{-1} is the average wall loss rate. Thus, one expects a polarization of order 10^{-7} after waiting one bottle lifetime; this is not too far from what could be detected with current techniques. We can derive the optimum time to wait assuming shot noise as in section 7.2.2:

$$S/N = (P(T)\sqrt{N(T)})\sqrt{t/T} = (A_z T/\tau)N_0^{1/2}e^{-T/2\tau}\sqrt{t/T} \qquad (7.66)$$

where $t \gg T$ is the total observation time for many measurements. This has a maximum when $T = \tau$, giving a signal-to-noise ratio of $0.6 A_z N_0^{1/2}\sqrt{n}$ where n is the total number of measurements; for $N_0 = 10^6$, we need 3×10^8 measurements to achieve $S/N = 1$. At $\tau = 50$ s, this gives 10^{10} s, or 440 years, so this technique is not so useful. One could consider some other geometries, such as detecting the penetration depth as a function of helicity, although such experiments would suffer the same problems as the usual beam experiments.

8

UCN scattering

8.1 GENERAL REMARKS

In comparison to the applications discussed in the last chapter UCN scattering
is only just learning to take its first steps. We begin our discussion with
some general remarks on neutron scattering. The cross section for neutron
scattering from a system of atoms is given by equation (A2.11) as

$$\frac{\mathrm{d}^2\sigma}{\mathrm{d}\Omega\,\mathrm{d}\omega} = a_{\mathrm{coh}}^2 \frac{k_{\mathrm{f}}}{k_{\mathrm{i}}} S_{\mathrm{coh}}\left(\boldsymbol{Q}, \omega\right) \tag{8.1}$$

with a similar expression for the incoherent cross section. This is a product of
terms depending on the neutron's properties $(k_{\mathrm{f}}/k_{\mathrm{i}})$, the system's properties
$S_{\mathrm{coh}}\left(\boldsymbol{Q}, \omega\right)$ and a term $\left(a_{\mathrm{coh}}^2\right)$ characterizing the interaction between them.
The influence of the system on the scattering depends only on the momen-
tum $(\hbar\boldsymbol{Q})$ and energy $(\hbar\omega)$ transfers between the system and the neutron,
independent of the incoming neutron momentum and energy. In Appendix
A5 we present an argument originally suggested by Maier-Leibnitz (1966)
which shows that for a measurement at fixed $\boldsymbol{Q}$ (small compared with the
incoming neutron $\boldsymbol{k}$), ω (small compared with the incoming neutron energy)
and fixed $\Delta Q/Q$ and $\Delta\omega/\omega$ (the relative precisions of the Q and ω determination) one can gain intensity by going to longer incident wavelengths,
provided that one makes use of the entire phase space allowed by the desired
ω, Q, $\Delta Q/Q$ and $\Delta\omega/\omega$. As a result of this we expect UCN scattering to
be beneficial at relatively small values of Q and ω. Turning to Appendix A2
sections A2.2.2 and A2.2.3 we see that for a spherical object of radius R we
require $QR \leq \pi$ to have significant coherent scattering (equation (A2.51))
and similarly $\omega\tau \leq \pi$ for motions with a correlation time τ. Thus UCN scattering, expected to be useful at small Q and ω, should be interesting for the
study of slow motions of large objects and one is led to the domain of large
molecules—polymers and biological molecules.

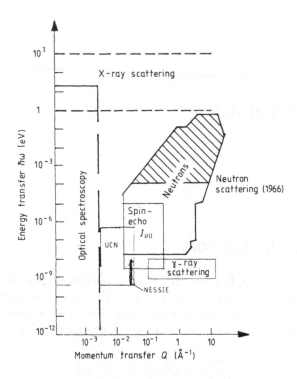

Figure 8.1   Regions of ω–Q space accessible to different scattering techniques. Progress in neutron scattering since 1966 is indicated. After Egelstaff (1967).

In figure 8.1 we show a plot of accessible regions in ω and Q. Such a plot was first introduced by Egelstaff (1967) and the shaded area shows the situation in neutron scattering at that time. The impressive improvement in neutron scattering techniques since that time is evident. The square labelled spin-echo refers to an instrument developed by Mezei (1972, see also Mezei 1980), using the precession of the neutron spin in a magnetic field to measure the time-of-flight and hence the velocity of the neutrons before and after the scattering. A short description of this and many other neutron scattering instruments is given by Bée (1988). It is seen from the figure that UCN scattering is expected to fill the gap in Q between present neutron techniques and light scattering techniques while taking the neutron techniques to lower values of ω. However the only existing UCN scattering spectrometer, NESSIE, described in Chapter 6 operates over a limited Q region at approximately 3×10^{-2} Å^{-1}. At the present time there are several projects aimed at reducing the ω–Q limits of the spin-echo technique. There is an instrument at Saclay which is said to provide a factor of two improvement in both limits (Bée 1988), and a project at the ILL which will work with 25 Å incident neutrons and which is expected to result in significantly lower Q values

1988). In addition there is a new type of spin-echo (neutron resonance spin-echo, NRSE) technique under development (Dubbers *et al* 1989a, Keller *et al* 1990), which might provide some improvements over the existing technique. However at the moment the NESSIE instrument provides the best available ω resolution. One can also consider the possiblity of a UCN spin-echo based on the NRSE idea.

In addition to considering the accessible ω–Q region there are some other properties of neutron scattering which are useful to the study of large molecules. In (A2.47) we showed that the coherent neutron scattering depends only on the variations of density (more precisely the scattering length density) with respect to that of the surrounding medium. The coherent scattering length for protons (-3.74 fm) and deuterons (6.67 fm) are so different that mixtures of H_2O and D_2O (scattering length of oxygen = 5.8 fm) covering a wide range of scattering length densities can be produced. Using this technique combined with selective deuteration of different parts of a large molecule one can adjust the relative scattering length densities so that the neutron scattering only takes place from a determined section of a molecule. Different parts of a molecule can be studied separately.

However, there is a difficulty with the application of this technique to UCN scattering. Hydrogen has an incoherent cross section of 80 b (1 b = 10^{-24} cm^2) which is more than an order of magnitude larger than that of any other nucleus. As a result of the factor (k_f/k_i) in equation (A2.12) the scattering cross section for a given $\omega \sim k_B T$ will be much larger for UCN than for faster neutrons, and the first experiments have been made only with solutions in pure D_2O. This incoherent scattering of UCN at relatively large ω can swamp the desired low ω coherent scattering. The problem requires further detailed study.

Incoherent scattering is also a problem for conventional small angle scattering. In order to overcome this Nierhaus *et al* (1982), (see also Stuhrmann 1982, May 1982) developed a technique where an entire biological molecule (in this case a ribosomal subunit of *E. coli*) can be contrast matched with a pure D_2O solution, thus eliminating the incoherent scattering from the protons in the solvent. The part of the molecule, e.g. a protein, that one desires to study can be replaced with a natural (protonated) protein, allowing measurements to be made on very dilute solutions. The possible applications of this technique to UCN scattering remain an open question.

8.2   QUASI-ELASTIC UCN SCATTERING FROM BIOLOGICAL MOLECULES

In this section we can only touch on a few of the main issues in the studies of biological molecules by physical methods, and the literature we cite can only represent the smallest fraction of the work that has been done in this field. Chapter 10 of Bée (1988) offers a more detailed review.

The forces between atoms in proteins and other biological molecules are little different from those between atoms in the usual solids and liquids. Protein dynamics simulations (McCammon and Harvey 1987) show frequency spectra with the majority of modes around 10^{12} Hz. However NMR and Mössbauer measurements give evidence for much slower motions. These slower motions are the results of strong overdamping of the vibrational motions by the viscous medium (Parak and Knapp 1984, Knapp *et al* 1983). The general problem has been lucidly described by Bée (1988):

> Biomolecules work in an aqueous, essentially isothermal environment. Many classes of biomolecules are continuously being assembled and broken down in complex chains of biochemical reactions which are catalysed by a plethora of highly specific enzymes and associated co-factors. Although structurally well-defined during most of the time, they must be able to undergo small-scale or large-scale isomerisations and to associate or dissociate easily in response to the presence of other molecules, changes of ionic milieu, external stimuli, etc. Biomolecular systems therefore are usually metastable in a strict thermodynamic sense, and at the molecular level this is reflected in a rich spectrum of low-frequency fluctuations and collective processes. Interactions at and between active sites are of particular interest since their structural and dynamical properties govern a variety of molecular recognition, reaction and transduction processes. The way biological macromolecules have been adapted or tuned, through evolution, to exploit the coupling or competition between cooperative and dissipative modes of motion in such processes is a key problem.

According to Sackmann (1985):

> One of the most interesting phenomena of molecular biology is the recognition between macromolecules, e.g. between an enzyme and its reaction partner, between an anti-body and its anti-gen or between a hormone (e.g. insulin) and its receptor. This recognition process is closely connected with the three-dimensional structure of the macromolecules, which are often built up from domains fulfilling special functions. The anti-bodies and the insulin receptors (for example) possess a Y-shaped structure. The vertical part can be used to anchor the molecule in a cell membrane, while the arms are the real recognition regions. Very probably, it is the motion of the arms which plays a decisive role in the recognition process itself. While the molecular structure, at least of crystallizable anti-bodies is well known from X-ray structure analysis very little is known about the relative motion of the domains. Here the quasi-elastic neutron scattering appears to be the most universal method for studying such processes. This holds

especially for the scattering of Ultra-Cold Neutrons.

As an example of what might be expected from a developed UCN scattering technique, we describe a measurement made with neutron spin-echo (Alpert *et al* 1985) on solutions of pig immunoglobulin G in 100% D_2O. Figure 8.2 shows the model used for the structure of the molecule based on X-ray crystallographic studies. The results of the spin-echo measurements are shown in figure 8.3 where $\Gamma/Q^2 = D_{\mathrm{eff}}$ ($\Gamma=$ linewidth of quasi-elastic line, see Appendix A2.2) is plotted as a function of Q for solutions of different concentrations. The model calculations presented by the authors represent a highly sophisticated application of the ideas explained in Appendix A2.2. Both the arm waving model and a rigid rotation model gave a reasonable fit to the data but the rigid rotation model required a much faster rotation than can be reasonably expected on the basis of the structure shown in figure 8.2. The dotted cone in that figure has an angle of 50° which gave the best fit to the measured data. The arm-waving model gave a relaxation time for the motion of approximately 1200 ns.

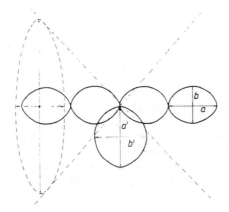

Figure 8.2   Vertical cross section of the design model used for calculations of the neutron scattering of pig immunoglobin. The broken lines represent the axonometric view of the cone inside which the motion of Fab arms was permitted at $\beta_{\mathrm{max}} = 50°$ (Alpert *et al* 1985).

The first experiments using the UCN spectrometer NESSIE, were carried out at the FRM reactor in Munich (thermal flux $\approx 10^{13}$ n cm^{-2} s^{-1}) which used a turbine (Steyerl 1975), see Chapter 3, to produce the UCN. Measurements were made on a lipid bilayer in solution in D_2O (Pfeiffer *et al* 1988).

Figure 8.4 shows the observed linewidth (corrected for instrument resolution) as a function of Q^2. The lowest Q point, $Q = 3 \times 10^{-2}$ Å^{-1}, was measured with UCN, the other points with neutron spin-echo. The UCN-measured linewidth was found to be temperature-dependent. The authors

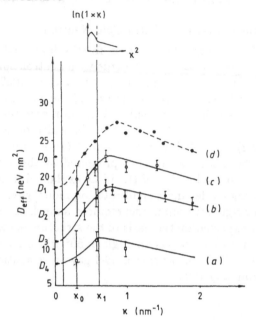

Figure 8.3   Experimental values of D_{eff} as a function of κ. Deuterium oxide solutions of immunoglobulin G dissolved at various concentrations in 0.1 M sodium chloride and 50 mM sodium phosphate, pH 6.6. Curves: (a), 7.33% (w/v) immunoglobulin G with 18% (w/v) sucrose; (b), 7.33 (w/v) immunoglobulin G; (c), 3.67% (w/v) immunoglobulin G; (d), $D_{\text{eff}}(\kappa)$ extrapolated to zero concentration of the immunoglobulin G, κ_0 is the reciprocal of R_g. D_0, translational diffusion coefficient of immunoglobulin G calculated from sedimentation data; D_1, translational diffusion coefficient of immunoglobulin G at zero concentration in the deuterium oxide solvent; D_2, D_3 and D_4 stand for the calculated translational diffusion coefficients for curves (c), (b), and (a), respectively, taking into account the viscosity of the solution (Alpert *et al* 1985).

considered three possible dynamic processes to account for the measured linewidths: (i) lateral diffusion of lipid molecules, (ii) diffusion of small vesicles; and (iii) surface undulations of the bilayers. (ii) would require extremely small vesicles, while (iii) would result in a much broader linewidth with a Q^3 dependence. (i) provides the best explanation of the results yielding a value for D ($\sim 2 \times 10^{-7}$ cm^2 s^{-1}) in good agreement with that obtained by other techniques. It should be emphasized that the UCN spectrometer is now installed at the ILL UCN source where the flux is about 3×10^3 times greater than at the FRM. Thus we can expect that UCN scattering will begin to play a role in further increasing the ω–Q region available for study.

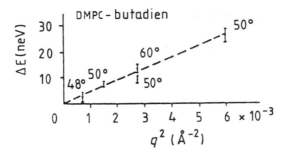

Figure 8.4 Measurement of linewidth as a function of the square of the scattering vector q. The sample is a bilayer dispersion of a 1:1 mixture of dimyristoyl-phosphatidyl-choline (DMPC) and the photopolymerized butadiene phospholipid in pure D_2O (45% by weight). The ΔE values for $q^2 > 3 \times 10^{-3}$ Å^{-2} are measured with the spin-echo spectrometer and the value with the smallest momentum transfer with the gravity spectrometer. The measuring temperatures are given in the figure (Pfeiffer *et al* 1988).

8.3 ELASTIC SCATTERING OF UCN

As the early UCN sources were very weak one looked for applications which were feasible with low counting rates. Steyerl and his co-workers carried out an intensive programme of pioneering studies to demonstrate the usefulness of total cross section measurements for VCN and UCN.

8.3.1 Scattering from homogeneous substances

As diffraction from crystals for neutrons with wavelengths $\lambda > 2a$, where a is the largest lattice spacing in the crystal, is impossible, the only effect of coherent elastic scattering is in the refraction which we discussed in detail in Chapter 2. As discussed in section 2.4.2 absorption and inelastic scattering cross sections should vary as $1/v'$ where $v' = \sqrt{2/m(E-V)}$ is the velocity in the medium taking account of the potential energy (or index of refraction) in the medium.

As a result of this dependence these processes will dominate the total cross section for VCN. This could be useful to measure absorption in rare isotopes or generally to make high accuracy measurements of absorption cross sections. The inelastic scattering is temperature-dependent and the temperature dependence of the total cross section can yield information about the phonon spectrum from (2.111). Calculations similar to those in section 3.5.4 for a Debye spectrum can be made. At low temperatures the inelastic scattering is negligible and we measure the absorption cross section. Steyerl and Vonach (1972) have measured the total cross section for gold, aluminium, copper, glass, mica and air, using time-of-flight in a vertical extraction set-up (section 6.1). Figure 8.5 shows the results for gold plotted against v'.

The broken line shows what would be expected if the cross section depended on $1/v$.

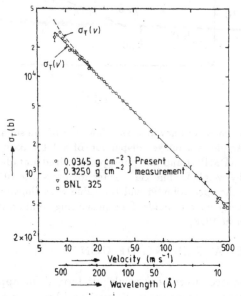

Figure 8.5   Total cross section of two gold foils against neutron velocity *in vacuo* (v) and inside the sample (v'); at temperatures 80 and 299 K (Steyerl and Vonach 1972).

By measuring the total cross section for aluminium at temperatures of 298, 95.5 and 33 K the authors extracted a Debye temperature $\Theta = 389 \pm 50$ K in reasonable agreement with other determinations.

The copper measurements showed some deviations from a $1/v'$ law presumably due to inhomogeneities caused by the cold-rolling treatment given to the samples. Glass and air showed $1/v'$ dependences ($v' = v$ for air) while mica showed deviations.

Dilg and Mannhart (1973) reported a further set of similar measurements but confined to higher incident neutron energies where the difference between v and v' was not apparent. The results reported for the absorption cross sections of scandium, vanadium, copper and rhodium were accurate to about 0.5%.

8.3.2   Scattering from static inhomogeneities

As shown in equation (A2.47) density inhomogeneities can contribute to the elastic scattering of UCN. As emphasized by Steyerl the total elastic scattering cross section

$$\sigma(k) = \int d\Omega \frac{d\sigma}{d\Omega} = \int d\Omega\, S(Q)$$

$$= \frac{2\pi a^2}{k^2} \int_0^{2k} S(Q)Q\, dQ \qquad (8.2)$$

using equations (A1.5) and (2.109), (the maximum momentum transfer from a neutron of momentum k is $Q = 2k$, backscattering) for small k contains the same information as $d\sigma/d\Omega \propto S(Q)$ for small angles (small Q). $S(Q)$ can, in principle, be recovered by differentiation of (8.2) but this is not necessary as the information can be recovered by an elegant technique due to Steyerl (Lermer and Steyerl 1976, Lengsfeld and Steyel 1977). In Appendix A2.2.3 we have taken scatterers in the form of solid spheres as a model. For $kR \gg 1$, with R the size of the spheres, the upper limit in (8.2) may be taken as infinite, the integral is then a constant and

$$\sigma(k) \propto \frac{1}{k^2}. \qquad (8.3)$$

For higher k the angular size of the detector, θ_1, becomes important since all the scatterings for which $Q < k\theta_1$, (in-scattering) result in the scattered neutrons hitting the detector. When k is large enough that the in-scattering includes all Q values except for the high Q tail

$$\sigma(k) = \frac{1}{k^4}. \qquad (8.4)$$

The intersection of the extrapolated behaviours (8.3) and (8.4) is shown in Appendix A2.2.3 to occur at

$$k_1 = \frac{1}{R\theta_1}. \qquad (8.5)$$

Figure 8.6 shows the results (Lengsfeld and Steyerl 1977) for a suspension of SiO_2 spheres of $R = 65$–70 Å in a mixture of D_2O and approximately 5% H_2O after correcting for the scattering of the liquid. The full line represents the fit generated by the function $g(x)$ in Appendix A2.2.3 corrected for the finite width of the incident beam.

Lermer and Steyerl (1976) have applied this technique to the study of ferromagnetic domains and domain walls. The analysis assumed the domain walls were regions in the shape of discs of diameter $d \approx$ (the domain size) and thickness t. While the analysis is more complicated than that given in Appendix A2.2.3 as it involves coupling with the neutron spin to the magnetic field fluctuation i.e. $a_{coh}\rho(r)$ is replaced by $-\mu \cdot B(r)$ (equation (2.3)), the results are essentially the same as in that appendix.

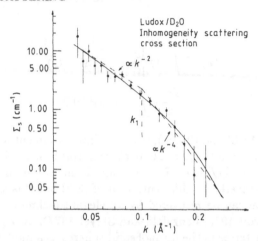

Figure 8.6   Macroscopic cross section Σ due to neutron scattering by the SiO_2-particles alone. The points which were obtained from the measured data by subtracting $A+B/k$ are represented well by the full fit curve of the form $C/k\ 2 \cdot g(kr\theta_1)$. This function shows a k^{-2} behaviour at small k-values and a k^{-4} asymptote at larger k (broken lines). The intersection point k_1 of the asymptotes is a convenient measure of the particle radius r, see Appendix A2.2.3 (Lengsfeld and Steyerl 1977).

For both the domains and the walls the low energy cross sections ($\sim 1/k^2$) change to a $1/k^4$ dependence at higher incident k. In the case of the domains themselves the crossover, k_1, (equation (A2.64)) is determined by $R = d/2$, i.e. the domain size, while for the domain walls it is determined essentially by the thickness t. Figure 8.7(a) shows the measured cross section $\sigma(k)$ plotted against neutron velocity ($k = 1.59 \times 10^{-3}\ v\ ms^{-1}\ Å^{-1}$) for nickel. The scattering above the usual $1/v$ law is seen to vanish as $H \gtrsim 100$ G indicating the disappearance of the domain walls as the material saturates. Figure 8.7(b) shows the $1/v^2$ and $1/v^4$ extrapolations for cobalt and the intersection point yielding $t = 225$ Å. From the size of the $1/k^4$ term it is possible to deduce the domain size; $d = 3\ \mu$m in cobalt. Thus the break, k_1, for the domains would come at a much lower neutron energy than that of the domain wall scattering.

This technique is like the inverse of conventional small-angle scattering. At high k the entire $S(Q)$ distribution is concentrated at very small angles and scattered into the detector. As k decreases the $S(Q)$ distribution spreads out to larger angles, the constant angle subtended by the detector acting as a window which rejects the larger Q values.

Since the pioneering work at a relatively weak source, this interesting technique has not yet been taken up at the more intense neutron sources, perhaps because the conventional small angle scattering instruments are so well developed and there is more interest in developing the inelastic scattering applications of UCN (see previous section).

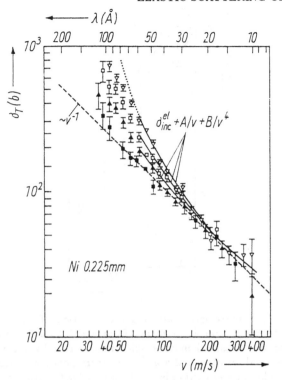

Figure 8.7 (a) Total experimental cross section σ_T per atom for nickel at different magnetic field strengths, plotted against the neutron velocity and wavelength (corrected for refraction within the sample). The $1/v$ contribution due to nuclear capture and thermal inelastic scattering (broken curve) is indicated. The full curves are fit curves taking account of incoherent elastic scattering (σ_{inc}^{el}), of absorption ($\sim 1/v$) and of the asymptotic behaviour of scattering on Bloch walls and domains as a whole ($\sim v^{-4}$ at large v). In the range $v < v_0 \approx 60$ m s^{-1} the data deviate noticably from this higher-energy fit curve (which is extrapolated as a dotted curve for $H = 0$ (∇)), as expected for Bloch wall scattering. The empty squares are for $H = 30$ Oe; $\blacktriangle$ 70 Oe; the full squares are for 100 Oe. (Lermer and Steyerl 1976).

However we should point out that the minimum Q available on a small angle scattering instrument at the ILL, $Q_{min} \approx 1 \times 10^{-4}$ Å^{-1} (ILL 1988), is reached at the relatively large angle of 10^{-2} rad for the 10 m s^{-1} neutrons which are at the peak of the turbine output spectrum (figure 3.8).

Binder (1971a,b) carried out a search for physical situations where measurement of total cross sections for UCN may be of interest. He suggested: (i) the diffraction of UCN from the periodic lattice formed by magnetic vortices in type II superconductors; (ii) scattering from spin waves in ferromagnetic materials; and (iii) scattering from the fluctuations in a ferromagnet near the critical transition. In all cases he calculated the total cross section to be

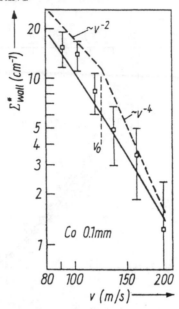

Figure 8.7   (*b*) Macroscopic scattering cross section of Bloch walls Σ^*_{wall} in cobalt in the low-energy transition region from a v^{-4} to a v^{-2} variation. Σ^*_{wall} is obtained from the measured macroscopic cross section by subtracting the $1/v$ contribution due to absorption and a v^{-4} term corresponding to scattering on the domains as a whole. The full curve is a calculation for the transition region. The intersection point at v_0 of the v^{-4} and v^{-2} asymptotes (broken lines) is a measure of Bloch wall thickness (Lermer and Steyerl 1976).

expected for UCN.

The treatment of the spin waves is quite similar to our treatment of phonons in Appendix A3, a major difference being that the spin waves ('magnons') satisfy a dispersion relation

$$\omega = \omega_0 + \alpha Q^2. \tag{8.6}$$

In the case of critical scattering one can (crudely) consider $S(Q,\omega)$ as being made up of the Fourier transform of the spin correlation function

$$\lim_{R\to\infty} \langle S_0 \cdot S_R \rangle \to \frac{1}{R} e^{-R/R_0} \tag{8.7}$$

which Fourier transforms into

$$\chi(Q) \propto \frac{1}{Q^2 + 1/(R_0)^2} \tag{8.8}$$

where the correlation length $R_0 \longrightarrow \infty$ as $T \longrightarrow T_c$. For small energy transfers

$$S(Q,\omega) \sim \chi(Q) = \frac{\Gamma(Q)}{\omega^2 + \Gamma^2(Q)} \tag{8.9}$$

with

$$\Gamma(Q) = \Lambda Q^2 \qquad \text{for } Q \ll 1/R_0 \qquad (8.10)$$

(compare equation (A2.45)), (Λ is called the spin diffusion constant) going over to a faster Q dependence for $Q \gtrsim 1/R_0$. The total cross section at $v \lesssim 10$ m s^{-1} shows a rapid rise as $T \longrightarrow T_c$.

8.4   UP-SCATTERING OF UCN

In addition to the types of scattering measurements mentioned in sections 8.2 and 8.3 there is another type of scattering which might be useful in certain circumstances, up-scattering of UCN. By this we mean inelastic scattering where the energy transfer is large compared with the original UCN energy.

Up-scattered UCN were first detected by Stoika et al (1978) who surrounded a UCN storage chamber with a detector in order to show that the rate of up-scattering could account for the unexpectedly high loss rate of UCN from storage vessels (Chapter 5).

We wish to turn our attention to measurements of the energy spectra of the up-scattered neutrons. As a means of measuring $S(Q, \omega)$ this has the disadvantage that only values of ω, Q along the 'free neutron dispersion curve'

$$\omega = \frac{\hbar Q^2}{2m} \qquad (8.11)$$

can be measured. The advantages of the method are based on the special properties of UCN, e.g. their penetration into surfaces for a distance of approximately 100 Å could allow studies of inelastic processes originating in this region, intermediate between the first few monolayers accessible to normal methods of surface physics and the true bulk matter. In addition large cross sections due to the $1/v$ dependence and the long mean free paths available for stored UCN in the absence of up-scattering, i.e. 5 m s^{-1} × 200 s= 10^3 m, indicate that one should be able to measure very weak scattering processes.

8.4.1   Up-scattering of UCN by superfluid helium-4

In section 3.5.3 we referred to some measurements of the neutrons upscattered by superfluid helium (figure 3.16) These experiments measured the total up-scattering rate, not the energy spectrum of the up-scattered neutrons. In this section we want to give some brief remarks on what can be learned by measurements of this spectrum which are currently in progress.

The physics of superfluid ^{4}He has been the subject of intense study for many decades (Wilks 1967) and there have been many neutron scattering studies of this material. Cowley and Woods (1971) and Woods and Cowley

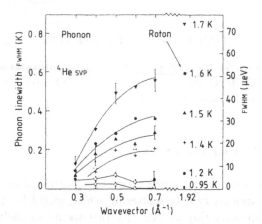

Figure 8.8 Intrinsic fullwidths for rotons and low q phonons in liquid ^{4}He at SVP for temperatures up to 1.7 K (Mezei and Stirling 1983).

(1973) provide comprehensive reviews while Svensson (1988) reviews the more recent work.

Figure 3.14 shows the dispersion curve of liquid helium and the 'free neutron dispersion curve'. The spectrum of up-scattered UCN will be proportional to $S(Q,\omega)$ along this latter curve. In figure 8.8 we show the measured linewidths (measured by spin-echo for the narrowest lines) as a function of Q and temperature T. We see that at 0.95 K the linewidths are just at the limit of what was observable (Mezei and Stirling 1983). More recent spin-echo measurements have shown a somewhat improved resolution (Mezei 1990).

The linewidths (inversely proportional to phonon lifetime) are determined by interactions between the phonons. These interactions can be understood by considering Landau's Hamiltonian density of a liquid (Landau and Khalatnikov 1949, see also Wilks 1967)

$$\mathcal{H} = \frac{m_{\text{He}}}{2} v \cdot \rho v + \varepsilon(\rho) \tag{8.12}$$

where $\varepsilon(\rho)$ is the internal energy of the fluid and

$$\rho = \rho_0 + \rho' \tag{8.13}$$

with ρ_0 the average density and ρ' the density fluctuations which will be treated as a quantized field and expanded as a sum of phonon creation and annihilation operators. $\varepsilon(\rho)$ can be expanded in a power series about ρ_0

$$\varepsilon(\rho) = \varepsilon_0 + \left(\frac{\partial \varepsilon}{\partial \rho}\right)_0 \rho' + \frac{1}{2!}\left(\frac{\partial^2 \varepsilon}{\partial \rho^2}\right)_0 (\rho')^2 + \frac{1}{3!}\left(\frac{\partial^3 \varepsilon}{\partial \rho^3}\right)_0 (\rho')^3. \tag{8.14}$$

Since $\varepsilon(\rho)$ is a minimum at ρ_0 the term $\propto \rho'$ is zero. The next term can be calculated from

$$\delta U = -\tfrac{1}{2}(\delta P)(\delta V) = \frac{1}{2}\left(\frac{1}{\kappa}\frac{\rho'}{\rho}\right)\left(\frac{\rho'N}{\rho^2}\right) = \frac{V}{2\kappa}\left(\frac{\rho'}{\rho_0}\right)^2 \qquad (8.15)$$

since $\rho = N/V$ and $\delta\rho = (-N/V^2)\delta V$. Then

$$\delta\varepsilon = \delta U/V = \frac{1}{2\kappa}\left(\frac{\rho'}{\rho_0}\right)^2 = \frac{m_{He}c^2}{2\rho_0}(\rho')^2 \qquad (8.16)$$

using $c^2 = (\kappa m_{He}\rho)^{-1}$, $c =$ velocity of sound and $\kappa =$ compressibility. Thus comparing (8.14) and (8.16) we identify

$$\frac{\partial^2\varepsilon}{\partial\rho^2} = \frac{m_{He}c^2}{\rho} \qquad \text{and} \qquad \frac{\partial^3\varepsilon}{\partial\rho^3} = m_{He}\frac{\partial}{\partial\rho}\left(\frac{c^2}{\rho}\right) = \frac{c^2}{\rho_0^2}(2u_0 - 1) \quad (8.17)$$

where the Gruneisen constant, u_0, is given by

$$u_0 = \frac{\rho_0}{c}\frac{\partial c}{\partial\rho} = 2.84$$

based on measurements of the velocity of sound (Abraham et al 1970).

The Landau quantum hydrodynamics proceeds by treating $\rho'(r)$ as a quantized field, expanding in the phonon creation and annihilation operators, a_k, a_k^+,

$$\rho'(r) = \sqrt{\frac{\rho_0}{2m_{He}cV}}\sum_k \sqrt{k}\left(a_k e^{ik\cdot r} + a_k^+ e^{-ik\cdot r}\right) \qquad (8.18)$$

satisfying the usual commutation relations for harmonic oscillator raising and lowering operators, with a similar expansion for the velocity v' (note $v_0 = 0$). Substitution of (8.18) into (8.12) and (8.14) results in a second-order Hamiltonian

$$H^{(2)} = \frac{\hbar c}{2}\sum_k k\left(a_k a_k^+ + a_k^+ a_k\right) = \hbar c\sum_k k\left(a_k^+ a_k + \tfrac{1}{2}\right) \qquad (8.19)$$

i.e. a sum of harmonic oscillator Hamiltonians with eigenvalues $\hbar ck\left(n_k + \tfrac{1}{2}\right)$ which can be interpreted as a state consisting of n_k phonons of wavenumber k. The kinetic energy term in (8.12) results in a third-order term

$$\frac{m_{He}}{2}v\cdot\rho'v \qquad (8.20)$$

which is added to the third-order term in (8.14) using (8.17). These third-order terms as well as the higher order terms in (8.14) mean that the eigenstates of $H^{(2)}$ (phonon states) are not eigenstates of the complete Hamiltonian and hence make transitions into other states. The third-order terms $H^{(3)}$ produce processes such as one phonon decaying into two, or two phonons merging into one (these are called three-phonon processes). The higher order terms $H^{(4)}$, $H^{(5)}$ etc lead to more complex processes. The higher order processes are less important than the contribution of $H^{(3)}$ which dominates where three-phonon processes are allowed by the conservation laws. The result is that the phonons have finite lifetimes and linewidths. Maris (1977) gives a review of the field of phonon interactions in ^{4}He. To apply this model to neutron scattering we write the interaction between a neutron and helium as (2.46)

$$V_{\text{n–He}} = \frac{2\pi\hbar^2}{m} a\rho(\boldsymbol{r}) \tag{8.21}$$

where the fluctuating part of $\rho(\boldsymbol{r})$ is given by (8.18). We can use this formalism to calculate the up-scattering rate for a one-phonon absorption process (up-scattering) according to the time-dependent perturbation theory

$$
\begin{aligned}
\frac{1}{\tau} &= N v_{\text{n}} \int \sigma(E_{\text{u}} \longrightarrow E_{\text{f}})\, \mathrm{d}E_{\text{f}} \\
&= \frac{2\pi}{\hbar} \int \int |\langle f, \boldsymbol{k}_{\text{f}} = \boldsymbol{Q}\, |V_{\text{n–He}}|\, \boldsymbol{k}_{\text{i}} = 0, \text{i}\rangle|^2 \\
&\quad \times \delta\left[\hbar c \left(k_1' - \frac{Q^2}{k^*}\right)\right] \frac{V\mathrm{d}^3 k_1}{(2\pi)^3} \frac{V\mathrm{d}^3 Q}{(2\pi)^3}
\end{aligned}
\tag{8.22}
$$

where k^* is defined by equation (3.27) and $\hbar c k_1'$ is the energy of a phonon of wavenumber k_1 ($k_1' = k_1$ for a linear dispersion relation). The matrix element is taken between the initial state i of the helium and a neutron at rest ($k_{\text{i}} = 0$) and the final state f of the helium (one phonon less than the initial state) and a neutron of wavenumber $k_{\text{f}} = \boldsymbol{Q}$. Using (8.18) and (8.21) we find

$$\frac{1}{\tau} = N\sigma \left(\frac{\hbar k^*}{m}\right) \left(\frac{E^*}{m_{\text{He}}c^2}\right) \mathrm{e}^{-E^*/k_{\text{B}}T} \tag{8.23}$$

with $\sigma = 4\pi a_{\text{coh}}^2$, the coherent scattering cross section and $E^* = \hbar c k^*$. Comparing with (3.35) we see that $(E^*/m_{\text{He}}c^2)$ in (8.23) is replaced by $\alpha S(k^*)$ in (3.45). $E^*/m_{\text{He}}c^2 \approx 0.4$ and $\alpha S(k^*) \approx 0.3$ using the measured value of $S(k^*)$. However this measured value includes both one-phonon and multi-phonon scattering (Cowley and Woods 1971, Woods and Cowley 1973). Thus the one-phonon scattering intensity predicted by (8.23) is distributed between one-phonon and multi-phonon scattering in the real liquid.

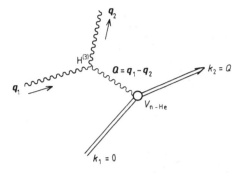

Figure 8.9 Main contribution to the up-scattering of UCN in ^{4}He at low temperatures.

We can calculate more complicated processes by replacing the matrix element in (8.22) by, say, the second-order matrix element involving $V_{\text{n-He}}$ (8.21) and the third-order phonon interaction $H^{(3)}$ as indicated by the diagram in figure 8.9. This is the dominant process below the dispersion curve, i.e. along the path (8.11) measured by UCN up-scattering. The two-phonon kinematics can be represented by the diagram in figure 8.10. With

$$Q = k_1 - k_2 \qquad (8.24)$$

we have

$$|k_1 - k_2| \leq Q \leq k_1 + k_2 \qquad (8.25)$$

where the left-hand side holds for the angle between k_1 and k_2, $\theta_{12} = 0$ and the right-hand side for $\theta_{12} = \pi$. For fixed Q (8.25) defines the shaded area in figure 8.10. From conservation of energy we have

$$k_1' - k_2' = \frac{Q^2}{k^*} \qquad (8.26)$$

which is plotted as the curve marked 'energy conservation'. The up-scattering cross section for a given Q is determined by the portion of this curve which lies in the shaded region.

Thus we see that measurement of the energy spectrum should provide information on the three-phonon interaction and its energy dependence down to temperatures inaccessible by other scattering methods. Details of the calculations are given by Golub (1979). The total up-scattering rate (integrated over Q) is proportional to T^7. Figure 8.11 shows the results of measurements of the total up-scattering rate. The broken line shows the results calculated for the two-phonon process previously described, while the full line includes the one-phonon and roton scattering (Golub *et al* 1983). The discrepancies may be due to our poor understanding of the three-phonon interaction.

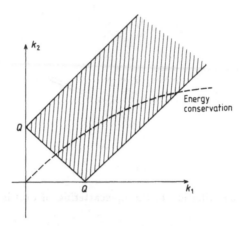

Figure 8.10 Conservation of momentum and energy for two-phonon up-scattering of UCN. k_1 is the incident phonon momentum; k_2 is the final phonon momentum. Conservation of momentum (8.26) defines the shaded region. Conservation of energy (8.27) defines the broken line.

Measurements of the spectrum of the up-scattered neutrons are in progress. A temperature dependence of the spectrum has been detected but the preliminary results are not reliable enough to justify any definite statement. It is hoped that more reliable results will be available shortly.

8.5 QUASI-ELASTIC SCATTERING OF UCN

In section 6.3.4 we have mentioned the possibility of measuring quasi-elastic scattering of UCN by placing a sample in a UCN storage vessel and measuring the spread in energy of an initially mono-energetic UCN sample by means of the 'gravitational monochromator'. This would provide sensitivity to values of $\omega \sim 10^{-11}$ eV which would be a uniquely small value for neutron scattering. In this case there would be absolutely no information concerning Q values and care would have to be taken to eliminate the incoherent scattering problem mentioned in section 8.1.

It remains to be seen if there are any cases where this technique can be interesting.

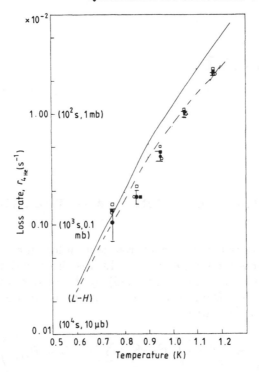

Figure 8.11   Loss rate due to the interaction of UCN with superfluid ^{4}He as a function of temperature. The numbers in brackets on the vertical scale give the corresponding storage times and total cross section (for 4.6 m s^{-1} UCN), respectively. The broken line shows the results for the two-phonon scattering process calculated using Landau's Hamiltonian (L–H). The full lines show the total loss rate. Different points have been corrected for wall losses by different methods. Their spread indicates the uncertainties involved (Golub *et al* 1983).

Appendix A1

A1.1 COHERENT SCATTERING AND THE STRUCTURE FACTOR $S(Q)$

Equation (2.36) gives the scattering cross section for an assembly of atoms. To obtain the cross section per atom, which is the way data are normally presented, it is necessary to divide (2.36) by the number of atoms present in the scatterer, N. Doing this and concentrating on the first (coherent scattering) term we can write:

$$\left(\frac{\mathrm{d}\sigma}{\mathrm{d}\Omega}\right)_{\mathrm{coh}} = \frac{a_{\mathrm{coh}}^2}{N} \left\langle \left| \int \mathrm{d}^3 r\, e^{i\mathbf{Q}\cdot\mathbf{r}} \sum_i \delta^{(3)}(\mathbf{r} - \mathbf{R}_i) \right|^2 \right\rangle \qquad (A1.1)$$

where the brackets refer to an average over the distribution of the positions R_i of the scattering atoms. This is the form we would obtain if we refrained from evaluating the integral as we did in going from (2.28) to (2.29). We rewrite (A1.1) as

$$\left(\frac{\mathrm{d}\sigma}{\mathrm{d}\Omega}\right)_{\mathrm{coh}}$$

$$= \frac{a_{\mathrm{coh}}^2}{N} \left\langle \int \mathrm{d}^3 r\, e^{i\mathbf{Q}\cdot\mathbf{r}} \sum_i \delta^{(3)}(\mathbf{r} - \mathbf{R}_i) \int \mathrm{d}^3 r'\, e^{-i\mathbf{Q}\cdot\mathbf{r}'} \sum_j \delta^{(3)}(\mathbf{r}' - \mathbf{R}_j) \right\rangle$$

$$\qquad (A1.2)$$

which can be transformed into

$$\left(\frac{\mathrm{d}\sigma}{\mathrm{d}\Omega}\right)_{\mathrm{coh}} = \frac{a_{\mathrm{coh}}^2}{N} \int \mathrm{d}^3 r_0\, e^{i\mathbf{Q}\cdot\mathbf{r}_0} \left\langle \sum_{i,j} \delta^{(3)}(\mathbf{r}_0 + \mathbf{R}_j - \mathbf{R}_i) \right\rangle \qquad (A1.3)$$

by substituting $\mathbf{r} - \mathbf{r}' = \mathbf{r}_0$. By separating the terms with $i = j$ from the double sum we obtain

$$\left(\frac{\mathrm{d}\sigma}{\mathrm{d}\Omega}\right)_{\mathrm{coh}} = a_{\mathrm{coh}}^2 \int \mathrm{d}^3 r_0\, e^{i\mathbf{Q}\cdot\mathbf{r}_0} \left(\delta^{(3)}(\mathbf{r}_0) + \frac{1}{N} \sum_{i \neq j} \left\langle \delta^{(3)}(\mathbf{r}_0 + \mathbf{R}_j - \mathbf{R}_i) \right\rangle \right).$$

$$\qquad (A1.4)$$

We can understand the significance of the second term in (A1.4) by first considering that we keep j fixed and noting that the δ-function gives a contribution every time $r_0 = R_i - R_j$, that is every time there is a particle (i) displaced by r_0 from the jth particle. Since the jth particle is not special in any way the contribution from the sum over j will be simply N times the contribution of the jth particle. Hence the double sum of the δ-function will be $Ng(r_0)$ where $g(r_0)$ is the probability of finding a particle at a position r_0 relative to a given particle and (A1.4) can be rewritten as

$$\left(\frac{\mathrm{d}\sigma}{\mathrm{d}\Omega}\right)_{\text{coh}} = a_{\text{coh}}^2 \int \mathrm{d}^3 r_0 e^{i\boldsymbol{Q} \cdot \boldsymbol{r}_0} \left[\delta^{(3)}\left(\boldsymbol{r}_0\right) + g\left(\boldsymbol{r}_0\right)\right]$$

$$\equiv a_{\text{coh}}^2 S\left(\boldsymbol{Q}\right). \tag{A1.5}$$

The integral, called the structure factor $S(Q)$, is simply one plus the Fourier transform of $g\left(r_0\right)$. The structure factor, or its Fourier transform, represents all the information about the static distribution of atoms in the scattering system that can be determined by scattering. It can be measured by X-ray as well as neutron scattering.

Appendix A2

A2.1 NEUTRON SCATTERING FROM A COLLECTION OF MOVING NUCLEI—THE SCATTERING LAW

When neutrons are scattered from a rigid array of nuclei the energy of the scattering system cannot change and so, by conservation of energy, the energy of the neutron cannot change either. Such scattering, where the neutron energy does not change, is conventionally referred to as elastic scattering. When we consider scattering from a real system in which the nuclei are able to move, there is a possibility of the scattering system exchanging energy with the scattered neutron. In this section we present the formalism (largely due to Van Hove (1954)) by which this type of scattering is usually discussed.

In Appendix A1 we saw that the elastic scattering was determined by the structure factor, $S(Q)$, which was related to the Fourier transform of the pair distribution function. In this section we will see that the inelastic scattering cross section is related to a function $S(Q, \omega)$ ($\hbar\omega$ is the energy lost by the neutron during the scattering) called the scattering law and given by the Fourier transform of a function $G(r, t)$ which is called the density correlation function and which represents the probability of finding a particle at position r at time t given that there was a particle at the origin ($r = 0$) at time $t = 0$.

We again consider the initial and final states of the neutron to be given by plane waves:

$$\psi_{i,f} = \frac{e^{i k_{i,f} \cdot r}}{L^{3/2}} \qquad \text{with energies} \quad \epsilon_{i,f} = \frac{\hbar^2 k_{i,f}^2}{2m} \qquad (A2.1)$$

where we have now normalized the neutron states to a volume L^3. We will designate the states of the scattering system by quantum numbers n_i, n_f with energies E_i and E_f. Thus the initial and final states of the entire system (neutron plus scatterer) will be designated

$$|\text{i}\rangle = \frac{e^{i k_i \cdot r}}{L^{3/2}} |n_i\rangle \qquad |\text{f}\rangle = \frac{e^{i k_f \cdot r}}{L^{3/2}} |n_f\rangle . \qquad (A2.2)$$

As in (2.28) the interaction between neutron and scattering system will be taken as:

$$V(r, R_i) = \frac{2\pi\hbar^2}{m} \sum_i a_i \delta^{(3)}(r - R_i).$$ (A2.3)

We again calculate the cross section by dividing the transition probability per second by the incident flux. The latter quantity is obtained from (A2.1) as follows

$$\text{incident flux} = \frac{1}{L^3} \cdot \frac{\hbar k_i}{m}$$ (A2.4)

and we will calculate the transition rate due to (A2.3) using perturbation theory. The standard result of perturbation theory (Golden Rule—Schiff 1968), in the case where there is a smooth distribution of final states (in our case the neutron energies form a continuum), is that the transition rate is given by

$$dP = \frac{2\pi}{\hbar} |\langle f|V|i\rangle|^2 \rho(\epsilon_f) \delta(E_f + \epsilon_f - E_i - \epsilon_i)$$ (A2.5)

where $\rho(\epsilon_f) d\epsilon_f$ is the number of states with final energies between ϵ_f and $\epsilon_f + d\epsilon_f$ and is given by

$$\rho(\epsilon_f) = \frac{mL^3}{8\pi^3\hbar^2} k_f \, d\Omega_f$$ (A2.6)

for final states whose momentum lies in the element of solid angle $d\Omega_f$. Using (A2.2), (A2.3) and (A2.5) and dividing by (A2.4) it is easy to see that the cross section per unit solid angle per unit final neutron energy—which is called the double differential cross section—is given by

$$\frac{d^2\sigma}{d\Omega \, d\epsilon_f} = \frac{k_f}{k_i} \left| \left\langle n_f \left| \sum_i a_i e^{iQ \cdot R_i} \right| n_i \right\rangle \right|^2 \delta(E_f - E_i - \hbar\omega)$$ (A2.7)

where $\hbar\omega = \epsilon_i - \epsilon_f$ is the energy lost by the neutron during the scattering. This is the cross section for a scattering process in which the scattering system makes a transition from the state $|n_i\rangle$ to the state $|n_f\rangle$. In an actual scattering situation the system is likely to be in thermal equilibrium which means it will be distributed over a range of initial states. We take

$$P_{n_i} = (1/Z)\exp[-E_i/k_\mathrm{B}T]$$

where k_B is Boltzman's constant and Z is a normalizing constant, as the probability of finding the system in a particular initial state $|n_i\rangle$ and the final states of the system will be distributed over a range of states $|n_f\rangle$. Therefore the actual scattering cross section will be given by summing (A2.7) over all possible final states and averaging over the initial states. Thus the cross section will be (we again divide by N to obtain the cross section per atom).

$$\frac{d^2\sigma}{d\Omega\,d\epsilon_f} = \frac{1}{N}\cdot\frac{k_f}{k_i}\sum_{n_f,n_i}P_{n_i}\left|\sum_i a_i\left\langle n_f\left|e^{i\boldsymbol{Q}\cdot\boldsymbol{R}_i}\right|n_i\right\rangle\right|^2\delta\left(E_f-E_i-\hbar\omega\right)\quad(A2.8)$$

which we rewrite as

$$\frac{d^2\sigma}{d\Omega\,d\epsilon_f} = \frac{1}{N}\cdot\frac{k_f}{k_i}\sum_{n_f,n_i}P_{n_i}\sum_{i,j}\overline{a_i^*a_j}\left\langle n_i\left|e^{-i\boldsymbol{Q}\cdot\boldsymbol{R}_i}\right|n_f\right\rangle$$
$$\times\left\langle n_f\left|e^{i\boldsymbol{Q}\cdot\boldsymbol{R}_j}\right|n_i\right\rangle\delta\left(E_f-E_i-\hbar\omega\right)\qquad(A2.9)$$

where, as in (2.31), the bar refers to averaging over the spin and isotope distribution of the nuclei. Using (2.35) we can separate (A2.9) into a coherent and incoherent part:

$$\frac{d^2\sigma}{d\Omega\,d\epsilon_f} = \left(\frac{d^2\sigma}{d\Omega\,d\epsilon_f}\right)_{\text{coh}} + \left(\frac{d^2\sigma}{d\Omega\,d\epsilon_f}\right)_{\text{inc}}\qquad(A2.10)$$

where

$$\left(\frac{d^2\sigma}{d\Omega\,d\omega}\right)_{\text{coh}} = \frac{\hbar}{N}\frac{k_f}{k_i}\sum_{n_f,n_i}P_{n_i}a_{\text{coh}}^2\left|\sum_i\left\langle n_f\left|e^{i\boldsymbol{Q}\cdot\boldsymbol{R}_i}\right|n_i\right\rangle\right|^2\delta\left(E_f-E_i-\hbar\omega\right)$$
$$\equiv a_{\text{coh}}^2\frac{k_f}{k_i}S_{\text{coh}}\left(\boldsymbol{Q},\omega\right)\qquad(A2.11)$$

and

$$\left(\frac{d^2\sigma}{d\Omega\,d\omega}\right)_{\text{inc}} = \frac{\hbar}{N}\frac{k_f}{k_i}a_{\text{inc}}^2\sum_{n_f,n_i}P_{n_i}\sum_i\left|\left\langle n_f\left|e^{i\boldsymbol{Q}\cdot\boldsymbol{R}_i}\right|n_i\right\rangle\right|^2\delta\left(E_f-E_i-\hbar\omega\right)$$
$$= a_{\text{inc}}^2\frac{k_f}{k_i}S_{\text{inc}}\left(\boldsymbol{Q},\omega\right)\qquad(A2.12)$$

where we have written $d\epsilon_f = \hbar d\omega$ (ϵ_i is taken as fixed). We are interested in the Fourier transforms of $S_{\text{coh}}\left(\boldsymbol{Q},\omega\right)$ and $S_{\text{inc}}\left(\boldsymbol{Q},\omega\right)$ which we denote by $G\left(\boldsymbol{r},t\right)$ and $G_s\left(\boldsymbol{r},t\right)$ respectively.

$$G\left(\boldsymbol{r},t\right) = \frac{1}{(2\pi)^3}\int\int e^{i(\omega t-\boldsymbol{Q}\cdot\boldsymbol{r})}S_{\text{coh}}\left(\boldsymbol{Q},\omega\right)d^3Q\,d\omega\qquad(A2.13)$$

which is from (A2.11)

$$G\left(\boldsymbol{r},t\right) = \frac{1}{(2\pi)^3}\frac{1}{N}\sum_{n_f,n_i}P_{n_i}\int d^3Q\sum_{i,j}e^{-i\boldsymbol{Q}\cdot\boldsymbol{r}}\left\langle n_i\left|e^{-i\boldsymbol{Q}\cdot\boldsymbol{R}_i}\right|n_f\right\rangle$$
$$\times\left\langle n_f\left|e^{iE_ft}e^{i\boldsymbol{Q}\cdot\boldsymbol{R}_j}e^{-iE_it}\right|n_i\right\rangle.\qquad(A2.14)$$

Note that we have carried out the ω-integration by using the properties of the energy conservation δ-function, and omitted a factor of $\hbar$ from the energy exponents.

We recall that we can describe time-dependent situations in quantum mechanics by means of the 'Heisenberg picture' in which the quantum mechanical operators are time-dependent and the quantum states are independent of time. The time-dependent form of an operator R_j, is given by

$$R_j(t) = e^{iHt/\hbar} R_j(0) e^{-iHt/\hbar} \qquad (A2.15)$$

where H is the Hamiltonian of the system and $R_j(0)$ is the operator at $t = 0$. By expanding the exponential $e^{i\mathbf{Q} \cdot \mathbf{R}_j(t)}$ in a power series and applying (A2.15) to each term we obtain

$$e^{i\mathbf{Q} \cdot \mathbf{R}_j(t)} = e^{iHt/\hbar} e^{i\mathbf{Q} \cdot \mathbf{R}_j(0)} e^{-iHt/\hbar} \qquad (A2.16)$$

which is the operator in the second matrix element in (A2.14).

Note

$$H |n_{i,f}\rangle = E_{i,f} |n_{i,f}\rangle \ . \qquad (A2.17)$$

Thus

$$G(r,t) = \frac{1}{(2\pi)^3 N} \sum_{n_i} P_{n_i} \sum_{i,j} \int d^3Q \, e^{-i\mathbf{Q} \cdot \mathbf{r}} \left\langle n_i \left| e^{-i\mathbf{Q} \cdot \mathbf{R}_i(0)} e^{i\mathbf{Q} \cdot \mathbf{R}_j(t)} \right| n_i \right\rangle \qquad (A2.18)$$

and by a similar argument:

$$G_s(r,t) = \frac{1}{(2\pi)^3 N} \sum_{n_i} P_{n_i} \sum_{i} \int d^3Q \, e^{-i\mathbf{Q} \cdot \mathbf{r}} \left\langle n_i \left| e^{-i\mathbf{Q} \cdot \mathbf{R}_i(0)} e^{i\mathbf{Q} \cdot \mathbf{R}_i(t)} \right| n_i \right\rangle . \qquad (A2.19)$$

Now (A2.18) and (A2.19) are the Fourier transforms of a product of two operators. As is well known the Fourier transform of a product of two functions (A and B) is given by the convolution of the Fourier transforms of A and B (Champeney 1973). We also know that, for example

$$\frac{1}{(2\pi)^3} \int d^3Q e^{-i\mathbf{Q} \cdot \mathbf{r}} e^{-i\mathbf{Q} \cdot \mathbf{R}_i(0)} = \delta^{(3)}(r + \mathbf{R}_i(0)) \qquad (A2.20)$$

so we can see that

$$G(r,t) = \frac{1}{N} \sum_{i} P_{n_i} \sum_{i,j} \int d^3r' \left\langle n_i \left| \delta^{(3)}(r + \mathbf{R}_i(0) - r') \, \delta^{(3)}(r' - \mathbf{R}_j(t)) \right| n_i \right\rangle \qquad (A2.21)$$

$$G_s(r,t) = \frac{1}{N} \sum_{n_i} P_{n_i} \sum_{i} \int d^3r' \left\langle n_i \left| \delta^{(3)}(r + \mathbf{R}_i(0) - r') \, \delta^{(3)}(r' - \mathbf{R}_i(t)) \right| n_i \right\rangle . \qquad (A2.22)$$

If the $R_i(t)$ were classical quantities, instead of operators, we could evaluate the integrals to obtain:

$$G(r,t) = \frac{1}{N} \sum_{n_i} P_{n_i} \sum_{i,j} \left\langle n_i \left| \delta^{(3)} (r + R_i(0) - R_j(t)) \right| n_i \right\rangle \qquad \text{(A2.23)}$$

with a similar result for $G_s(r,t)$.

By the same argument as was given following (A1.4) we see that this is the probability density of finding an atom at position r at time t given that there was an atom at the origin ($r = 0$) at time $t = 0$. Similarly $G_s(r,t)$ is the probability density of finding an atom at time t, given the *same atom* was at the origin at $t = 0$. The function $G(r,t)$ is called the space–time pair correlation function while $G_s(r,t)$ is called the space–time self-correlation function. In quantum mechanics these correlation functions, strictly speaking do not exist, since $R_i(0)$ and $R_i(t)$ do not commute and are not measurable simultaneously. Thus in general it is necessary to work with (A2.21) and (A2.22).

For $t = 0$ all the $R_i(0)$ commute so we can write from (A2.21):

$$G(r,0) = \frac{1}{N} \sum_{n_i} P_{n_i} \sum_{i,j} \left\langle n_i \left| \delta^{(3)} (r + R_i(0) - R_j(0)) \right| n_i \right\rangle \qquad \text{(A2.24)}$$

and separating into terms with $i = j$ and $i \neq j$ we obtain:

$$G(r,0) = \delta^{(3)}(r) + \frac{1}{N} \sum_{n_i} P_{n_i} \sum_{i \neq j} \delta^{(3)}(r + R_i(0) - R_j(0))$$

$$= \delta^{(3)}(r) + g(r) \qquad \text{(A2.25)}$$

from (A1.4) and (A1.5). Thus $G(r,0)$ is the spatial Fourier transform of the structure factor $S(Q)$.

Since for any function of time the integral of the Fourier transform over all ω is equal to the function evaluated at $t = 0$ it is easy to see that

$$\int S(Q,\omega)\, d\omega = S(Q) \qquad \text{(A2.26)}$$

a result known as the zero-moment sum rule.

In general it is not possible to calculate $S(Q,\omega)$ from first principles but its general behaviour can be understood in physical terms and discussion of scattering measurements in terms of $G(r,t)$ often leads to increased physical insight. If r_0 is the correlation length and τ the correlation time of $G(r,t)$, i.e. values of r and t for which $G(r,t)$ shows significant variations, then since $G(r,t)$ and $S(Q,\omega)$ form a Fourier transform pair the scattering will be significant for $Q \sim 1/r_0$ and $\omega \sim 1/\tau$.

By making use of the fact that:

$$\langle n_f | e^{iQ \cdot R} | n_i \rangle = (\langle n_i | e^{-iQ \cdot R} | n_f \rangle)^* \tag{A2.27}$$

we can prove that

$$S(-Q, -\omega) = e^{-\hbar\omega/k_B T} S(Q, \omega) \tag{A2.28}$$

if the scattering system is in thermal equilibrium. This is called the principle of detailed balance. From (A2.11) and (A2.28) it follows that

$$\epsilon_i e^{-\epsilon_i/k_B T} \sigma(\epsilon_i \longrightarrow \epsilon_f, \Omega_i \longrightarrow \Omega_f) = \epsilon_f e^{-\epsilon_f/k_B T} \sigma(\epsilon_f \longrightarrow \epsilon_i, \Omega_f \longrightarrow \Omega_i) \tag{A2.29}$$

where $\sigma(\epsilon_i \longrightarrow \epsilon_f, \Omega_i \longrightarrow \Omega_f)$ is the cross section for a neutron with energy ϵ_i travelling in the direction Ω_i finishing up with an energy ϵ_f travelling in the direction Ω_f as a result of the scattering.

Since the Maxwell–Boltzmann flux distribution is given by

$$\Phi(\epsilon)\, d\epsilon = \Phi_0 \frac{\epsilon}{k_B T} e^{-\epsilon/k_B T} \frac{d\epsilon}{k_B T} \tag{A2.30}$$

where Φ_0 is the total flux, we see the significance of (A2.29). The number of transitions per second made by neutrons from energy $\epsilon_i \longrightarrow \epsilon_f$, direction $\Omega_i \longrightarrow \Omega_f$ is exactly balanced by the number of neutrons per second going from $\epsilon_f \longrightarrow \epsilon_i, \Omega_f \longrightarrow \Omega_i$ if the neutrons are in thermal equilibrium in a moderator at temperature T. Thus when the neutrons are in thermal equilibrium with the moderator the balance is 'detailed', i.e. every transition takes place at the same rate as its inverse.

We will have occasion to make use of these properties of $S(Q, \omega)$ and the scattering cross section in our discussions of the scattering and production of UCN.

A2.2   MODELS OF $G(r,t)$

A2.2.1   Diffusion

As a simple model we consider a classical liquid whose density obeys the diffusion equation. This should be valid for liquids and gases for times large compared with the collision time and distances large compared with the inter-atomic distance.

From (A2.14) we see that

$$G^*(r,t) = G(-r, -t) \tag{A2.31}$$

with a similar expression for $G_s(r, t)$.

From (A2.13) this is only the condition that $S(Q, \omega)$ and the scattering cross sections be real. The diffusion equation follows from the equation of continuity (conservation of matter (2.58))

$$\frac{\partial \rho}{\partial t} + \nabla \cdot j = 0 \qquad (A2.32)$$

and the diffusion approximation

$$j = -D\nabla\rho \qquad (A2.33)$$

where D is the diffusion constant. Thus

$$\frac{\partial \rho}{\partial t} - D\nabla^2\rho = 0. \qquad (A2.34)$$

To find the self-correlation function $G_s(r, t)$ we solve (A2.34) with the initial condition of a particle at the origin at $t = 0$

$$\frac{\partial g_s(r, t)}{\partial t} - D\nabla^2 g_s(r, t) = \delta^{(3)}(r)\,\delta(t). \qquad (A2.35)$$

Putting

$$g_s(r, t) = \frac{1}{(2\pi)^3} \int d^3Q e^{iQ \cdot r} \gamma(Q, t) \qquad (A2.36)$$

we obtain

$$\frac{\partial \gamma(Q, t)}{\partial t} + DQ^2\gamma(Q, t) = \delta(t) \qquad (A2.37$$

with a solution

$$\gamma(Q, t) = e^{-DQ^2 t}\Theta(t) \qquad (A2.38)$$

where Θ is the unit step function. Substituting in (A2.36) we find

$$g_s(r, t) = \frac{1}{(4\pi Dt)^{3/2}} e^{-r^2/4Dt}\Theta(t). \qquad (A2.39)$$

This is the solution of the problem with the initial condition $g_s(r, t) = 0$ for $t \leq 0$.

Now $G_s(r, t)$ must satisfy condition (A2.31) so it is necessary to symmetrize (A2.39) with respect to the time to obtain

$$G_s(r, t) = \frac{1}{(4\pi D|t|)^{3/2}} e^{-r^2/4|t|D}. \qquad (A2.40)$$

Writing

$$I_s(Q,t) = \int d^3 r \, e^{iQ \cdot r} G_s(r,t) \tag{A2.41}$$

($I_s(Q,t)$ is called the intermediate scattering function) we find

$$I_s(Q,t) = e^{-DQ^2|t|} \tag{A2.42}$$

and

$$S(Q,\omega) = \frac{1}{2\pi} \int_{-\infty}^{\infty} dt \, e^{-i\omega t} e^{-DQ^2|t|} \tag{A2.43}$$

$$= \frac{1}{2\pi} 2\mathcal{R}_e \int_{0}^{\infty} dt \, e^{-i\omega t} e^{-DQ^2 t} \tag{A2.44}$$

$$S(Q,\omega) = \frac{1}{\pi} \frac{DQ^2}{\left(\omega^2 + (DQ^2)^2\right)} \tag{A2.45}$$

i.e. a Lorentzian in ω with width DQ^2.

This is a basic result in scattering theory and was first obtained by Vineyard (1958). As we said it should hold for small Q (large distances) and small ω (large times). It serves as the starting point for more accurate treatments.

For example replacing the width of the Gaussian in (A2.40) $4D|t|$ by $2\Gamma(t)$ leads to the 'Gaussian approximation' where more complicated kinds of diffusion can be represented by various forms for $\Gamma(t)$ (Turchin 1965). In liquid Na at 800 K DQ^2 in (A2.45) must be replaced by $DQ^2(1 - aQ)$ with $a = 0.15$ Å(Morkel et al 1987).

A2.2.2   Small motions of large objects

In many large molecules of biological interest there are motions which involve large pieces of a molecule moving over relatively small distances. An example is the motions in the myoglobin molecule which are involved in the opening of a channel allowing an oxygen molecule to be captured or emitted (Parak and Knapp 1984).

In this section we give a simple classical model of the neutron scattering for such a system. As a model we take a spherical shell of radius R, thickness ϵ of density ρ surrounded by an infinite medium of density ρ_0 with $\delta\rho = \rho - \rho_0$. The centre of the shell is located at a position $r = c$ and motion of the sphere is represented by taking c as a function of time.

We start with (2.93) for the static case

$$f(\theta) = a \int d^3 r \, \rho(r) e^{iQ \cdot r} \tag{A2.46}$$

and apply it to this case. We have

$$
\begin{aligned}
\frac{f(\theta)}{a} &= \int_{\text{shell}} \mathrm{d}^3r \rho(r)\, e^{iQ \cdot r} + \int_{\text{outside shell}} \mathrm{d}^3r\, \rho_0(r)\, e^{iQ \cdot r} \\
&= \int_{\text{allspace}} \mathrm{d}^3r\, \rho_0(r)\, e^{iQ \cdot r} + \int_{\text{shell}} \mathrm{d}^3r \delta\rho(r)\, e^{iQ \cdot r} \\
&= (2\pi)^3 \delta^{(3)}(Q) + \int_{\text{shell}} \mathrm{d}^3r\, \delta\rho(r)\, e^{iQ \cdot r}.
\end{aligned}
\tag{A2.47}
$$

Thus we see that in a uniform medium (first term) there is no scattering for $Q \neq 0$ and the scattering occurs only on inhomogeneities.

We now extend (A2.47) to the time-dependent case

$$
f(Q,\omega) = a \int \mathrm{d}^3r\, \delta\rho(r,t)\, e^{i(Q \cdot r - \omega t)}\, \mathrm{d}t
\tag{A2.48}
$$

carrying out the integration over d^3r for the spherical shell centred at $r = c(t)$ we have (see equation (A2.52))

$$
f(Q,\omega) = a\delta\rho \left(4\pi R^2 \epsilon\right) \int \mathrm{d}t\, e^{-i\omega t}\, \frac{\sin QR}{QR} e^{iQ \cdot c(t)}
\tag{A2.49}
$$

where we have used the property of Fourier transforms that if $F(Q)$ is the Fourier transform of $f(r)$ then $e^{iQ \cdot c} F(Q)$ is the Fourier transform of $f(r - c)$. Taking $c(t) = c_0 \cos \omega_0 t$ the time integral in (A2.49) can be evaluated using

$$
J_n(z) = \frac{i^{-n}}{2\pi} \int_0^{2\pi} e^{iz \cos \varphi} e^{-in\varphi}\, \mathrm{d}\varphi
\tag{A2.50}
$$

Champeney (1973), so that the result is

$$
\begin{aligned}
f(Q,\omega) &\propto a\delta\rho \frac{R^2 \epsilon}{\omega_0} \left(\frac{\sin QR}{QR}\right) \sum_{n=-\infty}^{\infty} (i)^{-n} J_n\left(Qc_0 \cos \theta\right) \delta\left(\omega - n\omega_0\right) \\
&\propto a\delta\rho \frac{R^2 \epsilon}{\omega_0} \left(\frac{\sin QR}{QR}\right) \sum_{n=-\infty}^{\infty} \frac{(i)^{-n}}{n!} \left(\frac{Qc_0 \cos \theta}{2}\right)^n \delta\left(\omega - n\omega_0\right)
\end{aligned}
$$

$$
\tag{A2.51}
$$

where θ is the angle between Q and c_0 and the last step is made under the assumption that $Qc_0 \ll 1$.

$$
\left(\text{Note that } J_n(x) = \frac{1}{n!}\left(\frac{x}{2}\right)^n \qquad \text{for } x \ll 1. \right)
$$

There are two things to note about (A2.51). First the Q dependence is such that the scattering is significant only for $Q \lesssim \pi/R$; that is the Q range is determined by the size of the moving object, not by the amplitude of the motion. Second, remembering we must take $|f(Q, \omega)|^2$ to get a scattering cross section we see that the term for $n = 1$ is very similar to the one for phonon scattering (A3.27) with the cross section proportional to $(\mathbf{Q} \cdot \boldsymbol{\xi})^2$ and an energy-conserving δ-function. Thus we have outlined a semi-classical derivation of the phonon-scattering cross section.

A2.2.3 Scattering from static spherical inhomogeneities

We proceed as in the previous section starting with equation (A2.48). In general the three-dimensional Fourier transform of a function of $r = |\mathbf{r}|$ can be calculated as

$$F(Q) = \int d^3 r\, e^{-i\mathbf{Q} \cdot \mathbf{r}} f(r) = 2\pi \int_0^\pi \sin\theta\, d\theta\, e^{-iQr\cos\theta} \int r^2 dr f(r)$$

$$= 4\pi \int f(r) \frac{\sin Qr}{Qr} r^2 dr. \tag{A2.52}$$

For a spherical region of radius R, and uniform density difference to the surrounding medium $\delta\rho$ we have, from (A2.47) and (A2.52)

$$\frac{f(\theta)}{a} = \frac{4\pi\delta\rho}{Q} \int_0^R r \sin Qr\, dr = 4\pi R^3 a\delta\rho \left(\frac{\sin QR}{(QR)^3} - \frac{\cos QR}{(QR)^2} \right)$$

$$= 4\pi R^3 a\delta\rho \left(\frac{j_1(QR)}{QR} \right) \tag{A2.53}$$

where $j_1(x)$ is the spherical Bessel function of order 1

$$j_1(x) = \frac{\sin x}{x^2} - \frac{\cos x}{x} \tag{A2.54}$$

(see Abramowitz and Stegun (1964) for the properties of spherical Bessel functions). Then

$$S(Q) = \left| \frac{f(\theta)}{a} \right|^2 = (4\pi R^3 \delta\rho)^2 \left(\frac{j_1(QR)}{QR} \right)^2. \tag{A2.55}$$

Using (A1.5) and (2.109) we have

$$\sigma_{\text{tot}} = \int d\Omega \left(\frac{d\sigma}{d\Omega} \right) = \frac{2\pi a^2}{k^2} \int S(Q) Q\, dQ \tag{A2.56}$$

where we took $k_i = k_f = k$. Substituting (A2.55) we find

$$\sigma_{tot} = \frac{2\pi a^2}{k^2} \left(4\pi R^2 \delta\rho\right)^2 \int_{k\theta_1}^{2k} \frac{[j_1(QR)]^2}{Q} \, dQ \qquad (A2.57)$$

where the lower limit on Q is determined by the angle subtended by the detector (θ_1)—neutrons scattered through a smaller angle hit the detector and are not counted as scattered in total cross section measurements ('in-scattering'). For large enough k (i.e. $kR \gg 1$) we can take the upper limit as ∞. Substituting $x = QR$, the integral in (A2.57) is (Lengsfeld 1977)

$$\int \frac{[j_1(x)]^2}{x} \, dx = -\tfrac{1}{4} \left[j_0^2(x) + j_1^2(x)\right] \qquad (A2.58)$$

as can be seen by differentiating the right-hand side using

$$\frac{dj_0(x)}{dx} = -j_1(x)$$

$$\frac{dj_1(x)}{dx} = j_0(x) - \frac{2}{x} j_1(x) \qquad (A2.59)$$

(Abramowitz and Stegun 1964). Using (A2.54) and

$$j_0(x) = \frac{\sin x}{x} \qquad (A2.60)$$

we have

$$g(x) \equiv j_0^2(x) + j_1^2(x) = \frac{1}{x^2} + \frac{\sin^2 x}{x^4} - \frac{2\sin x \cos x}{x^3} \qquad (A2.61)$$

Thus

$$g(0) = 1 \qquad \lim_{x \to \infty} g(x) = \frac{1}{x^2}$$

so that (A2.57) gives

$$\sigma_{tot} = \frac{\pi a^2 \left(4\pi R^2 \delta\rho\right)^2}{2k^2} \qquad (A2.62)$$

i.e. $\propto 1/k^2$ when 'in-scattering' is negligible ($\theta_1 \longrightarrow 0$) and

$$\sigma_{tot} = \frac{\pi a^2}{2k^4 \theta_1^2} (4\pi R\delta\rho)^2 \qquad (A2.63)$$

i.e. $\propto 1/k^4$ when k is large so that 'in-scattering' dominates and the detector measures only the scattering which is in the high Q tail. Comparing the asymptotic forms (A2.62) and (A2.63) we see that the intersection of their extrapolation occurs when

$$k_1 = \frac{1}{R\theta_1} \qquad (A2.64)$$

and hence can serve as a measure of the size of the scattering objects.

Appendix A3

A3.1 PHONONS—COHERENT AND INCOHERENT INELASTIC SCATTERING

We will now turn to a brief discussion of phonons in solids and their influence on neutron scattering. For further details the reader is referred to one of the comprehensive texts on neutron scattering (Turchin 1965, Lovesey 1984). Our main goals in this discussion are to emphasize the physics involved in coherent and incoherent inelastic neutron scattering from phonons and to present some results which we will use in the text.

For simplicity we will confine the discussion to ideal crystals with only one atom per unit cell. We write the position R_i of the ith atom in the crystal as

$$R_i = \rho_i + u_i(t) \tag{A3.1}$$

where ρ_i represents the equilibrium position of the ith atom in the crystal and $u_i(t)$ represents the motion of the atom about its equilibrium position. It is a basic assumption of our approach that $u_i \ll \rho_i$.

The kinetic energy of the crystal is the sum of the kinetic energies of all the atoms in the crystal

$$T = \sum_i \frac{m_i}{2} (\dot{u}_i)^2 \tag{A3.2}$$

and the potential energy is, in general, a function, $V(R_i)$, of all the R_i. We take the zero of potential energy to be at the equilibrium position ($R_i = \rho_i$ for all i) and consider that the potential energy is expanded in a power series in u_i. The leading term in this expansion can be written as

$$V = \tfrac{1}{2} \sum_{i,j} \sum_{\mu,\nu} \alpha_{i,j}^{\mu\nu} u_i^\mu u_j^\nu \tag{A3.3}$$

where the $\alpha_{i,j}^{\mu,\nu}$ are called the force constants of the crystal. The superscripts μ and ν label the three components of the vector u_i. Much of the work with neutron scattering is devoted to studying these force constants in different materials.

The neglect of higher order terms in the potential energy, (A3.3), is called the harmonic approximation. We will see that the Hamiltonian of the system (given by the sum of (A3.2) and (A3.3)) can be written as the sum of a set of harmonic oscillator Hamiltonians which, when quantized, will each have an energy $\hbar\omega(n + \frac{1}{2})$ where n is a quantum number. We will speak of the system in an eigenstate with quantum number n as containing n phonons of frequency ω. The neglected higher order terms in the potential energy will result in these phonon eigenstates not being exact eigenstates of the actual Hamiltonian. Thus the system will not remain in a given phonon state for an infinite time. We describe this by saying that the neglected terms result in interactions between the phonons with the result that each phonon has a finite lifetime. In addition these higher order terms result in the thermal expansion of solids and the fact that thermal conductivities are finite. These phonon energies are also temperature-dependent due to these terms.

The equations of motion of the system follow from (A3.2) and (A3.3):

$$m_i \ddot{u}_i^\mu = -\sum_{j,\nu} \alpha_{i,j}^{\mu\nu} u_j^\nu. \tag{A3.4}$$

For a crystal of N atoms there are $3N$ such equations of motion. We see that these equations are coupled—that is the motion of u_i^μ depends in general on all the other u_j^ν except that we expect the coupling to diminish as the atoms i and j are further apart. To simplify these equations we introduce the 'normal modes', that is we introduce new coordinates—linear combinations of the u_i—so that the equation of motion of each of the new coordinates only depends on the new coordinate itself. This is always possible in the case of equations like (A3.4) (Goldstein 1950).

To search for the normal modes we assume the u_i can be written (for the case that all the atoms have mass, $m_i = M$):

$$u_i(t) = \frac{\gamma(q)}{\sqrt{MN}} e^{i(q \cdot \rho_i - \omega t)} \tag{A3.5}$$

and substitute this into (A3.4). After some manipulation this reduces to

$$\omega^2 \gamma^\mu(q) = \sum_\nu \lambda^{\mu\nu}(q) \gamma^\nu(q) \tag{A3.6}$$

where the $\lambda(q)$ are linear combinations of $\alpha_{i,j}$. Equation (A3.6) represents three equations, one for each value of μ, and since these equations are linear homogeneous equations for the γ^μ they will only have a solution if the 3×3 determinant:

$$\det \left\| \omega^2 \delta_{\mu\nu} - \lambda^{\mu\nu}(q) \right\| = 0 \tag{A3.7}$$

which has three solutions for ω for each q . We denote these solutions as $\omega_s(q)$, $s = 1, 2, 3$. Once the frequencies $\omega_s(q)$ are known one can solve (A3.6) for the amplitudes $\gamma_s(q)$. The general solution of equations (A3.4) will be a superposition of terms like (A3.5).

$$u_i(t) = \sum_{s,q} \frac{A_{s,q}(t)}{\sqrt{N}} \frac{\gamma_s(q)}{\sqrt{M}} e^{iq \cdot \rho_i}. \tag{A3.8}$$

Substitution of (A3.8) into (A3.2) and (A3.3) yields the Hamiltonian as a function of the $A_{s,q}$

$$H = T + V = \tfrac{1}{2} \sum_{s,q} \left(\dot{A}_{s,q} \dot{A}^*_{s,q} + \omega_s^2(q) A_{s,q} A^*_{s,q} \right) \tag{A3.9}$$

which we recognize as a sum of harmonic oscillator Hamiltonians. Proceeding as in the quantization of the simple harmonic oscillator we introduce the harmonic oscillator raising, $a^+_{s,q}$, and lowering, $a_{s,q}$, operators by setting

$$A_{s,q} = \sqrt{\frac{\hbar}{2\omega_s(q)}} \left[a_{s,q} + a^+_{s,q} \right] . \tag{A3.10}$$

Substituting this into (A3.9) gives

$$H = \sum_{s,q} \hbar \omega_s(q) \left[a^+_{s,q} a_{s,q} + \tfrac{1}{2} \right] . \tag{A3.11}$$

The eigenvalues of $\left(a^+_{s,q} a_{s,q} \right)$ are integers, $n_{s,q}$, and

$$\begin{aligned} a^+_{s,q} | n_{s,q} \rangle &= \sqrt{n_{s,q} + 1} \, | n_{s,q} + 1 \rangle \\ a_{s,q} | n_{s,q} \rangle &= \sqrt{n_{s,q}} \, | n_{s,q} - 1 \rangle \end{aligned} \tag{A3.12}$$

where the $|n_{s,q}\rangle$ are the eigenstates of $\left(a^+_{s,q} a_{s,q} \right)$. Since the energy is a sum of terms, $\hbar \omega_s(q)$, multiplied by integers we speak of the system as if it was composed of excitations each with energy $\hbar \omega_s(q)$ and we call these excitations phonons. $n_{s,q}$ is then the number of (s, q) phonons present in the system. Since a and a^+ change the numbers of phonons we refer to them as phonon annihilation and creation operators. By substituting (A3.10) into (A3.8) we see that we can write

$$u_i(t) = \sum_{s,q} \left[\xi_i(s, q) a_{s,q} + \xi_i^*(s, q) a^+_{s,q} \right] \tag{A3.13}$$

where

$$\xi_i(s, q) = \sqrt{\frac{\hbar}{2 N M \omega_s(q)}} \gamma_s(q) e^{iq \cdot \rho_i} \tag{A3.14}$$

that is the displacements $u_i(t)$ are now quantum mechanical operators.

In order to calculate the scattering cross sections we return to (A2.11) and (A2.12). In the matrix element we will use the expression for $\boldsymbol{R}_i$ given by (A3.1) and A3.13):

$$\langle n_{\mathrm{f}} \left| e^{i\boldsymbol{Q}\,\cdot\,\boldsymbol{R}_i} \right| n_{\mathrm{i}} \rangle = e^{i\boldsymbol{Q}\cdot\boldsymbol{\rho}_i}$$
$$\times \prod_{s,q} \langle n_{\mathrm{f}} \left| \exp\left\{ i\boldsymbol{Q}\cdot \left[\boldsymbol{\xi}_i\,(s,q)\, a_{s,q} + \boldsymbol{\xi}_i^*\,(s,q)\, a_{s,q}^+ \right] \right\} \right| n_{\mathrm{i}} \rangle$$

$$(A3.15)$$

and expand the operator in a series

$$\exp\left\{ i\boldsymbol{Q}\cdot \left[\boldsymbol{\xi}_i\,(s,q)\, a_{s,q} + \boldsymbol{\xi}_i^*\,(s,q)\, a_{s,q}^+ \right] \right\} = 1 + i\boldsymbol{Q}\cdot \left[\boldsymbol{\xi}_i\,(s,q)\, a + \boldsymbol{\xi}_i^*\,(s,q)\, a^+ \right]$$
$$- \tfrac{1}{2} \left[(\boldsymbol{Q}\cdot\boldsymbol{\xi}_i)^2\, a^2 + (\boldsymbol{Q}\cdot\boldsymbol{\xi}_i^*)^2\, (a^+)^2 \right.$$
$$\left. + |\boldsymbol{Q}\cdot\boldsymbol{\xi}_i|^2\, (aa^+ + a^+a) \right] + \ldots$$

$$(A3.16)$$

where we have omitted the subscripts.

We see that the second term changes the phonon number by one while the third term changes the phonon number by two or zero. Thus the second term will result in scattering processes in which a single phonon is emitted or absorbed while the third term will result in processes in which either two phonons are emitted or absorbed or the phonon number does not change. Concentrating, for the moment, on processes of the latter type—these correspond to elastic scattering—we see that

$$\beta = \langle n_{\mathrm{f}} \left| e^{i\boldsymbol{Q}\,\cdot\,\boldsymbol{R}_i} \right| n_{\mathrm{i}} \rangle$$
$$= e^{i\boldsymbol{Q}\cdot\boldsymbol{\rho}_i} \prod_{s,q} \left[1 - \tfrac{1}{2} |\boldsymbol{Q}\cdot\boldsymbol{\xi}_i|^2\, (2n_{s,q} + 1) \right] \qquad (A3.17)$$

(the eigenvalue of $(aa^+ + a^+a)$ is $(2n + 1)$ from (A3.12)).

Since the second term in the square bracket is much less than one we can write

$$\beta_i = e^{i\boldsymbol{Q}\,\cdot\,\boldsymbol{\rho}_i} e^{-W(Q)} \qquad (A3.18)$$

where

$$W(Q) = \tfrac{1}{2} \sum_{s,q} \left[|Q \cdot \xi_i|^2 \left(2n_{s,q} + 1\right) \right]$$

$$= \tfrac{1}{2} \sum_{s,q} \frac{\hbar}{2M\omega_s(q)N} |Q \cdot \gamma(s,q)|^2 \left(2n_{s,q} + 1\right). \qquad (A3.19)$$

If we introduce

$$g(\omega) = \frac{1}{3N} \sum_{s,q} \delta\left(\omega - \omega_s(q)\right) \qquad (A3.20)$$

we see that $g(\omega)$ is normalized, i.e.

$$\int d\omega \, g(\omega) = 1 \qquad (A3.21)$$

and that $g(\omega) \, d\omega$ represents the probability of finding a state with energy given by $\hbar\omega$ where ω is in the range ω to $\omega + d\omega$. Since in thermal equilibrium

$$\langle 2n + 1 \rangle = \coth\left(\frac{\hbar\omega}{2k_B T}\right) \qquad (A3.22)$$

we can write (A3.19) as:

$$W(Q) = \frac{3\hbar}{4M} \int d\omega \frac{g(\omega)}{\omega} \left\langle |Q \cdot \gamma|^2 \right\rangle \coth\left(\frac{\hbar\omega}{2k_B T}\right) \qquad (A3.23)$$

where the average of $|Q \cdot \gamma|^2$ is taken over a surface in Q with constant ω and $e^{-2W(Q)}$ is called the Debye–Waller factor and represents the reduction in scattering caused by the vibrations of the atoms around their equilibrium positions. It is important in X-ray and Mössbauer scattering as well as in neutron scattering.

A3.1.1   One-phonon coherent inelastic scattering

For inelastic scattering we concentrate on the one-phonon case so only the second term in (A3.16) will contribute. The states $|n_i\rangle$, $|n_f\rangle$ will differ in that the quantum numbers of one particular phonon mode (s, q) will differ by ± 1. The operators for all the other modes (except s, q) will be diagonal and since the removal of a single mode from the sum (A3.19) will have no perceptible effect there will be a contribution $e^{-W(Q)}$ from all these diagonal modes. Hence (A3.15) becomes, using (A3.16), (A3.12) and (A3.14):

$$\beta_i = \langle n_f \left| e^{i\boldsymbol{Q} \cdot \boldsymbol{R}_i} \right| n_i \rangle = e^{-W(Q)} \left(\frac{\hbar}{2MN\omega_s(\boldsymbol{q})} \right)^{1/2} i\boldsymbol{Q} \cdot \boldsymbol{\gamma}(s, \boldsymbol{q})$$

$$\times \begin{cases} e^{i\boldsymbol{\rho}_i \cdot (\boldsymbol{Q}+\boldsymbol{q})} \sqrt{n_{s,q}} \\[2mm] e^{i\boldsymbol{\rho}_i \cdot (\boldsymbol{Q}-\boldsymbol{q})} \sqrt{n_{s,q}+1} \end{cases} \tag{A3.24}$$

for $\begin{cases} \text{absorption} \\ \text{emission} \end{cases}$ of a phonon.

For coherent scattering we require $|\sum_i \beta_i|^2$ (see (A2.11)) which is proportional to

$$\alpha_{\pm} = \frac{1}{N} \left| \sum_i e^{i\boldsymbol{\rho}_i \cdot (\boldsymbol{Q} \pm \boldsymbol{q})} \right|^2 . \tag{A3.25}$$

Now we can see that if, in a perfect crystal, the exponent in (A3.25) is an integer multiple of 2π for every ρ_i the sum will be very large. If this condition does not hold the sum will tend to zero. It turns out (see Turchin 1965, ch 3) that we can write

$$\alpha_{\pm} = \frac{(2\pi)^3}{B} \delta^{(3)} (\boldsymbol{Q} \pm \boldsymbol{q} - \boldsymbol{\tau}) \tag{A3.26}$$

where $\boldsymbol{\tau}$ is a reciprocal lattice vector and B is the volume of a unit cell in the crystal. If $\boldsymbol{\tau}$ were absent from (A3.26) the δ-function would represent the condition of conservation of momentum. The presence of $\boldsymbol{\tau}$ means that the neutron can be diffracted by the periodic crystal lattice during the inelastic scattering process. Using (A3.24)–(A3.26) in (A2.11) we finally obtain the coherent one-phonon inelastic scattering cross section:

$$\left(\frac{d^2\sigma}{d\Omega\, d\omega} \right) = a_{\text{coh}}^2 \left(\frac{k_f}{k_i} \right) e^{-2W(Q)} \frac{(2\pi)^3 \hbar}{NB2M} \sum_{s,q} \frac{|\boldsymbol{Q} \cdot \boldsymbol{\gamma}(s, \boldsymbol{q})|^2}{\omega_s(\boldsymbol{q})}$$

$$\times \begin{cases} \dfrac{\delta^{(3)}(\boldsymbol{Q}+\boldsymbol{q}-\boldsymbol{\tau})\delta(\hbar\omega_{s,q}+\hbar\omega)}{(e^{\hbar\omega/k_BT}-1)} \\[4mm] \dfrac{\delta^{(3)}(\boldsymbol{Q}-\boldsymbol{q}-\boldsymbol{\tau})\delta(\hbar\omega_{s,q}-\hbar\omega)}{(1-e^{-\hbar\omega/k_BT})} \end{cases} \tag{A3.27}$$

for $\begin{cases} \text{absorption} \\ \text{emission} \end{cases}$ of a phonon.

The presence of the two δ-functions indicates that in one-phonon coherent inelastic scattering, energy and momentum are both conserved. As a result of this the scattering cross section will show peaks at values of ω, Q satisfying

$$\omega = \pm\omega_s\left(Q\right) = \frac{\hbar}{2m}\left(k_i^2 - k_f^2\right) \qquad (A3.28)$$

although these peaks will be broadened to a finite width because of the anharmonic forces which we have neglected. Thus coherent inelastic scattering can be used to measure the functional dependence $\omega_s\left(q\right)$ called the dispersion curve— and there is a vast amount of research effort devoted to measuring dispersion curves in various materials and relating these curves to the force constants in the material. At small Q we have

$$\omega_s = c_s Q \qquad (A3.29)$$

these are the acoustic modes—the phonons representing sound waves. The Debye model assumes that (A3.29) holds up to some maximum value of Q. In the following discusson we shall make this assumption and concentrate on a single mode (single value of s).

From (2.30) we can write

$$Q = \sqrt{k_i^2 + k_f^2 - 2k_i k_f \cos\theta} \qquad (A3.30)$$

where θ is the angle between $\boldsymbol{k}_i$ and $\boldsymbol{k}_f$. Thus Q lies in the range

$$|k_i - k_f| \leq Q \leq k_i + k_f \qquad (A3.31)$$

corresponding to θ lying between $\theta = 0$ and $\theta = \pi$. Using (A3.29) and (A3.31) we can plot ω against k_f for different values of θ and for fixed k_i. Every point within the shaded region of figure A3.1 corresponds to some value of θ. On the same graph we can plot the magnitude of ω from (A3.28). Scattering is allowed for those values of k_f where the parabolae (A3.28) lie within the shaded region. For the case shown, corresponding to the initial neutron velocity, $v_i = \hbar k_i/m > c_s$ both phonon emission ($k_f < k_i$) and phonon absorption ($k_f > k_i$) are possible. In the case when this condition does not hold only phonon absorption is possible. We will have occasion to refer to this figure in our discussions of the scattering of UCN.

A3.1.2   Incoherent one-phonon inelastic scattering

For incoherent inelastic scattering we see from (A2.12) and (A3.24) that the cross section depends on

$$\alpha' = \sum_i |\beta_i|^2 = \sum_i \frac{\hbar}{2MN\omega_s\left(q\right)} \mathrm{e}^{-2W(Q)} |Q \cdot \boldsymbol{\gamma}_s\left(q\right)|^2 \left\{ \begin{array}{c} n_s\left(q\right) \\ \\ n_s\left(q\right) + 1 \end{array} \right\}.$$

$$(A3.32)$$

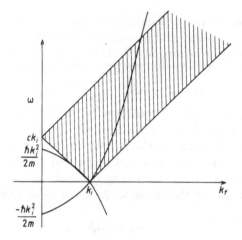

Figure A3.1   One-phonon coherent scattering. Energy transfer ω against final neutron wavenumber, k_f, for fixed incident neutron wavenumber k_i. (A3.29) and (A3.31) allow all points within the shaded region. Conservation of energy (A3.28) gives the parabola. Adapted from Turchin (1965).

We see that because of the form of (A3.32) the exponentials have disappeared from the result so that there will be no momentum conservation δ-functions (A3.26) in the incoherent scattering cross section. Thus incoherent scattering does not conserve momentum with an individual phonon in the way coherent scattering does. This is because, as we have seen in the discussion of correlation functions, incoherent scattering essentially takes place from individual atoms. During incoherent scattering the impulse transmitted from the neutron to the crystal displaces a single atom which then excites a superposition of phonon waves travelling in different directions. This does not require a discrete momentum to be transferred from the neutron as would be the case in the excitation of a single phonon.

Putting (A3.32) into (A2.12) we obtain the result for the incoherent scattering cross section:

$$\left(\frac{\mathrm{d}^2\sigma}{\mathrm{d}\Omega\,\mathrm{d}\omega}\right)_{\text{inc}} = a_{\text{inc}}^2 \frac{k_f}{k_i} \sum_{s,q} \frac{\hbar}{2MN\omega_s(q)} e^{-2W(Q)} |Q \cdot \gamma_s(q)|^2$$

$$\times \left\{ \begin{array}{l} \langle n_s(q)\rangle\, \delta\left(\hbar\omega_{s,q} + \hbar\omega\right) \\[2mm] \langle n_s(q) + 1\rangle\, \delta\left(\hbar\omega_{s,q} - \hbar\omega\right) \end{array} \right\} \qquad (A3.33)$$

where, again, the brackets $\langle\ldots\rangle$ refer to thermal averaging.

As in our discussion of the Debye–Waller factor, $W(Q)$, we can replace the summation over s, q by an integral

$$\sum_{s,q} \longrightarrow 3N \int d\omega'\, g(\omega') \tag{A3.34}$$

where $g(w)$ is given by (A3.20), and we evaluate the integral by means of the δ functions in (A3.33). Further, since

$$\frac{\langle n_s(q)+1\rangle}{\omega_{s,q}} = \frac{1}{\omega_{s,q}\left(1-e^{-\hbar\omega_{s,q}/k_BT}\right)} = \psi(\omega_{s,q})$$

$$\frac{\langle n_s(q)\rangle}{\omega_{s,q}} = \frac{1}{\omega_{s,q}\left(e^{\hbar\omega_{s,q}/k_BT}-1\right)} = \psi(-\omega_{s,q}) \tag{A3.35}$$

we can write the result in a form which applies to both phonon emission and absorption (dropping the subscripts on $\omega_{s,q}$).

$$\left(\frac{d^2\sigma}{d\Omega\,d\omega}\right)_{inc} = a_{inc}^2 \frac{k_f}{k_i} e^{-2W(Q)} |Q\cdot\gamma|_{ave}^2 \frac{3\hbar}{2M} g(\omega)\psi(\omega_{s,q}). \tag{A3.36}$$

Since all the factors in (A3.36) are known except for $g(\omega)$ one-phonon incoherent inelastic scattering can be used for studying the spectral density of normal modes, $g(\omega)$, in many materials including liquids.

As we shall see the coherent scattering of UCN is usually small because of the restrictions imposed by the momentum conservation conditions (A3.1) and incoherent inelastic scattering is usually the most important.

For applications involving integrals over the double-differential cross sections, (A3.27) or (A3.36), it is sufficient to use what is called the 'incoherent approximation'. This consists of using (A3.36) with a_{inc}^2 replaced by $a_{inc}^2 + a_{coh}^2$ as the total double-differential cross section (Turchin 1965).

Appendix A4

A4.1 BOUNDARY MATRICES

In this appendix we present another matrix method for calculating reflection from complex surfaces. This method has more complicated matrix elements but a simpler relation between the matrix elements and the reflection and transmission amplitudes than those of section 2.4.4. These matrices are based on relations between the coefficients of the two linearly independent solutions within a given layer. We assume the particles are incident from the left of the surface and number the layers with numbers increasing to the right.

Writing the wavefunction in the nth layer as (2.77) with coefficients A_n, B_n and denoting the wavevector in the region between (z_n, z_{n+1}) as k_n and that in the region (z_{n-1}, z_n) as k_{n-1}, we have as the boundary conditions at z_n

$$A_{n-1}e^{ik_{n-1}z_n} + B_{n-1}e^{-ik_{n-1}z_n} = A_n e^{ik_n z_n} + B_n e^{-ik_n z_n} \quad \text{(A4.1)}$$

$$k_{n-1}\left(A_{n-1}e^{ik_{n-1}z_n} - B_{n-1}e^{-ik_{n-1}z_n}\right) = k_n\left(A_n e^{ik_n z_n} - B_n e^{-ik_n z_n}\right)$$

$$\text{(A4.2)}$$

or solving for A_n and B_n we find

$$\begin{pmatrix} A_n \\ B_n \end{pmatrix} = \frac{1}{2}\begin{bmatrix} (1+\gamma_n)\,e^{i(k_{n-1}-k_n)z_n} & (1-\gamma_n)\,e^{-i(k_{n-1}+k_n)z_n} \\ (1-\gamma_n)\,e^{i(k_{n-1}+k_n)z_n} & (1+\gamma_n)\,e^{-i(k_{n-1}-k_n)z_n} \end{bmatrix} \cdot \begin{pmatrix} A_{n-1} \\ B_{n-1} \end{pmatrix}$$

$$\text{(A4.3)}$$

$$\equiv \bar{M}_n \begin{pmatrix} A_{n-1} \\ B_{n-1} \end{pmatrix}$$

with $\gamma_n = k_{n-1}/k_n$. Taking (2.48) for the wavefunction we have

$$\begin{pmatrix} A_0 \\ B_0 \end{pmatrix} = \begin{pmatrix} 1 \\ R \end{pmatrix} \quad \text{(A4.4)}$$

268

and taking (2.49) for the wavefunction inside the material, we have

$$\begin{pmatrix} T \\ 0 \end{pmatrix} = \bar{M} \begin{pmatrix} 1 \\ R \end{pmatrix} \tag{A4.5}$$

with

$$\bar{M} = \bar{M}_N \ldots \bar{M}_2 \cdot \bar{M}_1 = \begin{pmatrix} \bar{M}_{11} & \bar{M}_{12} \\ \bar{M}_{21} & \bar{M}_{22} \end{pmatrix}. \tag{A4.6}$$

From (A4.3) we have

$$\det \bar{M}_n = \gamma_n \tag{A4.7}$$

so that

$$\det \bar{M} = \frac{k_1}{k_2} \tag{A4.8}$$

and

$$R = \frac{-\bar{M}_{21}}{\bar{M}_{22}} \qquad T = \frac{k_1/k_2}{\bar{M}_{22}}. \tag{A4.9}$$

It is interesting to note that we can obtain the solution for particles travelling through the same barrier in the opposite direction by replacing (A4.5) by

$$\begin{pmatrix} \tilde{R} \\ 1 \end{pmatrix} = \bar{M} \cdot \begin{pmatrix} 0 \\ \tilde{T} \end{pmatrix} \tag{A4.10}$$

where $\tilde{T}$ and $\tilde{R}$ are the transmission and reflection amplitudes for this opposite direction (particles incident from the right) and $\bar{M}$ is the same as in (A4.6).

Thus we find

$$\tilde{T} = \frac{1}{\bar{M}_{22}} \qquad \tilde{R} = \frac{\bar{M}_{12}}{\bar{M}_{22}}. \tag{A4.11}$$

Thus

$$\frac{T}{\tilde{T}} = \frac{k_1}{k_2} \qquad \frac{R}{\tilde{R}} = \frac{-\bar{M}_{21}}{\bar{M}_{12}} \tag{A4.12}$$

for any barrier. Note that the matrices based on the linearly independent solutions (A4.3) each represent a single boundary while the matrices based on $\psi(x)$ and $\psi'(x)$ of section 2.4.4 each represent an entire layer.

Appendix A5

A5.1 PRECISION OF INELASTIC SCATTERING EXPERIMENTS FOR LOW ω AND Q

In this appendix we present an argument of Maier-Leibnitz (1966) which shows that for an inelastic scattering experiment at fixed ω, Q where $\hbar\omega \ll E_i$, $Q \ll k_i$ ($E_{i,f}$, $k_{i,f}$ represent the initial and final neutron energies and wavevectors) one can always gain intensity by going to slower incident neutrons.

As usual the momentum transfer

$$Q = k_f - k_i \tag{A5.1}$$

or

$$|Q|^2 = k_f^2 + k_i^2 - 2k_i k_f \cos\theta_s \tag{A5.2}$$

where

$$\theta_s = \theta_f - \theta_i = \text{ scattering angle.} \tag{A5.2a}$$

The energy transfer

$$\hbar\omega = \frac{\hbar^2}{2m} \left(k_f^2 - k_i^2 \right). \tag{A5.3}$$

From equation (A5.2) we have for the fractional uncertainty in momentum transfer

$$\frac{\Delta Q}{Q} = \left[R_f^2 \left(k_f^2 - k_i k_f \cos\theta_s \right)^2 + R_i^2 \left(k_i^2 - k_i k_f \cos\theta_s \right)^2 + \dots \right.$$
$$\left. + \left(k_i k_f \sin\theta_s \right)^2 \left(\Delta\theta_s \right)^2 \right]^{1/2} \times \frac{1}{Q^2} \tag{A5.4}$$

where $R_{f,i} = (\Delta k/k)_{f,i}$ is the velocity resolution for incoming and outgoing neutrons, and we have assumed the variations in k_i, k_f, θ_s are random and uncorrelated.

From (A5.2a)

$$\Delta\theta_s = \left[(\Delta\theta_f)^2 + (\Delta\theta_i)^2\right]^{1/2} \qquad (A5.5)$$

and from (A5.3)

$$\frac{\Delta\omega}{\omega} = \frac{\hbar}{m}\frac{\left[(k_i^2 R_i)^2 + (k_f^2 R_f)^2\right]^{1/2}}{\omega}. \qquad (A5.6)$$

We wish to consider an experiment at a given set of values of ω and Q (assumed to be small), and we ask how the experimental resolutions of input and output velocities and angles must vary with k_i, so as to keep $\Delta Q/Q$, $\Delta\omega/\omega$ (i.e. the accuracy with which the measured quantities are determined) constant:

$$R_{i,f} = \left(\frac{\Delta k}{k}\right)_{i,f} \propto \frac{1}{(k_{i,f})^2}$$

$$\Delta\theta_i \approx \Delta\theta_s \qquad \text{i.e. } \Delta\theta_i \propto \frac{1}{k_i k_f \sin\theta_s} \propto \frac{1}{k_i^2 \sin\theta_s} \qquad (A5.7)$$

assuming $\hbar\omega \ll \hbar k_{f,i}^2/2m$ i.e. $k_i \approx k_f$.

The intensity of neutrons available at a target is

$$I_i \propto k_i^4 R_i (\Delta\theta_i)^2 \qquad (A5.8)$$

assuming a Maxwell distribution and neglecting the exponential term which will be close to unity. Using (A5.7), equation (A5.8) becomes ($\theta_s \ll 1$)

$$I_i \propto \frac{1}{k_i^2 \theta_s^2} \qquad (A5.9)$$

but for a fixed Q, small θ_s and $k_i = k_f$, equation (A5.2) gives $k_i^2 \propto 1/\theta_s^2$. Therefore for a fixed Q, ω, $\Delta Q/Q$ and $\Delta\omega/\omega$ the intensity of neutrons at the target is independent of the incident neutron wavelength.

In an inelastic scattering experiment the counting rate, I, will be proportional to $I_i \, d\Omega_{det}$ i.e.

$$I \propto d\theta_{det} \, d\varphi_{det} \sin\theta_s \qquad (A5.10)$$

where φ is the azimuthal angle. Since $d\theta_{det} \approx d\theta_s$ using (A5.8)

$$I \propto d\varphi_{det}/k_i^2. \qquad (A5.11)$$

This means that with the above requirements it is more advantageous to work with longer wavelengths. This does also mean that any experiment at long wavelengths will be superior to any experiment at short wavelengths as there may be some physical restriction which prevents the long wavelength system from using the entire volume of phase space allowed by kinematic requirements. We have also assumed that at each incident wavelength the same fraction of incident neutrons will be isotropically scattered.

Appendix A6

A6.1 SCATTERING FROM ROUGH SURFACES

In this appendix we will calculate the reflection of UCN from a rough surface, following Lekner (1987), using an extension of a method originally due to Rayleigh. The calculation is an alternative to those discussed in section 2.4.5 and allows an extension to time-dependent surfaces, i.e. surface waves on liquids. Lekner (1987) gives a detailed list of references.

We limit the discussion to a one-dimensional rough surface which we will represent by

$$\xi(x) = \sum_n e^{ingx} \alpha_n \tag{A6.1}$$

where $n = 0$ is excluded from this and all following sums. This assumes the surface is periodic with a period $2\pi/g$ and we will use some results of diffraction theory. Non-periodic surfaces can be studied by assuming they are repetitions of a large section (size L) of surface. In this case the Fourier coefficients in (A6.1) are given by the Fourier tranform of the non-periodic surface evaluated at $2\pi n/L$ (Champeney 1973).

We take the incident beam as

$$\psi_i = e^{i(k_x x + k_z z)} \tag{A6.2}$$

the specularly reflected beam as

$$\psi_0 = A_0 e^{i(k_x x - k_z z)} \tag{A6.3}$$

and the diffracted beam as

$$\psi_n = A_n e^{i(k_x^{(n)} x - k_z^{(n)} z)} \tag{A6.4}$$

where

$$k_x^2 + k_z^2 = k_0^2. \tag{A6.5}$$

The initial kinetic energy of the neutron is $\hbar^2 k_0^2/2m$ and

$$\left(k_x^{(n)}\right)^2 + \left(k_z^{(n)}\right)^2 = k_0^2 \qquad k_x^{(n)} = k_x + ng \qquad (A6.6)$$

is a well-known result of diffraction theory (Sears 1989).

The z-direction is taken as positive going into the material, positive $\xi(x)$ represents a valley on the surface. For $z < \xi$ we take

$$\psi = \psi_{\mathrm{i}} + \psi_0 + \sum_n \psi_n. \qquad (A6.7)$$

For the wave transmitted into the material we take

$$\tilde{\psi} = B_0 \mathrm{e}^{(\mathrm{i}k_x x - \kappa z)} + \sum_n B_n \mathrm{e}^{(\mathrm{i}k_x^{(n)} x - \kappa^{(n)} z)} \qquad (A6.8)$$

with

$$k_x^2 + k_{\mathrm{c}}^2 - \kappa^2 = k_0^2$$
$$\left(k_x^{(n)}\right)^2 + k_{\mathrm{c}}^2 - \left(\kappa^{(n)}\right)^2 = k_0^2 \qquad (A6.9)$$

and $\hbar^2 k_{\mathrm{c}}^2/2m = V$, the effective potential for the material. For $\xi = 0$, A_n and B_n are zero so that A_n and B_n are proportional to ξ in lowest order. We will apply the usual boundary conditions at the surface specified by $(x, \xi(x))$ i.e.

$$\psi = \tilde{\psi}$$
$$(\boldsymbol{\nabla}\psi \cdot \boldsymbol{n}) = \left(\boldsymbol{\nabla}\tilde{\psi} \cdot \boldsymbol{n}\right) \qquad (A6.10)$$

i.e. ψ and its derivative with respect to the surface normal are continuous. We have

$$(\boldsymbol{\nabla}\psi \cdot \boldsymbol{n}) = \frac{\partial\psi}{\partial x}\sin\theta + \frac{\partial\psi}{\partial z}\cos\theta = \cos\theta\left(\frac{\partial\psi}{\partial z} - \frac{\partial\psi}{\partial x}\frac{\mathrm{d}\xi}{\mathrm{d}x}\right) \qquad (A6.11)$$

or

$$\left(\frac{\partial\psi}{\partial z} - \frac{\partial\psi}{\partial x}\frac{\mathrm{d}\xi}{\mathrm{d}x}\right) = \text{continuous} \qquad (A6.12)$$

taking acount of the definition of the positive z-direction. Continuity of ψ at $(x, \xi(x))$ gives

$$e^{ik_x x} \left(e^{ik_z \xi} + A_0 e^{-ik_z \xi} + \sum_n A_n e^{i(ngx - k_z^{(n)} \xi)} \right.$$

$$\left. = B_0 e^{-\kappa \xi} + \sum_n B_n e^{ingx} e^{-\kappa^{(n)} \xi} \right) \qquad \text{(A6.13)}$$

where we used (A6.6). To first order in ξ

$$(1 + A_0) + ik_z \xi (1 - A_0) + \sum_n A_n e^{ingx} = B_0 (1 - \kappa \xi) + \sum_n B_n e^{ingx} \quad \text{(A6.14)}$$

and equating terms of the same order in ξ

$$1 + A_0 = B_0 \qquad \text{(A6.15)}$$

$$ik_z \xi (1 - A_0) + \sum_n A_n e^{ingx} = -\kappa \xi B_0 + \sum_n B_n e^{ingx}. \qquad \text{(A6.16)}$$

Using (A6.1) and equating the coefficients of e^{ingx} we have

$$ik_z \alpha_n (1 - A_0) + A_n = -\kappa B_0 \alpha_n + B_n$$

or

$$B_n - A_n = \alpha_n [ik_z (1 - A_0) + \kappa B_0]. \qquad \text{(A6.17)}$$

To apply condition (A6.12) we calculate

$$\frac{\partial}{\partial z} \left(\psi - \tilde{\psi} \right) = e^{ik_x x} \left(ik_z \left(e^{ik_z \xi} - A_0 e^{-ik_z \xi} \right) - \sum_n A_n ik_z^{(n)} e^{ingx} e^{-ik_z^{(n)} \xi} \right.$$

$$\left. + \kappa B_0 e^{-\kappa \xi} + \sum_n B_n \kappa^{(n)} e^{ingx} e^{-\kappa^{(n)} \xi} \right) \qquad \text{(A6.18)}$$

and

$$\frac{d\xi}{dx} \frac{\partial}{\partial x} \left(\psi - \tilde{\psi} \right) = e^{ik_x x} \left[ik_x \left(e^{ik_z \xi} + A_0 e^{-ik_z \xi} + \sum_n A_n e^{ingx - ik_z^{(n)} \xi} \right. \right.$$

$$\left. - B_0 e^{-\kappa \xi} - \sum_n B_n e^{ingx - \kappa^{(n)} \xi} \right)$$

$$\left. + \sum_n (ing) e^{ingx} \left(A_n e^{-ik_z^{(n)} \xi} - B_n e^{-\kappa^{(n)} \xi} \right) \right] \sum_n \alpha_n (ing) e^{ingx}.$$

$$\text{(A6.19)}$$

To zero order in ξ (A6.12) yields using (A6.18) and (A6.19)

$$ik_z(1 - A_0) + \kappa B_0 = 0 \qquad (A6.20)$$

while to first order in ξ we have

$$ik_z\left[ik_z\xi(1 + A_0)\right] + \sum_n A_n\left(-ik_z^{(n)}\right)e^{ingx}$$

$$-\kappa^2 B_0\xi + \sum_n B_n\kappa^{(n)}e^{ingx} = \left(\sum_n \alpha_n(ing)e^{ingx}\right)ik_x\eta$$

$$(A6.21)$$

where $\eta \equiv 1 + A_0 - B_0 = 0$ according to (A6.15). Substituting (A6.20) into (A6.17) yields $(A_n = B_n)$ and equating coefficients of e^{ingx} in (A6.21) we find

$$A_n = \frac{\alpha_n\left(\kappa^2 + k_z^2\right)B_0}{\left(-ik_z^{(n)} + \kappa^{(n)}\right)}. \qquad (A6.22)$$

From (A6.15) and (A6.20) we have

$$B_0 = \frac{2ik_z}{ik_z - \kappa} \qquad A_0 = \frac{ik_z + \kappa}{ik_z - \kappa} \qquad (A6.23)$$

or just the reflection and transmission coefficients for a flat surface, equations (2.52) and (2.53). From (A6.22) we have for the non-specularly reflected intensity

$$|A_n|^2 = 4k_z^2|\alpha_n|^2 \qquad (A6.24)$$

using (A6.5), (A6.6) and (A6.9). Note that $|\alpha_n|^2$, the square of the Fourier transform of the surface roughness function, is equal to the Fourier transform of the surface roughness correlation function (Champeney 1973).

This approach depends on the validity of the 'Rayleigh hypothesis', i.e. that (A6.7) is general enough to satisfy the boundary conditions on an arbitrary surface. While this is only strictly true in some cases, it can always be applied in a restricted least-squares sense. Lekner (1987) gives a more detailed discussion with references.

REFERENCES

Abraham B M, Eckstein Y, Ketterson J B, Kuchnir M and Roach P R 1970 *Phys. Rev.* A **1** 250

Abramowitz M and Stegun L A 1964 *Handbook of Mathematical Functions* (Washington DC: National Bureau of Standards)

Ageron P, de Beaucourt P, Harig H D, Lacaze A and Livolant M 1969 *Cryogenics* **9** 42–50

Ageron P, Mampe W, Bates J C and Pendlebury J M 1986 *Nucl. Instrum. Methods* A **249** 261

Ageron P, Mampe W, Golub R and Pendlebury J M 1978 *Phys. Lett.* **66A** 469

Ageron P, Mampe W and Kilvington A I 1985 *Z. Phys.* B **59** 261

Alpert Y, Cser L, Farago B, Franck F, Mezei F and Ostanevich Y M 1985 *Biopolymers* **24** 1769–84

Altarev I S *et al* 1980 *Nucl. Phys.* A **341** 269.

Altarev I S *et al* 1986a *Sov. Phys.-JETP Lett* **44** 344–8

—— 1986b *Sov. Phys.-JETP Lett* **44** 460–5

Amaldi E 1959 *Handbuch der Physik* XXXVIII/2

Aminoff C G, Larat C, Leduc M and Laloë F 1989 *Rev. Phys. Appl.* **24** 827

Anton F, Paul W, Mampe W, Paul L and Paul S 1989 *Nucl. Instrum. Methods* A **284** 101

Antonov A V, Isakov A I, Kazarnovskii M V and Solidov U E 1969a *Sov. Phys.-JETP Lett* **10** 241 (2)

Antonov A V, Vul D E, Kazarnovskii M V 1969b *Sov. Phys.-JETP Lett* **9** 180–3

Arif M, Deslattes R D, Dewey M S, Greene G L and Schroder I G 1989 *Nucl. Instrum. Methods* A **284** 216 (1)

Arzumanov S S, Masalovich S V, Strepetov A N and Frank A I 1984 *Sov. Phys.-JETP Lett.* **39** 590

—— 1986 *Sov. Phys.-JETP Lett.* **44** 271

Bates J C 1978 *Nucl. Instrum. Methods* **150** 261–72

—— 1983 *Nucl. Instrum. Methods* **216** 535

Bates J C and Roy S 1974 *Nucl. Instrum. Methods* **120** 369–70

Baumann J, Gahler R, Ioffe A I, Kalus J and Mampe W 1989 *Nucl. Instrum. Methods* A **284** 130

Baumann J, Gahler R, Kalus J and Mampe W 1988 *Phys. Rev.* D **37** 3107

Beckmann P and Spizzichino A 1963 *The Scattering of Electromagnetic Waves from Rough Surfaces* (Oxford: Pergamon)

Beckurts K H and Wirtz K 1964 *Neutron Physics* (Berlin: Springer)

Bée M 1988 *Quasi-elastic Neutron Scattering* (Bristol: Adam Hilger)

Berceanu I and Ignatovich V K 1973 *Vacuum* **23** 441–5

Berman A S 1965 *J. Appl. Phys.* **36** 3356

Berry M V 1984 *Proc. R. Soc.* A **392** 45

Bethe H A 1937 *Rev. Mod. Phys.* **9** 69

Bevington P R 1969 *Data Reduction and Error Analysis in the Physical Sciences* (New York: McGraw-Hill)

Bialynicki-Birula I I and Mycielski J 1976 *Ann. Phys.* **100** 62

Binder K 1971a *Z. Naturforsch.* **26a** 432–41

Binder K 1971b *Z. Ang. Phys.* **3** 178–83

278

Bitter T and Dubbers D 1987 *Phys. Rev. Lett.* **59** 251
Blokhintsev D I and Plakida N M 1977 *Phys. Status Solidi* b**82** 627-32
Bohm D 1951 *Quantum Theory* (New York: Prentice Hall)
Bonse U and Rauch H (eds) 1979 *Neutron Interferometry* (New York: Oxford University Press)
Borisov Yu V *et al* 1988 *Sov. Phys. Tech. Phys.* **33** 574
Born M and Wolf E 1959 *Principles of Optics* (London: Pergamon)
—— 1970 *Principles of Optics* 4th edn (Oxford: Peragmon)
Brown M, Golub R and Pendlebury J M 1975 *Vacuum* **25** 61-4 (2)
Brun T O, Carpenter J M, Krohn V E, Ringo G R, Cronin J W, Dombeck T W, Lynn J W and Werner S W 1980 *Phys. Lett.* A **75** 223-4
Bugeat J P and Mampe W 1979 *Z. Phys.* B **35** 273
Buras B and Giebultowicz T 1972 *Acta Crystallogr.* A **28** 151-3
Burnett S M 1982 *PhD Thesis* University of Sussex (unpublished)
Butterworth I, Egelstaff P A, London H and Webb F J 1957 *Philos. Mag.* **2** 917-27
Byrne J *et al* 1990 *Phys. Rev. Lett.* **65** 289
Carpenter J M 1977 *Nucl. Instrum. Methods* **145** 91
Casella R 1969 *Phys. Rev. Lett.* **22** 554
Chadwick J 1932 *Proc. R. Soc. London* A **136** 692
Champeney D C 1973 *Fourier Transforms and their Physical Applications* (London: Academic)
Christensen J H *et al* 1946 *Phys. Rev. Lett.* **13** 138
Cimmino A, Opat G I, Klein A G, Kaiser H, Werner S A, Arif M and Clothier R 1989 *Phys. Rev. Lett.* **63** 380
Clausing P 1932 *Ann. Phys. Lpz.* **14** 134
Cohen-Tannoudji C, Diu B and Laloe F *Quantum Mechanics* (New York: Wiley) 443
Commins E D and Bucksbaum P H 1980 *Ann. Rev. Nucl. Part. Sci.* **30** 1
Cowley R A and Woods A D B 1971 *Can. J. Phys.* **49** 177-99
Crampin N 1989 *PhD Thesis* University of Sussex (unpublished)
Davies H 1954 *Proc. IEE Monograph No.* **90** 209-14
Davis D H 1960 *J. Appl. Phys.* **31** 1169
De Groot M H 1975 *Probability and Statistics* (Ontario: Addison Wesley)
De Marcus W C 1961 *Rarified Gas Dynamics* ed L Talbot (New York: Academic) 161-7
Denegri D, Sadoulet B and Spiro M 1990 *Rev. Mod. Phys.* **62** 1
Desplanques B, Gönnenwein G and Mampe W 1984 *Workshop on Reactor Based Fundamental Physics J. Physique Supp. C.3 Colloque*
Dilg W and Mannhart W 1973 *Z. Phys.* **266** 157-60
Dombeck T W, Lynn J W, Werner S A, Brun T, Carpenter J, Krohn V and Ringo R 1979 *Nucl. Instrum. Methods* **165** 139-55
Doroshkevich A G 1962 *Sov. Phys.-JETP* **16** 56-7 (1)
Dubbers D 1989 *Nucl. Instrum. Methods* A **284** 22
Dubbers D, El-Muzeini P, Kessler M and Last J C 1989a *Nucl. Instrum. Methods* A **275** 294
Dubbers D, Mampe W, Schreckenbach K 1989b *Nucl. Instrum. Methods* A **284** 1
Dushman S 1922 *Production and Measurement of High Vacuum* (New York: General Electric Company)

Ebeling H 1990 *Institut Laue-Langevin Internal Report* 90EB07T
Eckert J 1983 *Physica* B **120** 25–30
Egelstaff P A (ed) 1965 *Thermal Neutron Scattering* (London: Academic)
—— 1967 *An Introduction to the Liquid State* (London: Academic)
Egelstaff P A and Poole M J 1969 *Experimental Neutron Thermalization* (Oxford: Pergamon)
Egorov A I, Lobashev V M, Nazarenko V A, Porsev G D and Serebov A P 1974 *Sov. J. Nucl. Phys.* **19** 147
Ellis J 1989 *Nucl. Instrum. Methods* A **284** 33
Erozolimskii B G 1975 *Sov. Phys.-Usp* **18** 377–86
—— 1989 *Nucl. Instrum. Methods* A **284** 89–93
Fermi E 1936 *Ricerca Scientifica* **7** 13–52
—— 1950 *Nuclear Physics* compiled by J Orear, A H Rosenfeld and R A Schluter, University of Chicago
Fermi E and Marshall L 1947 *Phys. Rev.* **71** 666–77
Fermi E and Zinn W N 1946 *Phys. Rev.* **70** 103
Fidecaro G *et al* 1985 *Phys. Lett.* B **156** 122
Foldy L L 1945 *Phys. Rev.* **67** 107–119
—— 1966 *Preludes in Theoretical Physics* ed A de Shalit, H Feshbach and L Van Hove (Amsterdam: North-Holland)
Fortson E N and Lewis L L 1984 *Phys. Rep.* **113** 289
Frank A I 1979a I V Kurchatov Inst. of Atomic Energy, preprint 3203
—— 1983 *Sov. Phys. Tech. Phys.* **28** 600
—— 1987 *Sov. Phys. Usp.* **30** 110
Frank I M 1972 *Priroda* **9** 29
—— 1975 *JINR, Dubna, Commun.* P **3** 8851
—— 1976 *JINR, Dubna, Commun.* P **3** 9846
—— 1979 *Comm. Joint Inst. for Nucl. Res. Dubna*, P **3** 12829 (in Russian)
Frank I M and Pacher P 1983 *Physica* B **120** 37–44
Friedburg H and Paul W 1951 *Naturwissenschaften* **38** 159 and *Z. Phys.* **130** 493
Gahler R, Klein A G and Zeilinger A 1981 *Phys. Rev.* A **23** 1611
Gerasimov A S, Ignatovich V K, Kazarnovsky M V 1973 *JINR, Dubna, Commun.* P **4** 6940
Gmal B 1981 *Dissertation, Techn.* Universität Munchen (unpublished)
Goldberger M L and Seitz F 1947 *Phys. Rev.* **71** 294–310
Goldstein H 1950 *Classical Mechanics* (New York: Addison-Wesley)
Golikov V V, Kozlov Z A, Kulkin L K, Pikelner L B, Rudenko V T and Sharapov E I 1971 *SINR. Dubna Commun.* P3-5736
Golikov V V, Luschikov V I and Shapiro F L 1973 *Sov. Phys.-JETP* **37** 41–4
Golikov V V and Taran Y U 1975 *Instrum. Expt. Tech.* **18** 36–9
Golub R 1972 *Phys. Lett.* A**38** 177–8
—— 1978 *Inst. Phys. Conf. Ser.* vol 42 104
—— 1979 *Phys. Lett.* **72A** 387
—— 1983 *J. Physique* **44** L321
—— 1984 *Nucl. Instrum. Methods* **226** 558–9
—— 1986 *Sov. Phys. Tech. Phys.* **31** 945
—— 1987 *Proc. 18th Int. Conf. on Low Temperature Physics, Pt. 3: Invited Papers* (Kyoto, Japan 1987) 2073

280

Golub R and Böning K 1981 *Report Jül, Spez.* 113/KfK 3175 Part III-C, Nuclear Research Institute, Jülich
—— 1983 *Z. Phys.* B **51** 187
Golub R, Felber S, Gähler R and Gutsmiedl E 1990 *Phys. Lett.* A **148**, 27
Golub R, Jewell C, Ageron P, Mampe W, Heckel B and Kilvington A I 1983 *Z. Phys.* B **51** 187
Golub R, Mampe W, Pendlebury J M and Ageron P 1979 *Scientific American* **240** 106 June, 1979
Golub R and Pendlebury J M 1972 *Contemp. Phys.* **13** 519–58
—— 1974 *Phys. Lett.* **50**A, 177
—— 1975 *Phys. Lett.* **53**A 133–5
—— 1977 *Phys. Lett.* **62**A 337–9
—— 1979 *Rep. Prog. Phys.* **42** 439–501
Golub R and Yoshiki H 1989 *Nuc. Phys.* A **501** 869–76
Golub R, Yoshiki H and Gahler R 1989 *Nucl. Instrum. Methods* A **284** 16
Greene G L (ed) 1986 *The Investigation of Fundamental Interactions with Cold Neutrons* NBS Special Publication 711 (Washington DC: NBS)
Greene G L, Ramsey N F, Mampe W, Pendlebury J M, Smith K F, Dress W D, Miller P D and Perrin P 1979 *Phys. Rev.* D **20** 2139
Groshev L V, Dvoretsky V N, Demidov A M, Panin Yu N, Luschikov V I, Pokotilovsky Yu N, Strelkov A V and Shapiro F L 1971 *Phys. Lett.* **34B** 293–5 (1)
Groshev L V *et al* 1973 *JINR, Dubna, Commun.* P3 7282
—— 1975 *JINR, Dubna, Commun.* P3
—— 1976 *JINR, Dubna, Commun.* P3 9534
Gudkov V P *et al* 1991 *Phys. Lett.* B to be published
Gurevich I I and Nemirovskii P E 1962 *Sov. Phys. JETP* **14** 838
Haddock R P, Salter, R M, Zeller M, Czirr J B and Nygren D R 1965 *Phys. Rev. Lett.* **14** 318–23
Hanson A O, Taschek R F and Williams J H 1949 *Rev. Mod. Phys.* **21** 635
Hari Dass N D 1976 *Phys. Rev. Lett.* **36** 393
Hauge E H and Støvneng J A 1989 *Rev. Mod. Phys.* **61** 917
He X, McKellar B and Pakvassa S 1990 *Int. J. Mod. Phys.* A **4** 5011
Heckel B, Ramsey N F, Green K, Greene G L, Gähler R, Schärpf O, Forte M, Dress W, Miller P, Golub R, Byrne J and Pendlebury J M *Phys. Lett.* **119B** 298–302
Heckel B R 1989 *Nucl. Instrum. Methods* A **284** 66
Hermann P, Steinhauser K A, Gahler R, Steyerl A and Mampe W 1985 *Phys. Rev. Lett.* **54** 1969
Ignatovich V K 1973 *JINR, Dubna, Commun.* P4 7055
—— 1974 *JINR, Dubna, Commun.* P4 7831
—— 1975a *Phys. Status Solidi* B **71** 477
—— 1975b *JINR, Dubna, Commun.* P4 9007
—— 1986 *Physics of Ultra-cold Neutrons* (Moscow: Nauka) (in Russian)
—— 1987 *JINR, Dubna, Commun.* P4 87402
—— 1990 *The Physics of Ultra-cold Neutrons* (Oxford: Clarendon) (English translation of the 1986 work)
Ignatovich V K and Luschikov V I 1975 *JINR, Dubna, Commun.* P3 8795

—— 1981 *JINP-P-4-81-77, Dubna, USSR*

Ignatovich V K and Satarov L M 1977 *Kuratchov Institute of Atomic Energy Moscow Report* IAE-2820

Ignatovich V K and Terekhov G I 1976 *JINR, Dubna, Commun.* P4 9567

—— 1977 *JINR, Dubna, Commun.* P4 10548

ILL 1988 *Guide to Neutron Facilities (Yellow Book)* Grenoble France: (Institut Laue–Langevin)

Ishikawa Y 1983 *Physica* B **120** 3–14

Jacobsson R 1965 *Progress in Optics* vol 5 ed E Wolf (Amsterdam: North Holland)

Jewell I J 1983 The interaction of UCN with superfluid ^{4}He, and a new possible super-thermal UCN source *PhD Thesis*, University of Lancaster

Kashoukeev N T, Stanev G A, Ianeva N B and Mirtchova D S 1975 *Nucl. Instrum. Methods* **126** 43–48

Kazarnovskii M V, Kuz'min V A and Shaposhnikov M E 1981 *Sov. Phys.-JETP Lett.* **34** 47

Keller T, Zimmermann P, Golub R and Gähler R 1990 *Physica* B **162** 327

Kennard E H 1938 *Kinetic Theory of Gases* (New York: McGraw-Hill)

Kharitonov A G, Nesvizhevsky V V, Serebrov A P, Taldaev R R, Varlamov V E, Vasilyev A V, Alfimento V P, Luschikov V I, Shvetsov V N and Strelkov A V 1989 *Nucl. Instrum. Methods* A **284** 98

Khriplovich I B and Pospelov M E 1989 *University of Minnesota Theoretical Physics Institute Report* TPI-MINN-89/34-T

Kilvington A I, Golub R, Mampe W and Ageron P 1987 *Phys. Lett.* **125A** 416

Knapp E W, Fischer S F and Parak F 1983 *J. Chem. Phys.* **78** 4701

Knudsen M 1909 *Ann. Phys. Lpz.* **28** 75

Kossel D 1948 *Optik* **3** 266–73

Kosvintsev Yu Yu, Kulagin E N, Kushnir Yu A, Morozov V I and Strelkov A V 1977 *Nucl. Instrum. Methods* **143** 133–7

Kosvinstev Yu Yu, Kushnir Y A and Morozov V I 1979 *Sov. Phys.-JETP Lett.* **50** 642

Kosvintsev Yu Yu, Kushnir Y A, Morosov V I, Stoika A D and Strelkov A V 1978 *Sov. Phys.-JETP Lett.* **28** 153

Kosvintsev Yu Yu, Morozov V I and Terekhov G I 1982 *Sov. Phys.-JETP Lett.* **9** 424

—— 1986 *Sov. Phys.-JETP Lett.* **44** 571

Kosvintsev Yu Yu, Terekhov G I, Morosov V I, Panin Yu. N, Rogov E V and Fomin A I 1987 *Proc. First Int. Conf. on Neutron Phys.* vol. 1 (*Kiev, 14–18 September 1987*) p 231

Kügler K J, Paul W and Trinks U 1978 *Phys. Lett.* **72B** 422

Kügler K J, Moritz K, Paul W and Trinks U 1985 *Nucl. Instrum. Methods* **228** 240

La Marche P H, Lanford W A and Golub R 1981 *Nucl. Instrum. Methods* **189** 535

Lamoreaux S K 1986 private communication to the ILL EDM group

Lamoreaux S K 1989 *Nucl. Instrum. Methods* A **284** 43

—— 1988 *Institute Laue–Langevin Internal Report* 88LA01T

—— 1989 *Nucl. Instrum. Methods* A **284** 43

Lamoreaux S K and Golub R 1989 private communication, to be published

—— 1989 *Nucl. Instrum. Methods* A **284** 43
Lamoreaux S K and Golub R 1989 private communication, to be published
Lamoreaux S K, Jacobs J P, Heckel B R, Raab F J and Fortson E N 1986 *Phys. Rev. Lett.* **57** 3125
—— 1987 *Phys. Rev. Lett.* **59** 2275
—— 1988 *Phys. Rev.* A **39** 1082
Landau L 1957 *Nucl. Phys.* **3** 127
Landau L D and Khalatnikov I M 1949 *Collected Papers of L D Landau* ed D ter Haar (Oxford: Pergamon 1965) pp 494–510
Landau L D and Lifschitz E M 1958 *Quantum Mechanics* (Oxford: Pergamon)
—— 1960 *Electrodynamics of Continuous Media* (Oxford: Pergamon)
Lander G H 1983 *Physica* B **120** 15–24
Lanford W and Golub R 1977 *Phys. Rev. Lett.* **39** 1509
Lax M 1951 *Rev. Mod. Phys.* **23** 287–310
—— 1952 Phys. Rev. **85** 621–9
Lee T D and Yang C N 1956 *Phys. Rev.* **104** 254
Leitner J and Okubo S 1964 *Phys. Rev.* B **136** 1542
Lekner J 1987 *Theory of Reflection* (Dordrecht: Martinus Nijhoff)
Lengsfeld W and Steyerl A 1977 *Z. Phys.* B **27** 117–20
Lermer R and Steyerl A 1976 *Phys. Status Solidi* A **33** 531–41
Lobashev V M, Porsev G D and Serebrov A P 1973 *Kostantinova* (Leningrad: Institute of Nuclear Physics) report no. 37
Lovesey S W 1984 *Theory of Neutron Scattering from Condensed Matter* vols 1 and 2 (Oxford: Oxford University Press)
Luschikov V I 1977 *Phys. Today* **30** (June) 42–51
Luschikov V I, Pokotilovsky Y N, Strelkov A V and Shapiro F L 1968 *Preprint P3-4127*, Joint Institute for Nuclear Research Dubna USSR
—— 1969 *Sov. Phys. JETP Lett.* **9** 23–6
—— 1976 *Proc. Int. Conf. on the Interaction of Neutrons with Nucleii, Lowell, Mass* ed E Sheldon (Userda: Technical Information Centre) pp 117–42
—— 1977 *Proc. Int. Symp. on Neutron Inelastic Scattering* (Vienna: IAEA) pp 39–52
Lynn J W, Miller W A, Dombeck T W, Ringo G R, Krohn V E and Freedman M S 1983 *Physica* B **120** 114–7
Maier-Leibnitz 1966 *Nukleonik* **8** 5
Maier-Leibnitz H and Springer T 1963 *J. Nucl. Energy* A/B **17** 217–25 (1)
Mampe W 1989 private communication
Mampe W, Ageron P, Bates J C, Pendlebury J M and Steyerl A 1989a *Nucl. Instrum. Methods* A **284** 111–5
—— 1989b *Phys. Rev. Lett.* **63** 593–6
Mampe W, Ageron P and Gahler R 1981 *Z. Phys.* B **45** 1
Maris H J 1977 *Rev. Mod. Phys.* **49** 341–59
Marshak R E 1947 *Phys. Rev.* **72** 47–50
Marshak R E *et al* 1969 *Theory of Weak Interactions in Particle Physics* (New York: Wiley)
Marshall W C and Lovesey S W 1971 *Theory of Thermal Neutron Scattering* (Oxford: Clarendon Press)
May R P 1982 *Proc. Conf. on the Neutron and its Applications* Cambridge

Maysenholder W 1976 *Nucl. Instrum. Methods* **137** 291

McCammon J A and Harvey S C 1987 *Dynamics of Proteins and Nuclear Acids* (Cambridge University Press)

McClintock P V E 1978 *Cryogenics* **18** 201

Messiah A 1966 *Quantum Mechanics* (New York: Wiley)

Mezei F 1972 *Z. Phys.* **255** 146

—— 1976 *Commun. Phys.* **1** 81

—— (ed) 1980 *Neutron Spin Echo* (Berlin: Springer)

—— 1990 private communication

Mezei F and Dagleish P A 1977 *Commun. Phys.* **2** 41

Mezei F and Stirling W G 1983 *25th Jubilee Conf. on Helium-4* ed J G M Armitrage (Singapore: World Scientific)

Miranda P C 1987 *PhD Thesis* University of Sussex (unpublished)

—— 1988 *J. Phys. D: Appl. Phys.* **21** 1326

Mohapatra R N 1989 *Nucl. Instrum. Methods* A **284** 1

Mohapatra R N and Marshak R E 1980 *Phys. Rev. Lett.* **44** 1316

Morkel Ch, Gronemeyer Ch and Gläser W 1987 *Phys. Rev. Lett.* **58** 1873

Morozov V I 1989 *Nucl. Instrum. Methods* A **284** 108

Morozov V I *et al* 1987 *Sov. Phys.-JETP Lett.* **46** 377

Morse P M and Feshbach H 1953 *Methods of Theoretical Physics* (New York: McGraw-Hill)

Namiot V A 1973 *Sov. Phys. Dokl.* **18** 481-2 (1974)

Nierhaus K H, Lietzke R, May R P, Volker N, Schulze H, Simpson K, Wurmbach P and Stuhrmann H B 1982 *Proc. Nat. Acad. Sci. (USA)* **80** 2889-93

Okun L B 1969 *Comm. Nucl. Part. Phys.* **3** 133-5

Okun L B and Rubbia C 1967 *Int. Conf. on Elementary Particles* (Heidelberg, 1967) p 338

Parak F and Knapp E W 1984 *Proc. Nat. Acad. Sci. USA* **81** 7088

Pauli W 1955 *Neils Bohr and the Development of Physics* (New York: Pergamon) ch 4

Pendlebury J M 1978 private communication

—— 1982 private communications

—— 1984 private communication

—— 1988a *Proc. Ninth Symposium on Grand Unification* (France, Aix-les-Bains)

—— 1988b private communication

—— 1990 private communication

Pendlebury J M and Richardson D J 1990 (to be published)

Pendlebury J M *et al* 1984 *Phys. Lett.* B **136** 327

Peshkin M and Ringo G R 1971 *Am. J. Phys.* **39** 324-7

Pfeiffer W, Schlossbauer G Koll W, Farago B, Steyerl A and Sackmann E 1988 *J. Physique* **49** 1077-82

Puglierlin G 1989 *Nucl. Instrum. Methods* A **284** 9

Purcell E M and Ramsey N F 1950 *Phys. Rev.* **78** 807

Puricia I I 1979 *Rev. Roum. Phys.* **24** 243

—— 1988 *Rev. Roum. Phys.* **33** 847

Rabi I I 1936 *Phys. Rev.* **49** 324-8

Ramsey N F 1956 *Molecular Beams* (London: Oxford University Press)

—— 1958 *Phys. Rev.* **109** 225

—— 1982 *Ann. Rev. Nucl. Part. Sci.* **32** 211

284

—— 1984 *Acta Physica Hungarica* **55** 117

Reif F 1965 *Fundamentals of Statistical and Thermal Physics* (New York: McGraw-Hill)

Rice S O 1951 *Commun. Pure Appl. Math.* **4** 351–78

Richardson D J 1985 Final Year Undergraduate Project, University of Sussex (unpublished)

—— 1989 *PhD Thesis, University of Sussex* (to be published)

Richardson D J and Golub R 1989 *Institut Laue–Langevin Proposal* 3-12-28

Richardson D J and Lamoreaux S K 1987 *Phys. Rev. Lett.* **61** 2050

—— 1989 *Nucl. Instrum. Methods* A **284** 192

Richardson D J, Pendlebury J M, Yaidzhiev P and Mampe W 1990 (accepted for *Nucl. Instrum. Methods* A)

Robson J M 1976 *Can. J. Phys.* **54** 1277

Robson J M and Winfield D 1972 *Phys. Lett.* **40B** 537–8

Rose M E 1942 *Phys. Rev.* **63** 111–120

Sackmann E 1985 *Das Konzept der Neuen Neutronenquelle der TUM* ed K Böning, W Gläser, E Steichele and A Steyerl, Garching

Sandars P G H and Lipworth E 1964 *Phys. Rev. Lett.* **13** 718

Scheckenhofer H and Steyerl A 1977 *Phys. Rev. Lett.* **39** 1310

—— 1981 *Nucl. Instrum. Methods* **179** 393

Schiff L I 1963 *Phys. Rev.* **132** 2194

—— 1968 *Quantum Mechanics* 3rd ed (New York: McGraw-Hill)

Schmelev A B *Sov. Phys. Usp.* 1972 **15** 173–83

Schramm D A 1989 *Nucl. Instrum. Methods* A **284** 84

Schubert K R *et al* 1970 *Phys. Lett.* **313** 662

Schutz G, Steyerl A and Mampe W 1980 *Phys. Rev. Lett.* **44** 1400

Sears V F 1989 *Neutron Optics* (Oxford: Oxford University Press)

Serebrov A P 1989 *Nucl. Instrum. Methods* A **284** 212

Shabilin E P 1982 *Sov. J. Nucl. Phys.* **36** 575

—— 1983 *Sov. Phys. Usp.* **26** 297

Shapere A and Wilczek F 1989 *Geometric Phases in Physics* (Singapore: World Scientific)

Shapiro F L 1968 *Sov. Phys. Usp.* **11** 345

—— 1972a *Proc. Int. Conf. on Nuclear Structure Study with Neutrons (Budapest)* ed J Ero and J Szucs (New York: Plenum) pp 259–84

—— 1972b *JINR, Dubna* P3-5544

Shapiro I S and Estulin I V 1956 *Sov. Phys.-JETP* **3** 345

Siebert H W 1989 *Nucl. Instrum. Methods* A **284** 94

Skachkova O S and Frank A I 1981 *JETP Lett.* **33** 203

Smith J H, Purcell E M and Ramsay N F 1957 *Phys. Rev.* **108** 120

Smith K F *et al* 1990 *Phys. Lett.* **234B** 191

Steckelmacher W 1966 *Vacuum* **16** 561

Steenstrup J and Buras B 1978 *Nucl. Instrum. Methods* **154** 549–55

Steigman G 1989 *Ann. NY Acad. Sci.* **578** 138

Steinhauser K A, Steyerl A, Scheckenhofer H and Malik S S 1980 *Phys. Rev. Lett.* **44** 1306

Steyerl A 1969 *Phys. Lett.* **29B** 33–5

—— 1972a *Z. Phys.* **252** 371–82

—— 1972b *Z. Phys.* **254** 169–88

—— 1972c *Nucl. Instrum. Methods* **101** 295–314

—— 1974 *Proc. 2nd Int. School on Neutron Physics, (Alushta), JINR, Dubna* vol 42 p 90

—— 1975 *Nucl. Instrum. Methods* **125** 461–9

—— 1977 *Springer Tracts in Modern Physics* vol 80 57–201

—— 1978 *Z. Phys.* B **30** 235

—— 1984 *J. Physique* **45** Coll. C3, Suppl. No. 3, C3 255

—— 1989a *Int. Conf. on Neutron Scattering* Grenoble: 1989

—— 1989b private communication

Steyerl A, Drexel W, Ebisawa T, Gutsmiedl E, Steinhauser K-A, Gahler R, Mampe W and Ageron P 1988b *Rev. Phys. Appl.* **23** 171

Steyerl A, Drexel W, Malik S S and Gutsmeidl E 1988a *Physica* B **151** 36–43

Steyerl A, Ebisawa T, Steinhauser K A and Utsuro Z 1981 *Z. Phys.* B **41** 283

Steyerl A, Gmal B, Steinhauser K-A, Achiwa N and Richter D 1983 *Z. Phys.* **50** 281

Steyerl A and Malik S S 1989 *Nucl. Instrum. Methods* A **284** 200–207

Steyerl A, Malik S S, Steinhauser K A and Berger L 1979 *Z. Phys.* B **36** 109

Steyerl A, Nagel H, Schreiber F K, Steinhauser K A, Gähler R, Gläser W, Ageron P, Astruc J M, Drexel W, Gervais R and Mampe W 1986 *Phys. Lett.* **116A** 347–52

Steyerl A and Schutz G 1978 *Appl. Phys.* **17** 45

Steyerl A, Steinhauser K A, Malik S S and Achiwa N 1985 *J. Phys. D: Appl. Phys.* **18** 9

Steyerl A and Trustedt W D 1974 *Z. Phys.* **267** 379

Steyerl A and Vonach H 1972 *Z. Phys.* **250** 166

Stodolsky L 1982 *Nucl. Phys.* B **197** 213

Stoika A D, Strelkov A V and Hetzelt M 1978 *Z. Phys.* B **29** 349; (see also, Strclkov A V and Hetzelt M 1978 *Sov. Phys. JETP* **47** No. 1

Stuhrmann H B 1982 *Proc. Conf. on the Neutron and its Applications* Cambridge

Sumner T J 1977 *PhD Thesis*, University of Sussex (unpublished)

Sumner T J, Pendlebury J M and Smith K f 1987 *J. Phys. D: Appl. Phys.* **20** 1095

Svensson E C 1988 *Elementary Excitations in Quantum Fluids—Hiroshima Symposium August 1987* ed K Ohbayashi and M Watabe (Berlin: Springer) pp 59–95

Taylor A 1975 *PhD Thesis*, University of Sussex (unpublished)

Tolman R C 1938 *The Principles of Statistical Mechanics* (Oxford: Oxford University Press)

Trump J G and van De Graaff R J 1947 *J. Appl. Phys.* **18** 327

Turchin V F 1965 *Slow Neutrons* (Israel: Program for Scientific Translations)

Utsuro M 1983 *Nucl. Instrum. Methods* **213** 557

Utsuro M, Ebisawa T, Okumura K, Shirahoma S and Kawabata Y 1988 *Nucl. Instrum. Methods* A **270** 450–61

Van Hove L 1954 *Phys. Rev.* **95** 249–62

—— 1958 *Physica* **24** 404–8

Vineyard G H 1958 *Phys. Rev.* **110** 999

Vladimirskii V V 1961 *Sov. Phys. - JETP* **12** 740–6

Von Egidy T 1978 ed *Fundamental Physics with Reactor Neutrons and Neutrinos (Inst. Phys. Conf. Series 42)* (Bristol: Institute of Physics)

286

Von Smoluchowski M 1910 *Ann. Phys. Lpz.* **33** 1559

Webb F J 1963 *J. Nuclear Energy* A/B **17** 187–215

Wilkinson D 1982 *Nucl. Phys.* A **377** 474

Wilks J 1967 *The Properties of Liquid and Solid Helium* (Oxford: Clarendon)

Windsor L G 1981 *Pulsed Neutron Scattering* (London: Taylor and Francis)

Winfield D and Robson J M 1975 *Can. J. Phys.* **53** 667

Woods A D B and Cowley R A 1973 *Rep. Prog. Phys.* **36** 1135

Wu C S, Ambler E, Hayward R W, Hoppes D D and Hudson R P 1957 *Phys. Rev.* **105** 1413

Wu C S and Moszkowski S A 1966 *Beta Decay* (New York: Wiley)

Yamada S, Ebiwawa T, Achiwa N, Akiyoshi T and Okamoto S 1978 *Annual Report Research Reactor Institute Kyoto University* **11** 8–27

Yoshiki H, Ishimoto S and Utsuro M 1987 *Z. Phys.* B **67** 161–8

Yu Y Ch, Malik S S and Golub R 1986 *Z. Phys.* B **62** 137–42

Zeldovich Ya B 1959 *Sov. Phys. JETP*-**9** 1389–90

Subject Index

Author Index

Milton Keynes UK
Ingram Content Group UK Ltd.
UKHW021621071024
449327UK00020BA/1144